CULTURE AND BIOLOGY AT A CROSSROADS: THE MIDDLE PLEISTOCENE RECORD OF YARIMBURGAZ CAVE

(THRACE, TURKEY)

CULTURE AND BIOLOGY AT A CROSSROADS: THE MIDDLE PLEISTOCENE RECORD OF YARIMBURGAZ CAVE (THRACE, TURKEY)

EDITORS

F. Clark Howell †
Güven Arsebük
Steven L. Kuhn
Mihriban Özbaşaran
Mary C. Stiner

YAYINLARI

With great respect and gratitude to Clark *Hoca*,

The participators of Yarımburgaz
1988-1990

TABLE OF CONTENTS

LIST OF CONTRIBUTORS

Güven Arsebük (retired)
İstanbul Üniversitesi
Edebiyat Fakültesi
Arkeoloji Bölümü
Prehistorya Anabilim Dalı
Ordu Caddesi, 34134 Laleli-İstanbul, Turkey

Bonnie A. B. Blackwell
Department of Chemistry
Williams College
Williamstown, MA, 01267, USA

Joel I. B. Blickstein
RFK Science Research Institute
Glenwood Landing, NY, 11547, USA.

Mimi Divjak
RFK Science Research Institute
Glenwood Landing, NY, 11547, USA

William Farrand
Exhibit Museum of Natural History,
The University of Michigan
Ann Arbor, MI 48109-1079, USA

F. Clark Howell (deceased)
Laboratory for Human Evolutionary Studies
Museum of Vertebrate Zoology
University of California, Berkeley

Wighart v. Koenigswald
Institut für Paläontologie
Universität Bonn
Nussallee 8
D-53115 Bonn, Germany.

Steven L. Kuhn
School of Anthropology
University of Arizona
Tucson, AZ 85721-0030 USA

Christa Lindenau
Institut für Paläontologie
Universität Bonn
Nussallee 8
D-53115 Bonn, Germany.

Joyce Lundburg
Department of Geography
Carleton University
Ottawa, ON, K1S 5B6, Canada

Jill McMahon
Exhibit Museum of Natural History,
The University of Michigan
Ann Arbor, MI 48109-1079, USA

Mihriban Özbaşaran
İstanbul Üniversitesi
Edebiyat Fakültesi
Arkeoloji Bölümü
Prehistorya Anabilim Dalı
Ordu Caddesi, 34134 Laleli-İstanbul, Turkey

Henry P. Schwarcz
Department of Geology
McMaster University
Hamilton, ON, L8S 4M1, Canada.

Anne R. Skinner
Department of Chemistry
Williams College
Williamstown, MA, 01267, USA

Winfried Santel
Institut für Paläontologie
Universität Bonn
Nussallee 8
D-53115 Bonn, Germany

Mary C. Stiner
School of Anthropology
University of Arizona
Tucson, AZ 85721-0030 USA

Evangelia Tsoukala
Aristotle University
School of Geology
54124 Thessaloniki, Greece

PREFACE

The current volume has a long history. Most of the chapters by various specialists were first prepared around 1995-1996. Clark had always planned to write a concluding chapter and submit the entire manuscript for publication. Like so many projects, life intervened.

In the summer of 2006 Clark Howell asked Mary Stiner, Steve Kuhn and Mihriban Özbaşaran to take over the final push to publication of the Yarımburgaz monograph. He was in good spirits at the time, full of stories and plans, seemingly unaware of the illness that would soon claim his life. Within eight months Clark Howell was dead.

We decided to revive the Yarımburgaz monograph, in part as a tribute to Clark, but mainly because of the substantive contributions of many of the project's participants. Steve Kuhn, with the assistance of Tim White, retrieved the manuscript and figures from Clark's archives at The University of California, Berkeley. All authors were given the opportunity to revise and update their chapters and all of them did. The two sections originally prepared by Clark Howell, Chapters 1 and 8, have somewhat different histories. Chapter 1, covering the archaeological, environmental and historical contexts of the Yarımburgaz Paleolithic deposits, has been extensively revised and updated. Nonetheless, descriptions of the history of the excavations remain substantially as Clark had written them. In contrast, Chapter 8 is more or less exactly in Clark's words, although we have done some technical editing and changed the reference format. This chapter reflects the state of his thinking about the site and its vertebrate fauna as of about 1999. Readers will quickly note that the tone and style of Clark's chapter differ somewhat from those of other contributors to the volume. That was our intention. The way Clark Howell wrote was a direct reflection of his intensive and exhaustive style of research. Much as we might aspire to it, this style of scholarship is seldom matched today.

Project participants and contributors

A great many individuals have contributed to the excavations, analyses and publication of the work on the Pleistocene deposits in Yarımburgaz Cave. The excavation team consisted of geomorphologist Prof. Sırrı Erinç (Istanbul 1988-90), geologist Prof. W. R. Farrand (University of Michigan 1989-90); geochronologist Prof. H. P. Schwarcz (McMaster University, 1989); architect A. Koyunlu (Istanbul 1988-90); geologist/sedimentologist J. McMahon (University of Michigan, 1990); field supervisor Dr. M. Özbaşaran (Istanbul 1988-90); draftswoman J. Ogden (Walnut Creek, California 1988-90); computer specialist O. Tanındı, MA (Istanbul 1988-90); and photographer A. Boratav (Istanbul 1988-90). Student members of the excavation team were K. Adatepe MA (1988), Ş. Alptekin (1988-90), R. Aslan (1989-90), D. Büyükdora (1988-89), A. Demirtaş

(1988-90), N. Özbaşaran Dede, MA (1988-90), O. Denizci (1989), İ. Devir (1990), G. Egeli, MA (1988-90), B. Güzeldoğu (1988), B. Gürdil (1989-90), K. İnanç (1989-90), Ö. Işıltan (1988), A. Kayabal, MA (1988-90), K. Kartallıoğlu (1989-90), S. Kibaroğlu Küpeli, MA (1990), S. Zurikoğlu Par (1989), A. Taşkıran, MA (1988-89), A. Temel (1990), İ. Tığlı (1990). The Ministry of Culture was represented in 1988 and 1989 by Çiğdem Girgin (Archaeological Museums Istanbul), and in 1990 by Süleyman Eskalen (Hagia Sophia Museum, Istanbul).

The excavation program at Yarımburgaz Cave was largely supported by the National Geographic Society in the form of consecutive grants from the Committee on Research and Exploration to F.C.H. in 1988, 1989, 1990, and an initial visit in 1987 to examine materials from the excavations by M. Özdoğan of 1986. Further assistance (to G.A.) was afforded by Edebiyat Fakültesi, Istanbul Üniversitesi and Türkiye Turing ve Otomobil Kurumu. Post-excavation support (grants to F.C.H.) from the American Philosophical Society of Philadelphia and the Wenner-Gren Foundation for Anthropological Research in New York enabled processing of faunal and archeological residues, preparation of illustrations, and sedimentary determinations. The co-directors of the Yarımburgaz project are most appreciative of this assistance and also particularly grateful to all participants for their efforts in bringing the project to fruition. Finally, we thank Dr. Celal Şengör and Dr. Namık Çağatay for their careful review and extensive comments on sections of Chapter1.

All color photographs and many of the black and white images appearing in this monograph are the work of Ahmet Boratav.

LIST OF FIGURES

LIST OF TABLES

LIST OF PLATES

LIST OF COLOUR PLATES

CHAPTER 1

THE CAVES OF YARIMBURGAZ: GEOLOGICAL CONTEXT AND HISTORY OF INVESTIGATION

Güven Arsebük – F. Clark Howell – Steven L. Kuhn
William R. Farrand – Mihriban Özbaşaran – Mary C. Stiner

INTRODUCTION

The Yarımburgaz cave complex is located in eastern Thrace (Figure 1.1), approximately 22 km west of the city of Istanbul (Figure 1.2). Thrace, which constitutes the European portion of Turkey, represents the terminus of a large peninsula. This peninsula is 24,000 km^2 in area and bounded on the east by the the Black Sea, on the south by the Bosporus-Marmara Sea-Dardanelles system, and on the north and northeast by the Istıranca mountain chain (now called Yıldız Dağları), a continuation of the Rhodope mountains. The Thracian peninsula forms a land bridge between Europe and Asia (Anatolia). Although it has been interrupted by narrow waterways—the Bosporus and the Dardanelles—for parts of the Pleistocene and Holocene, it has long been viewed as a conduit for the flow of people and cultures between Europe and Asia.

The Yarımburgaz caves are situated ca. 1.5 km north of the northern shores of the modern lagoon of Küçükçekmece - the smaller of two local embayments of the Sea of Marmara. These features are flooded valleys that extend as submarine features into the sea (Erinç 1954). The twin openings of the Yarımburgaz cave complex (Figure 1.3) face south toward the Sazlıdere valley (Figures 1.4 and 1.5). The higher opening leads into the Upper Cave (UC), and the lower opening leads into the much more extensive Lower Cave (LC). Eocene limestone (Digis 1986: 8) exposed on both the eastern and western sides of the valley contain features developed erosionally by stream flow as well as by dissolution. The western side includes mainly large niches or rockshelters. The large openings of the Yarımburgaz caves occur on the eastern side of the valley.

Although the Yarımburgaz caves have known to science since the second half of the 19th century (e.g., Eyice 1978), recognition that the caves might have something to contribute to the Pleistocene prehistory of Turkey came only in 1964-1965 when Şevket Aziz Kansu (1963, 1972) opened several tests in conjunction with his exploration of a Chalcolithic occupation in the cave complex. In addition to artifacts from later periods, these *sondages* yielded limited fauna and Paleolithic artifacts of indeterminate age. The implications and potential of this archaeological discovery were not pursued for two decades. In 1986, Prof. Mehmet Özdoğan (Istanbul University), in the course of his archeological surveys in Thrace, turned his attention to the Chalcolithic remains at Yarımburgaz. His excavations in the cave's upper chamber demonstrated the presence of some Neolithic and limited Middle/Upper Paleolithic occupations below the Chalcolithic deposits (Figure 1.6). However, test soundings in the UC and adjacent areas of the LC revealed unexpected evidence of a much older and more substantial infilling with multiple Lower Paleolithic occupations (Özdoğan 1987a; Özdoğan and Koyunlu 1986). Artifacts

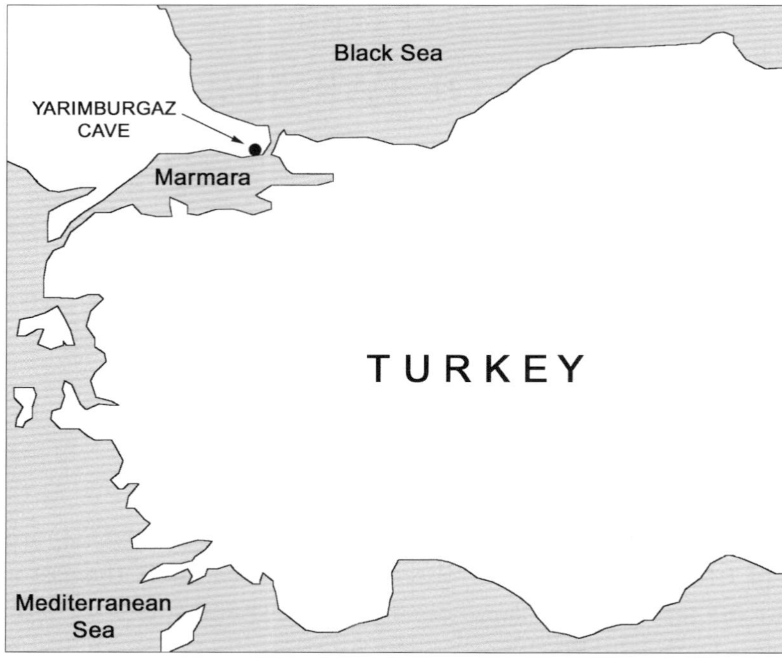

Fig. 1.1. Location of Yarımburgaz cave in western Turkey.

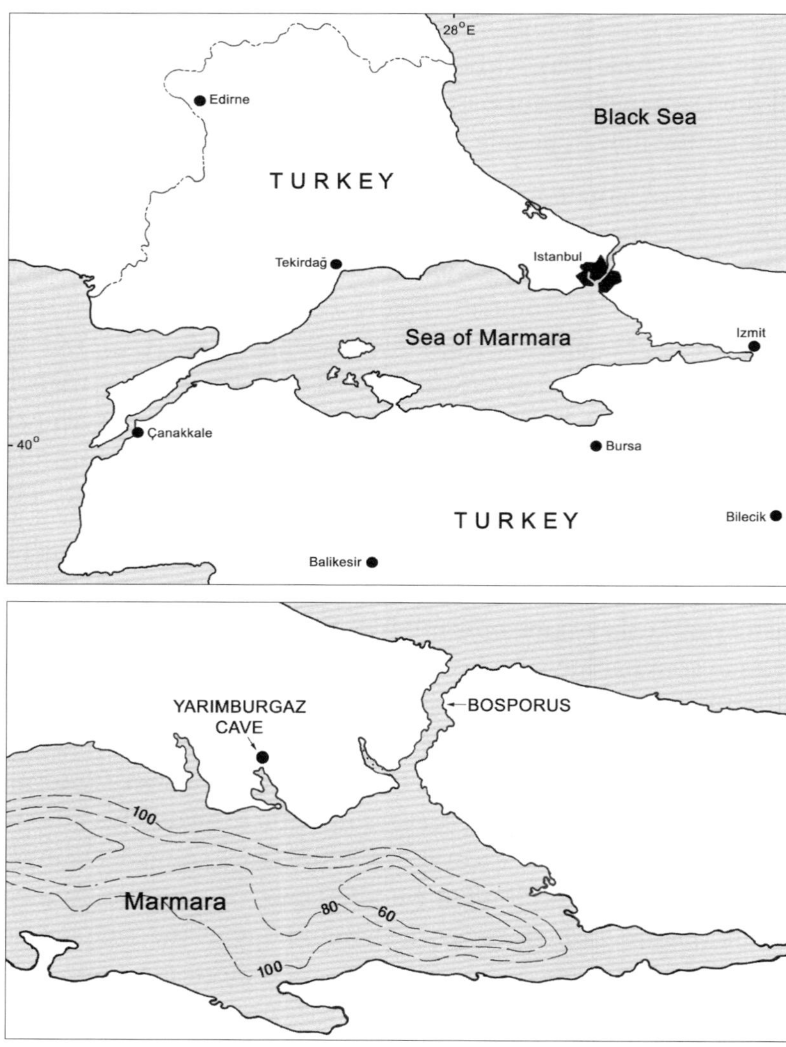

Fig. 1.2. Location of Yarımburgaz cave in relation to the modern Sea of Marmara and Bosporus, and general bathymetry of the Marmara (below).

Fig. 1.3. Entrances of the Upper and Lower Caves of Yarımburgaz.

Fig. 1.4. View of Sazlıdere Valley from just above Yarımburgaz cave (1990).

and faunal remains from these soundings were examined in 1987 by one of us (Howell [1988]), who recognized the necessity for an in-depth study of this unique situation. Three field seasons in 1988, 1989 and 1990 were devoted to an investigation of the site. These excavations yielded abundant and diverse vertebrate fossils in association with what are clearly human-produced artifacts (Howell and Arsebük 1989, 1990). This volume reports and synthesizes a wide range of geological, paleontological and archaeological findings from the 1988-1990 field seasons, along with limited information from Özdoğan's 1986 investigations.

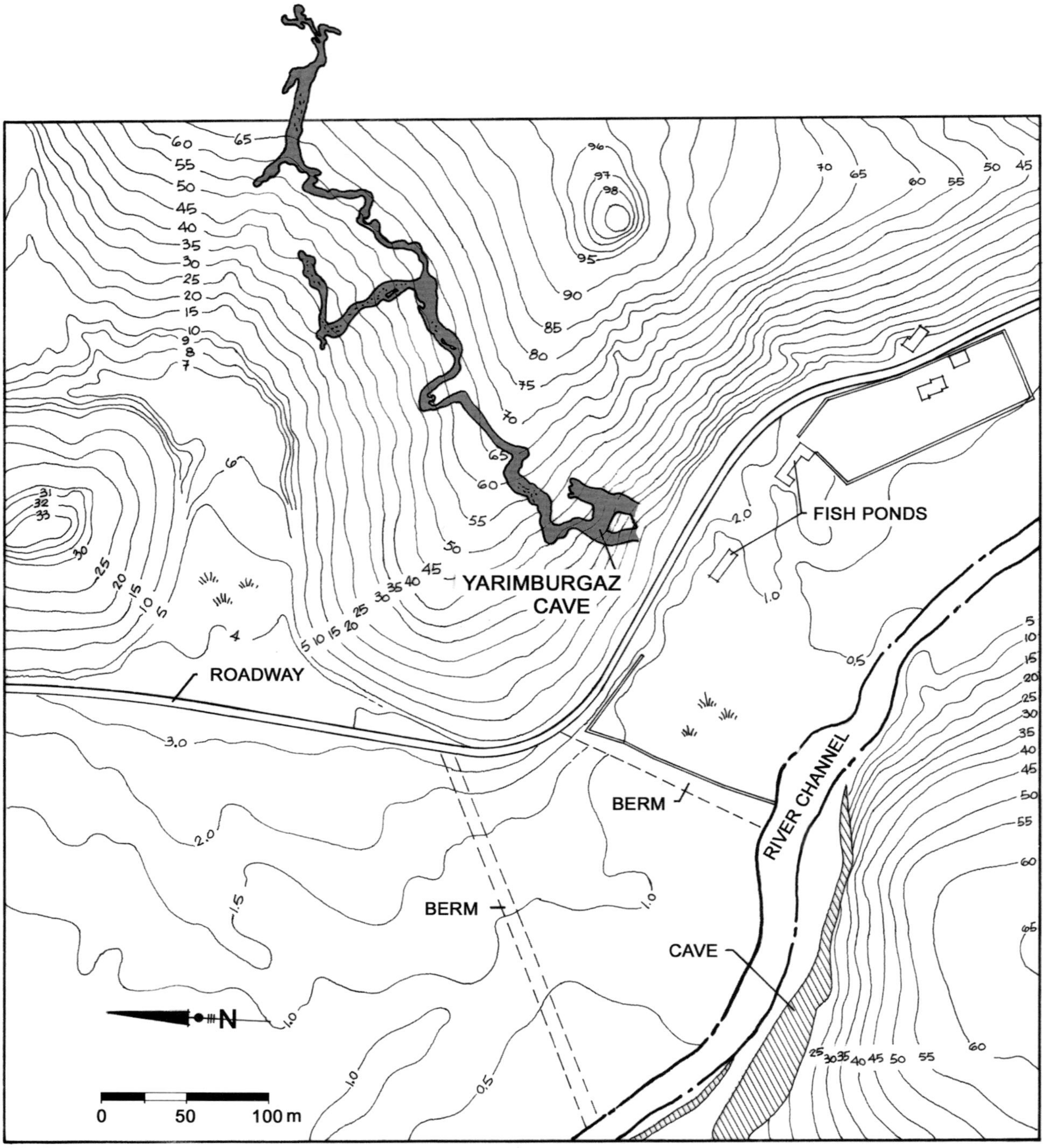

Fig. 1.5. Local topography of the Sazlıdere Valley and full extent of the (short) Upper and (long) Lower Chambers of Yarımburgaz cave in the bedrock.

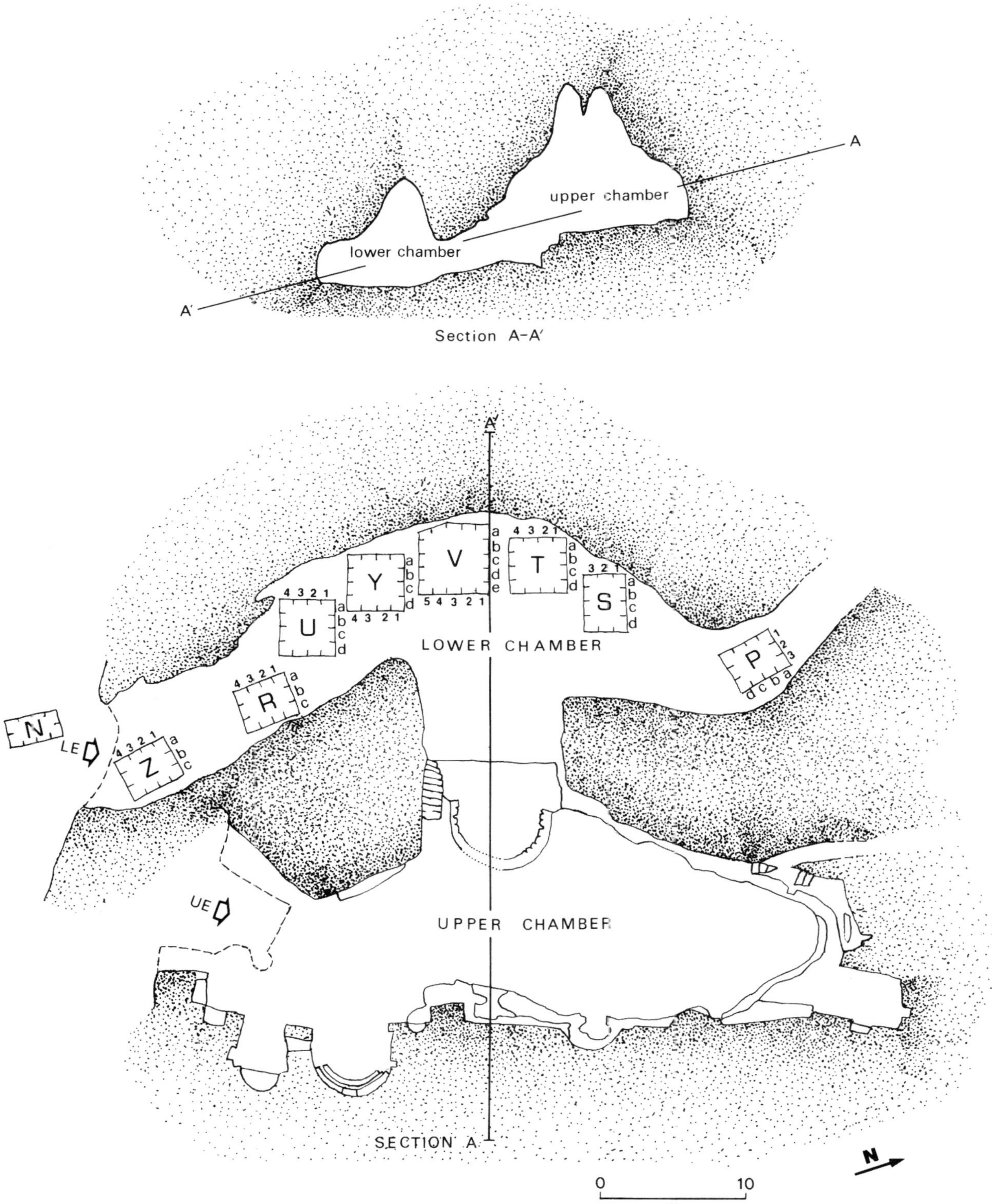

Fig. 1.6. Plan of the Yarımburgaz cave chamber complex, with E-W cross-section, showing locations of archaeological trenches or "squares" in the Lower Chamber. (UE) upper entrance; (LE) lower entrance. The 1986 sounding (A.71) was later expanded and renamed Square V-88. Squares V-88, Y-88 and Z-88 were excavated in 1989. Squares S-89, T-89 and U-89 were excavated in 1989. Squares N-90, P-90 and R-90 were excavated in the final year of the project (1990).

ARCHAEOLOGICAL CONTEXT

In 1988, when the most recent excavations of Yarımburgaz cave complex began, the great extent of Turkey (ca. 780,000 km^2) was still considered largely *terra incognita* in terms of Pleistocene prehistory. Turkey's contributions to Paleolithic archaeology and human evolution research seemed particularly limited considering the size of its territory and its position at the geographic interface between western Asia and Europe. Upper and Middle Paleolithic human occupations, probably dating to the later Pleistocene, had been documented in a few caves in the southern part of the country (Bostancı 1971; Şenyürek 1959; Şenyürek and Bostancı 1958). Middle and Lower Paleolithic surface finds had been recorded across Anatolia (the Asian parts of Turkey), though few were found in geological context (Kuhn 2002; Özdoğan 1984, 1987b; Taşkıran 2008 Yalçınkaya 1981). A notable and important exception was the large Karain cave complex, near Antalya, first excavated in the 1950s and 1960s by K. Kökten (reviewed by Yalçınkaya [1989]) and subsequently investigated more extensively by an international team led by I. Yalçınkaya (Otte et al. 1995a, b, 1998; Rink et al. 1994; Yalçınkaya et al. 1992). The deeply-stratified Karain caves, and the neighboring site of Öküzini (Otte et al. 1995c; Yalçınkaya et al. 1995), contain levels that span the later Middle Pleistocene to the Holocene. The Middle Paleolithic is particularly well represented at Karain.

The pace of Paleolithic research in Turkey has accelerated noticeably in the two decades since the Yarımburgaz project began. The first definitively *in situ* evidence for a Lower Pleistocene human presence has been discovered at the Dursunlu lignite quarry near Konya, in the central part of the country (Güleç et al. 1999, 2009). Researchers working at another central Anatolian site, Kaletepe Deresi 3, have exposed a sequence of Middle and Lower Paleolithic deposits dating to between 0.16 and 1.1 mya, including the first stratigraphically sealed Acheulean assemblages in Turkey (Slimak et al. 2004, 2007, 2008). A calvarium attributed to *Homo erectus* was discovered fortuitously in a travertine quarry near Denizli, on the Aegean coast (Kappelman et al. 2007), but unfortunately without definitive archaeological or geological associations. Research on the Middle and Upper Paleolithic continues at the Karain cave complex, as well as at sites in the Hatay region along the southern Mediterranean coast (Kuhn 2004a, 2004b; Kuhn et. al. 1999, 2009). In recent years many additional Middle and Lower Paleolithic localities have been identified through surface reconnaissance, particularly in southeastern Turkey but also in Thrace and western Anatolia (e.g., Dinçer and Slimak 2007; Şahin et al. 2009; Taşkıran 2008). The Upper Paleolithic currently is less well represented in western Turkey but has been identified in some surface sites (Runnels and Özdoğan 2001).

If anything, results from the past two decades of research have heightened the interest and relevance of the findings from the work conducted at Yarımburgaz between 1986 and 1990. Yarımburgaz remains as one of the earliest well-documented human sites in Turkey, thanks to the generally high level of preservation, the presence of a deep Middle Pleistocene stratigraphy, and spatially extensive deposits that allowed for the exposure of large areas within the Lower Cave. The distinctive aspects of the flaked-stone artifacts from the Lower Cave raise important questions about Middle Pleistocene hominin cultural geography, the distribution of the Acheulean in particular, and about the arbitrary definitional boundary between the Lower and Middle Paleolithic. Paradoxically, the artifacts co-occur in the deposits with cave bear bones and comparatively scant remains of other macromammals. The fact that the artifacts are closely associated spatially with Pleistocene bears—which apparently used the cave as a den—has significant implications for the ecology and behavior of both bears and hominids. Yarımburgaz thus also provides a paleontological resource of unique importance with respect to the Ursids as well as for other taxa.

PLEISTOCENE CONDITIONS IN THE MARMARA BASIN

The Marmara basin and its present continental sea is the dominant feature of the region around Yarımburgaz the caves. Conditions in the Marmara during the Pleistocene would have had significant impacts on the terrestrial as well as littoral environments experienced by the hominins who used the cave. The Marmara basin covers some 11,350 km^2 and is roughly 280 km long east to west. The maximum width at the center is 76 km, but the basin narrows sharply at its eastern and western ends. It consists of three deep sub-basins (>-1100 to -1250 m) separated by NW-trending ridges: from east to west these are the Çınarcık, central and Tekirdağ basins. Uplands isolated as islands occur in a number of locations, including the well-known Prince Islands near Istanbul and the much larger Marmara Island in the western part of the basins. The current configuration of the Marmara is strongly influenced by two complex fault systems. The Thrace-Eskişehir system formed some of the basin's earliest features, whereas the western extension of the North Anatolian Fault has been a more active force since the late Pliocene (Figure 1.7, Barka and Kadinsky-Cade 1988; Gökaşan et al. 2002; Okay et al. 2000; Şengör 1979; Şengör et al. 2005; Yaltırak 2002).

The Marmara connects to the Mediterranean (Aegean) Sea via the Dardanelles (Çanakkale strait) on the west, and to the Black Sea via the 31 km long Bosporus strait. The Dardanelles follows a valley that formed as a result of Pliocene tectonic activity (Yaltırak et al. 2000). The current sill depth is variously reported at between -60 and -80 m. The Bosporus has a more complex origin. The northern (Black Sea) part of the strait is of fluviatile origin, whereas the southern part is formed by a deeper basin that probably is the result of faulting (Gunnerson and Özturgut 1974; Gökaşan et al 1997). The sill at the northern end of the Bosporus is composed of Pleistocene sediments and is roughly -60 m deep. The present day sill at the southern end of the strait, composed of Quaternary sediment, is much shallower (ca. -30 m) (Gökaşan et al 1997; Meriç and Algan 2007) but the underlying bedrock sill is much deeper (Major et al. 2002).

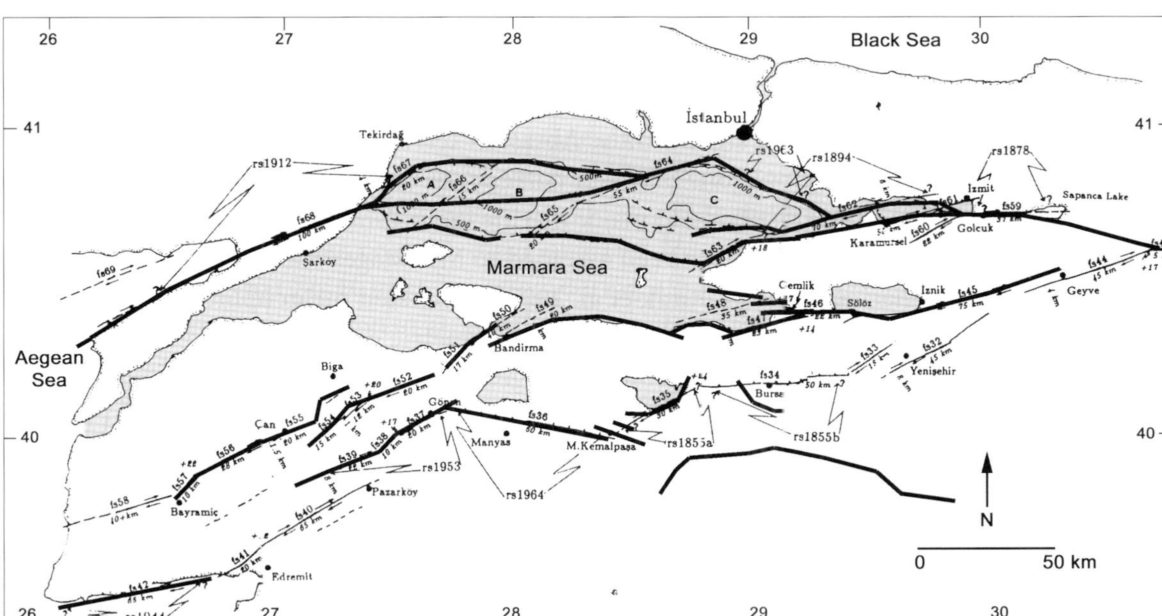

Fig. 1.7. Active fault segments in the western North Anatolian fault zone, Marmara Sea region. After Barka and Kadinsky-Cade (1988) and Okay et al. (2000). Faults illustrated are those that cut Miocene and younger sediments.

> "In fact, although the floristic composition may be very different from site to site depending on the history of individual taxa, the variations of the vegetation structure, indicated by the general trends of deforestation and afforestation, are generally related to climatic changes, at least until human activity significantly modified the landscape. Although temperature variations between glacials and interglacials have certainly had some influence on vegetation, the main driving factor for forest expansion all around the Mediterranean was water availability, which is at present the main limiting factor for Mediterranean vegetation."

Patterns of vegetation succession were generally similar between interglacials (e.g. Tzedakis et al. 2001, Tzedakis 2007), suggesting that plant species expanded repeatedly from persistent refugia in the eastern Mediterranean area. Periods of forest expansion in the interglacials are closely associated with the timing of the summer insolation peak (Watts 1988; Tzedakis 2005).

Unfortunately, few if any pollen records from Turkey, or eastern Thrace, extend back before 20 ky. The deep pollen record closest to the Yarımburgaz caves comes from Tenaghi Philippon in eastern Macedonia, a sequence now known to extend back to ca. 1.35 mya (van der Wiel et al. 1987a, 1987b; Tzedakis et al. 2006). The environment of the Tenaghi Philippon plain is not altogether similar to that of the Küçükçekmece valley, in that the former is situated at a similar elevation (ca 40m asl) but farther inland. However, the Tenaghi Philippon plain can provide a notion of the range of environments and scale of variation that might be expected during the Middle Pleistocene in the area of the Yarımburgaz caves. The vegetational succession at Tenaghi Philippon is closely tied to the global record of Pleistocene glacial/interglacial cycles (Tzedakis et al. 2006). This sequence documents periods of treeless steppe vegetation, presumed to represent full glacial conditions, interglacial environments oak forests with pistachio and chestnut and other thermophilic trees, and a range of intermediate vegetation types.

Deep sea coring in the Black Sea basin has been more limited (see Ross et al. 1978). Moreover, the paleoclimatic and paleoenvironmental implications of these cores have been difficult to untangle, due to problems of inter-core correlation, lack of a clear chronological framework, and poor pollen representation from widely spaced samples (see Paluska and Degens 1978; Schrader 1979; Degens and Stoffers 1980). The recent spate of research into the hypothetical early Holocene Black Sea flood has stimulated a prodigious amount of research (summarized in Yanko-Hombach et al. 2007, as well as two special issues of the journal *Quaternary International* [167/168, 2007 and 197(1-2) 2008] among other places), and the great majority of it refers to periods that long post-date the Paleolithic occupations of Yarımburgaz cave.

Zubakov (1988) synthesized the results of extensive Soviet investigations into the Pleistocene record of marine and adjacent continental (riverine and cover) deposits of the northern Black Sea-Azov Sea area. There it has been possible to correlate and integrate results on terrestrial and marine stratigraphy, geomagnetic polarity changes and secular variations, isotopic age determinations, salinity changes of marine waters and their molluscan faunas, and the relationships of alluvial-deltaic and loessic sediments and paleosols, the latter containing important vertebrate faunal assemblages. Since 1.1 ma, 8 salinity events and 7-10 cold freshwater events have been distinguished. The 8 salinity events are marked by the penetration of Mediterranean waters and mollusc species into the Black Sea. The 7-10 cold freshwater events involved inflow of Caspian Sea and proglacial-derived waters and attendant molluscan species into the Black Sea. Ages of these various events have been ascertained by TL and polarity datum planes. These units match quite well with Mediterranean MIS events and sapropel accumulations, as well as the number and ages of Eurasian glacial-interglacial cycles.

THE UPPER AND LOWER CAVES OF YARIMBURGAZ

Yarımburgaz cave comprises an Upper Chamber (UC) and a Lower Chamber (LC), each with a separate south-facing entrance (Figure 1.3). Named the "Upper Gallery" by some researchers, the UC is over 50 m in length and between nine and 15 m wide (Figure 1.5). The ceiling height reaches 10 m in some places. The surface of the UC deposits is situated at 18.5 m above modern sea level. The much more extensive LC system is more than 500 m long but averages only 5 m in width over much of its length. The LC bifurcates about 310 m from the entrance, and the northwestern arm continues approximately 130 m into the bedrock, while the southern fork reaches a further 200 m. The sill at the LC entrance is situated at 11.5 m above sea level. The outermost 60 m of the LC is a large gallery with a ceiling that slopes down toward the drip-line at the entrance.

Some 10 m inside the entrance to the LC the two outer chambers are joined by a horizontal passage (see Figures 1.5 and 1.6). The LC widens to about 10 m at this juncture. Daylight penetrates approximately 50 m inside the entrance. The bedrock floor of the LC was reached immediately outside the entrance in Square N-90. Bedrock appears on the surface at the point at which the two arms of LC split, suggesting that the floor of the cave forms a broad, U-shaped basin.

During the Roman and Byzantine Periods a large building complex was constructed both inside and outside Yarımburgaz cave, although evidence of these constructions is now badly disturbed. During the Byzantine period, the front of the UC was modified to form a chapel, and traces of a division wall remain on the ceiling. Other remnants of Byzantine modifications to the UC include a carved column with capital on a projection on one wall, three carved vaults in the ceiling, nine niches on the east wall, stairs, and an apse with altar in the middle. Traces of joists/supports in the walls suggest that the constructions near the mouth of the Upper Chamber originally had a second storey. The inner area was also used as a cemetery for a time (M. Özdoğan and A. Koyunlu 1986: 8).

On the south side of the passage between the chambers are stairs carved into the wall and another apse, probably also of Byzantine age. The architectural remains of the Byzantine period terminate at a rubble wall which ends at the passage between the two chambers. It thus appears that the LC was used little if at all during this period, although architectural features revealed in Trench N-90 at the LC entrance may relate to the enlargement of Byzantine constructions inside the UC.

The vicinity of Yarımburgaz, and even the caves themselves, have been the focus of some historical interest for well over a century. However, their archeological potential remained largely unexploited until quite recently. Semavi Eyice (1978) provides the most thorough documentation of the Küçükçekmece area from an historical perspective, a history that extends back into Graeco-Roman and Byzantine times through the Ottoman Period.

The cave complex of Yarımburgaz first attracted the attention of researchers in the second half of the 19th century, though primarily to the geological formation. Dr. Hammerschmidt (then known as *Miralay Macarlı Abdullah Bey*) from the Faculty of Medicine, University of Istanbul (*Mekteb-i Tıbbiye-i Şahane*), visited the cave in 1845. He subsequently published his observations in two articles (Abdullah-Bey 1869, 1870). Viquesnel (1868), in his treatment of the physical geography and geology of Thrace, had mentioned briefly the existence of the caves and their geological situation. This account also mentioned the human utilization and transformation of the upper chamber in Roman and Byzantine times and thus highlighted the particular historical interest of Yarımburgaz. Bousquet (1900-1901) provided a charming account of a day's excursion to the caves, and some details of the structural configurations in the UC. The caves received further brief notice from Harun Reşid Kocacan (1921).

Kansu and Kökten re-visited the research of Abdullah Bey and published a summary of it in 1972. Kansu and Kökten were the first to realize the cave's archeological importance, and they later conducted excavations there (Kansu 1972: 24-26). Abdullah Bey judged the age of the limestone hills in which the cave is located to be Miocene, and believed the formation of the caves took place in a much later period. He describes "going from the Upper Chamber down to the Lower Chamber, the *stalactites* on the cave wall, where the deep gallery starts..." (Kansu 1972: 25). Unfortunately, no such elements exist in the cave today. Prof. R. Hovasse, then on the Faculty of Sciences (Istanbul), and his colleagues Prof. R. Jeannel (Paris) and İbrahim Hakkı (Akyol), were the first to make soundings in the cave, their main aim being to locate the bedrock floor. Though many small trenches were opened in various locations, they were unable to reach the bedrock (Hovasse 1927: 405; Kansu 1972: 30). Hovasse (1927) published the first detailed account of the cave complex, including a map of the cave and its conformation, and a preliminary identification of the extant invertebrate fauna. He described in detail the boat paintings found on the wall of a chamber he named "Salon F" [inward of the LC system], which he felt gave important indications of the relation of the Sea of Marmara to the lagoon and river. In favorable conditions, large boats could come to within 200 m of the cave via the river that connects to the lagoon (Hovasse 1927: 407), the Marmara now being about 10 km from the cave.

Somewhat later Hubbard (1932) published an account of the upper and lower chambers of Yarımburgaz cave, including their karstic development and a map of their form, extent, and the locations of dripstone formations. Ernest Chaput (1936), in his exceptional monograph on the geology and geomorphology of Turkey, devotes a major chapter to the Marmara and its setting. This includes the geological situation of the Küçükçekmece, the surrounding limestone countryside, and the exposure of (upper) Miocene marine sediments, interbedded with continental deposits near the village of the same name (Nafiz and Malik 1933; Yalçınlar 1971, 1983). This important locality was subsequently demonstrated to be of early upper Miocene age (Benda et al. 1978; Chaput and Gillet 1939). The Yarımburgaz cave complex has also attracted the interest of speleologists, and some of its copepod fauna was reported by Lindberg (1952). It is curious to note that a volume on Turkish caves oriented toward speleology (Aygen 1984) concludes that Yarımburgaz "....has no outstanding features except that is very near to Istanbul."

The first systematic archaeological excavations in Yarımburgaz were carried out by Kökten and Kansu in 1959. They excavated a 2.5 x 1.5 m sounding adjacent to the northern wall of the LC down to 1.1 m in depth, and they were able to distinguish two layers. The uppermost layer, 0.50-0.60 m thick, yielded Classical pottery. The lower layer, distinguished by small pebbles and yellowish sand, included deer teeth, scrapers on flint flakes, broken pieces of hammerstones, together with fossilized bones and teeth of cave bear. Kansu's study of this material suggested "...a strong probability of Lower and Middle Paleolithic..." (Kansu 1967: 13; 1972: 277-278). This is the first report of prehistoric cultural residues from the site.

Together with N. Dolunay of the Archaeological Museums of Istanbul, Kökten and Kansu resumed their fieldwork in 1964-1965, on behalf of the Turkish Historical Society. They opened four trenches in the LC (named "B") and one in the UC ("A"). The first layer from the top (1.1 m thick) was composed of recent bat and sheep droppings; the second layer (0.60 m thick) had some concentrations of stones together with Byzantine pottery; the third layer (2.0 m thick) consisted of clay, yellow sand and coarsely flaked artifacts. The existence of a Paleolithic occupation was demonstrated by the presence of 'Mousterian' type points of either flint or quartz and the teeth, long bones and skull fragments of what they identified as "*Ursus spelaeus*" in this lower layer (Kökten 1963: 278; Kansu 1967:13). Although excavations continued to a depth of three meters, they did not reach bedrock.

Kökten and Kansu's excavations in the UC identified a Byzantine fill 0.45 m in thickness, followed by a second layer 2.0 m thick that included flint tools, a quantity of pottery and some bone implements. Of the pottery in the third level, "both sides are either black or dull buff in color, burnished and with grooved, fluted, dotted and encrusted decoration," indicating its Chalcolithic character. A stone celt and mortars were also recovered in this level. Below the Chalcolithic was a yellowish clayey sand layer, and bedrock appeared at 4.8 m below the surface (Kansu 1972: 23).

After two seasons of excavation by Kansu, Kökten and Dolunay, the site of Yarımburgaz suffered serious destruction and degradation. Contemporary building and construction activities near the cave and new interest of treasure seekers led to the loss of the stalagmites and stalactites (Ketin 1970), defacement of the apses and niches, general displacement of fills, and continued utilization of the outer chambers as sheep folds.

EXCAVATIONS IN 1986

In an effort to prevent further damage to the Yarımburgaz caves, the General Directorate of Antiquities and Museums of Turkey instigated salvage excavations in 1986 by a team from the University of Istanbul, Prehistory Department, under the direction of Prof. M. Özdoğan. Özdoğan began with excavation of nine trenches in the UC, four in the LC and three in the middle passage between the chambers (see also Farrand and McMahon 1997). The results showed that stratification in the UC was the result of topographical and environmental conditions during the natural formation process of the cave rather than from human activities. Flowing water eroded the cave floor in a 15 m wide depression in the inner half of the UC. At times this area was a pond or bog and at others it was totally dry. Prehistoric deposits were preserved in the UC only by virtue of the fact that this depression protected some sediments from the floor-leveling activities of the Byzantine period.

In all, Özdoğan identified 14 or 15 layers in UC in 1986. The stratigraphy of the UC infilling, originally determined by Özdoğan (1987a, 1990; Özdoğan and Koyunlu 1986), was re-examined by Farrand and McMahon (1997, and this volume). The uppermost layer in the UC dated to Roman and Byzantine periods, while the second and fifth layers preserved in the depression dated to Chalcolithic and Neolithic ages, respectively. The ^{14}C samples from the UC have given varied results: samples taken from the Chalcolithic layers 2, 3, and 4 were dated to 5930 ± 110 (GrN 15534), 6880 ± 90 (GrN 15528) and 7330 ± 60 (GrN 15529) B.P. Those from the underlying Neolithic layers 7, 7b, and 7b yielded ages of 7640 ± 90 (GrN 15533), 9190 ± 100 (GrN 15531) and 10,000±890 (Beta 42734) B.P. (M. Özdoğan, personal communication).

Pottery finds in these last cultural layers were important for demonstrating links between the Balkans and the Aegean and Anatolian cultures via Thrace. Although no architectural remains were connected with the pottery, the stratigraphy afforded an important local cultural sequence (M. Özdoğan 1989; 1990: 380). Moreover, the fine handmade, decorated Chalcolithic pottery of layers 2 and 4 (Özdoğan 1985), the Neolithic types of layer 5, and the pots, ground stones, celts and bone implements all suggest a continuous use of the UC from at least 5500 B.C. onward. Further indication of a Chalcolithic settlement in the LC was observed during our own 1990 excavation season. The same kind of pottery – out of stratigraphic context in this case – was found near the cave wall in Square P-90. These sherds were either washed in and deposited when parts of the Chalcolithic layers were eroded by flowing water, or they were discarded there as trash.

Layers 6 through 10 in the UC, attributed by Özdoğan to the Middle and Upper Paleolithic, were poor in archeological finds. Layer 8 yielded a ^{14}C age of 24,510±240 BP (Beta 42732), suggesting a rare Upper Paleolithic occupation for Turkey. While it is doubtful that intact Middle Paleolithic deposits are present in the UC, some Levallois flakes and sidescrapers were collected. An underlying sterile naturally stratified sand and gravel deposit, Layer 11, over 3.0 m thick, is now known to represent a former high (transgressive) sea level that flooded into the cave mouth. The earliest deposits excavated by Özdoğan in the UC, layers 13 through 15, were exposed in three squares—A.30, A.40 and A.39 (see Figure 2 in Farrand and McMahon 1997). These were assigned to Lower Paleolithic on the basis of a limited number of artifacts and some mammalian faunal remains; artifacts were most abundant in level 14. Mammal fossils from A.40 included *Ursus* vertebrae and rib fragments, other indeterminate rib and vertebrae fragments, and long bone diaphyses, some possibly *Bovine*, and some chiropteran postcrania. There is also a distal tibia of a large canid, tentatively referred to *Cuon* or possibly (?) *Xenocyon*. Square A.39 yielded some cranial and vertebral fragments of *Ursus* and an innominate fragment of an indeterminate bovid/cervid.

A preliminary study of the Paleolithic material from the LC excavations, notably that from Özdoğan's A.71 deep sounding, clearly demonstrated its Lower Paleolithic character (Özdoğan and Koyunlu 1986). This was further confirmed by F. C. Howell's examination of the lithic assemblage and associated faunal remains from the A.71 sounding in 1987. Square A.71 comprised layers 1 through 12 from top to bottom. Lower Paleolithic artifacts were found in levels 4 (12), 5 (119), 6 (194), 7 (25), 8 (23), 9 (8) and 10 (8). Mammal remains were found in layers 3, 5, 6, 7, 9 and 12, and they were most abundant in layers 5 and 6. The recognition of Yarımburgaz as a rare example of a stratified Paleolithic cave site in Turkish Thrace (Arsebük 1993, 1996) led to the extensive excavation program from 1988 to 1990, the results from which are reported in this volume.

1988 - 1990 EXCAVATIONS OF THE LOWER CAVE

A multi-national team jointly directed by Prof. Güven Arsebük and Prof. F. Clark Howell initiated the first of three consecutive seasons of fieldwork at Yarımburgaz in 1988. The project's aims were as follows:

1) Elucidation of the infilling of the LC, its stratigraphic succession, processes of accumulation, and linkages (if any) with external events/processes.
2) Employment of appropriate methods of age determination, in respect to both inorganic and organic dateable resources.
3) Maximum recovery of the vertebrate fossil record, including an intensive screening operation for micromammal taxa, in association with close stratigraphic and spatial control of the distribution of all residues.
4) Maximum recovery of the associated Lower Paleolithic assemblage(s) in their stratigraphic and spatial contexts and in respect to co-association(s) with organic residues.

Each of the three seasons of field work lasted approximately 40 days. The system of excavation followed methods normally used by the Prehistory Department of the University of Istanbul in the Keban Project Excavations in Eastern Anatolia between 1968-1974 and practices followed by Howell in extensive Lower Paleolithic excavations in Spain. A map of the LC produced by A. Koyunlu (in Özdoğan and Koyunlu 1986), with his and other modifications, was employed as a base for the situation of the excavation trenches or "squares". A set point on the north wall of the largest niche in the UC was selected as a permanent (0) vertical datum point. This elevation was transferred by transit to the LC and, successively, to all corner stakes of each excavation square. All feature records and their vertical elevations were calculated according to this set datum. Individual excavation trenches were normally 3 x 3 m or 3 x 4 m, and sub-divided to individual 1 m^2 blocks, and were gridded by alphabetic letters (E-W) and Arabic numbers (N-S). Each excavation square received a letter and year designation (e.g., Y-88). The distribution of squares is shown in Figure 1.6. Strata in each square were enumerated from level 1 downwards. Hence, level terms in each square are unique and not necessarily equivalent from square to square (see Farrand and McMahon for cross-correlations, this volume). All finds, including fossils and natural blocks, were mapped in plan at 1:20 scale (Figure 1.8). Information for each find was recorded on printed labels and computer forms. Descriptions and observations of sediment structure and of the context of finds were noted in daily field notebooks.

Dry sieving of the excavated sediment was achieved using a "Shaker" of the type designed by Dr. C. B. M. McBurney (Cambridge University) (Payne 1972: 49-51). The sieve was furnished with various graded screens - the smallest being 0.005 m^2 - and all of these were used consistently. All residues were then picked from the sieved material and boxed according to stratigraphic level and square. Except for surface soil and evidently disturbed areas, each cultural unit was either sieved completely or was sampled.

Fig. 1.8. 1988 excavation of Trench T (plan 6).

Fig. 1.9. View of the southeast corner of Trench V (1988).

Extensive use of computers was decided on early and a data base was designed for this purpose. An entry form was designed for all finds and a data base structure was set up by creating fields which took into account later query and sorting operations. This data base was updated daily. Along with find information, maps and small find drawings were scanned in as produced. The main data base was then updated and verified by specialists after the end of each season of excavations, when all finds were reexamined and further identified.

The results of each season were presented annually in a Symposium organized by the Ministry of Culture/Turkey, and preliminary results were published (in Turkish) in the Proceedings—*Kazı Sonuçları Toplantısı*—by the same institution (Arsebük, Howell and Özbaşaran 1990, 1991, 1992). Annual reports were also prepared for the National Geographic Society and other funding institutions.

The squares designated Z-88, Y-88 and V-88 (=A.71) were opened in the first excavation season; U-89, T-89, and S-89 were added the following year; and n-90, R-90 and P-90 in 1990. Geological, archaeological, and paleontological findings from these trenches are introduced below. The geological context is described in much greater detail in Chapter 2.

Excavations commenced in Square V-88 (5 x 5 m), layer 12, which was a deeper continuation of Özdoğan's 1986 A.71 trench (Figure 1.9). The layer yielded a substantial quantity of animal bones but few artifacts. The finds were labeled as layer 12 of square V-88, as that layer had already been documented in 1986 excavations (square A.71). In order to prevent confusion, the four subsequent (deeper) layers were denoted, from the top down, as a, b, c and d, respectively. They were differentiated according to textures, color and nature of the sediment. These lowermost layers all proved to be sterile. Work was then confined to the southern half of the square trench, measuring 2.5 x 5.0 m. When confirmed as being of only geological interest, the southern half was again subdivided and work continued in the southeast quarter of the original A.71 trench. This 2.5 x 2.5 m area constituted a deep sounding. Bedrock was not reached even though the excavation reached 5.0 m below the surface.

Square Y-88, (4 x 4 m) was opened close to the cave wall, to the south of V-88, with a 1 m baulk separating the two squares. This trench was excavated to a depth of 2.80 m below surface. Square Y-88 yielded diverse amounts of fossilized faunal remains, abundant microfauna and Lower Paleolithic stone artifacts from six levels (Figure 1.10). A zone of mixed fill about 1.5 m wide in the eastern half of the trench was identified as back-dirt from Kansu and Kökten's 1964-65 soundings.

In Y-88, the first layer (1a-b) below the surface sediment contained partially cemented material with abundant limestone fragments in its upper part. Elsewhere it consisted of various lenses of different colors and structural characteristics. Over 200 lithic artifacts and nearly 300 vertebrate

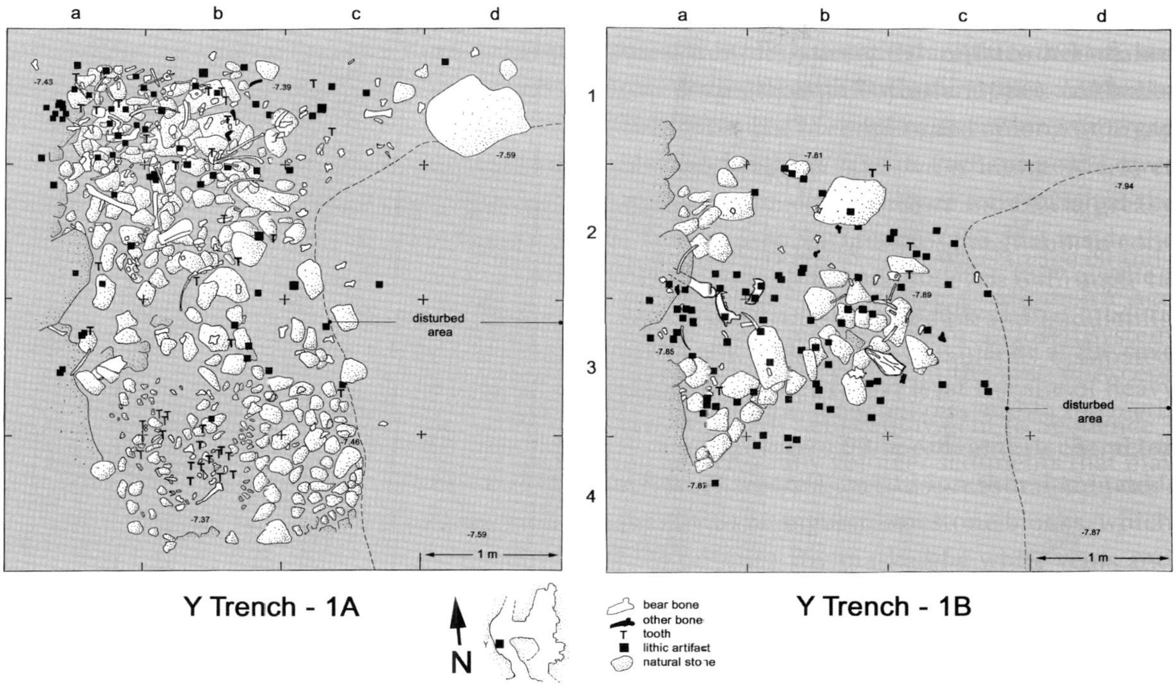

Fig. 1.10. Co-occurrences of plotted lithic artifacts and bones in Trench Y (1, a-b).

remains were found in layer 1. The second layer of Y-88 also had two sub-levels (2a and 2b): 2a provided more than 40 artifacts, but only 2 pieces were found in 2b. The third layer (3a, 3b) was structurally different from those above, being rich in small limestone pieces and localized stony debris: this layer yielded 81 artifacts in all. The fourth layer was sterile, and the fifth contained a substantial amount of microfauna, but scant macrofauna.

The central area of the LC in which squares V-88 and Y-88 are situated evidently represents the most intensively occupied part of the chamber. However, this "concentration area" continues both to the north and south, albeit with decreasing amounts of artifacts and fauna.

Square T-89 (4 x 4 m) (see Figure 1.8), located immediately north of V-88, yielded abundant artifacts and fauna, revealing that this area was also within the central concentration. The total number of stone artifacts from all layers amounted to more than 430. More than 900 identifiable bone specimens and fragments were collected as well. Six layers were distinguished in T-88. The first layer was rich in faunal remains but contained few lithic artifacts. The second layer contained abundant fallen blocks of limestone. Among these there was much fauna, but again few stone tools. Layers were 3 and 4 were much richer in artifacts; each of these layers had a thickness of about 0.5 m and together they provided some 430 artifacts and more than 600 faunal remains. The majority of the fauna consisted of *Ursus* bones, as was the case for all Paleolithic layers. The third layer was also particularly rich in microfaunal remains. The underlying 5th and 6th layers were purely geological in character: the 5th layer consisted of a cloddy soil with vertical cracks; the 6th was a reddish yellowish loam with seemingly cemented lenses.

Square S-89 (3 x 4 m) was opened immediately to the north of T-89, separated from it by a 1 m baulk and offset several m to the east. Under the mixed surface sediment, the first intact layer, which appeared in only part of square S-89, proved to be a continuation of layer 2 in T-89 and thus could be traced over a wider area. The third layer, however, was again of limited extent, while

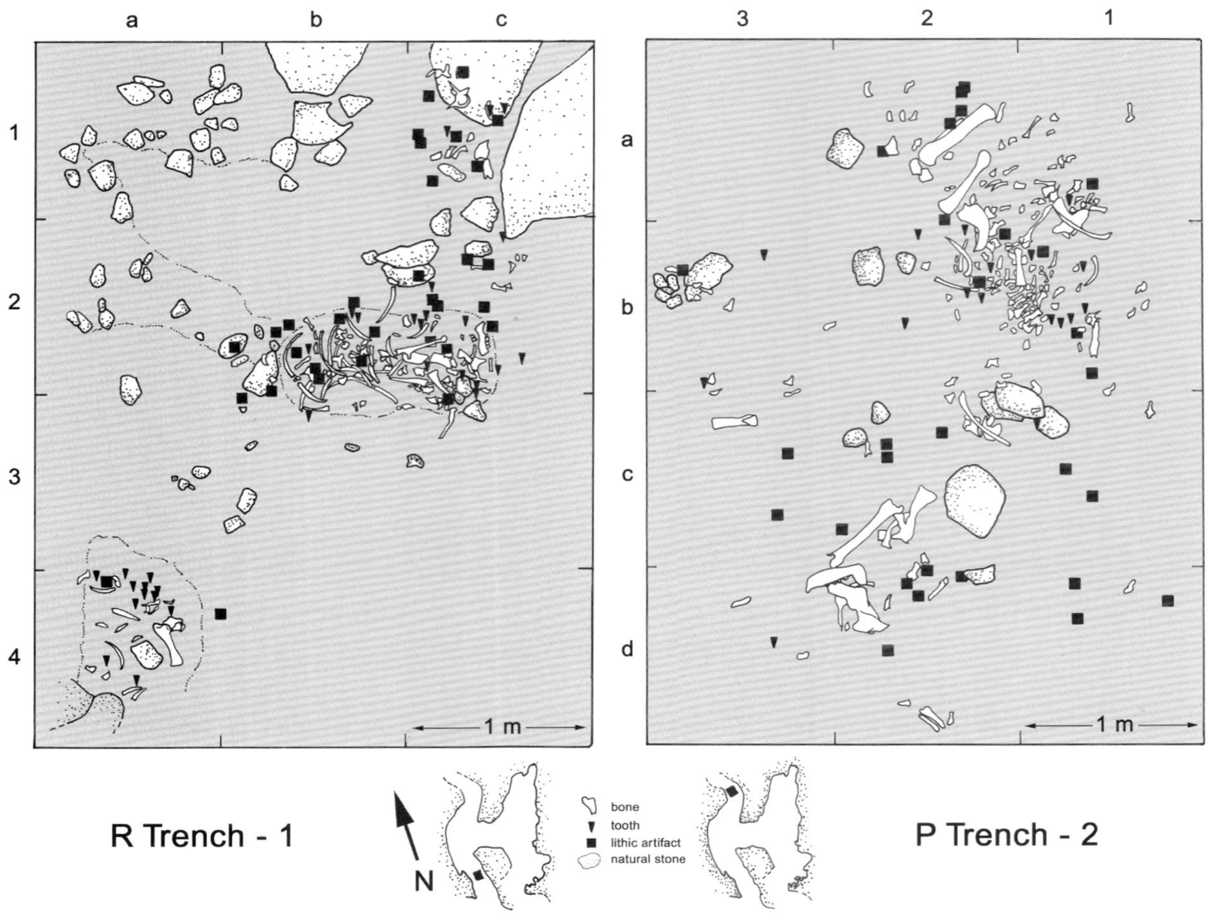

Fig. 1.11. Co-occurrences of plotted lithic artifacts and bones in Trenches P and R.

the fourth (as in square T-89, layers 3 and 4), yielded a substantial amount of fauna and lithics. The 5th and 6th layers were archeologically sterile. S-89 was closed at a depth of 2.0 m below the surface, without reaching bedrock.

Three seasons of excavations revealed that Paleolithic humans utilized most intensively the area along the western wall of the LC, within ca. 45 m of the entrance. The limits of this area, beginning about 10 m from the entrance (Square R-90), extended northwards to just south of the narrowest part of the gallery (and where Square P-90 was situated). According to the spatial distribution of the stone tools and animal bones, it can be argued that biological activities of all sorts were particularly concentrated in the area of the Y, V and T squares.

To establish further the borders of the apparent "intensively used/occupied area" of the LC, squares P-90 and R-90 were opened in 1990 (Figure 1.11). Square P-90, about 45 m from the entrance and largely beyond natural light, was opened to define the northern limits of occupation. Measuring 3 x 4 m, the deposits in P-90 were found to include four different cultural units. The first layer had a fill rich in fauna but was poor in stone artifacts. The second layer contained very hard, cloddy and stony sediment s with many faunal remains (200), but again few artifacts.The third and fourth layers, the latter examined only in a restricted area, were both sterile. Two artifacts found in the upper part of layer 3 were considered as probably intrusive (from layer 2).

Square U-89 (4 x 4 m), south of Y-88, was opened to clarify the southward extent of the occupation area. Below the surface sediment, a huge fallen roof block covered most of the square (Figure 1.12). The sediments deposited around this block were excavated in several levels. The entire 3 x 4 m area contained a large amount of both macro- and microfauna. Its further excavation was reluctantly abandoned owing to the impracticality of removing the massive block of limestone.

Square Z-88 (3 x 4 m), the first trench of 1988, was intentionally situated just inside the present drip-line, close to the entrance of the LC. The deepest point of excavation measured from the surface was -3.60 m. Contrary to expectations, undisturbed Paleolithic deposits were not reached here; most of the layers had been destroyed by building activity during the Byzantine period. All layers from the surface (0) through layer 3 contained mixed archaeological contents. Layer 4 and below had undisturbed sediments. The 8th and 12th layers had plentiful microvertebrate remains but the remainder were sterile.

The area of Square R-90 (3 x 4 m), between U-89 and Z-88, contained more abundant material than did the other adjacent squares (see Figure 1.11). R-90 was excavated to a depth of -2.30 m, through six layers. We identified several concentrations of finds, containing mostly bones and teeth. One such concentration, occupying about 1.0 square meter roughly in the middle of the square, contained almost 100 teeth and 290 bones, and nearly 170 stone artifacts. In layer 2 the stone artifacts outnumbered the faunal remains. A disconformity was encountered between layers 2 and 3, below which the number of finds decreased. Layer 4 was sterile. Layer 6 yielded large animal bones and microfauna but lacked artifacts.

Square N-90 (2 x 4 m), was opened at the end of the 1990 season, outside the cave in front of the entrance. The aim was to reach bedrock, which had never been attained within the cave. This square yielded no archeological finds of Pleistocene age. Beneath the surface soil the remains of a building was cleared, probably of the same age as the Byzantine chapel in the UC. The architectural features excavated possibly represented the staired pavement of a building, rising towards the LC. Six depositional units were observed in N-90, layer 1 being the fill just above

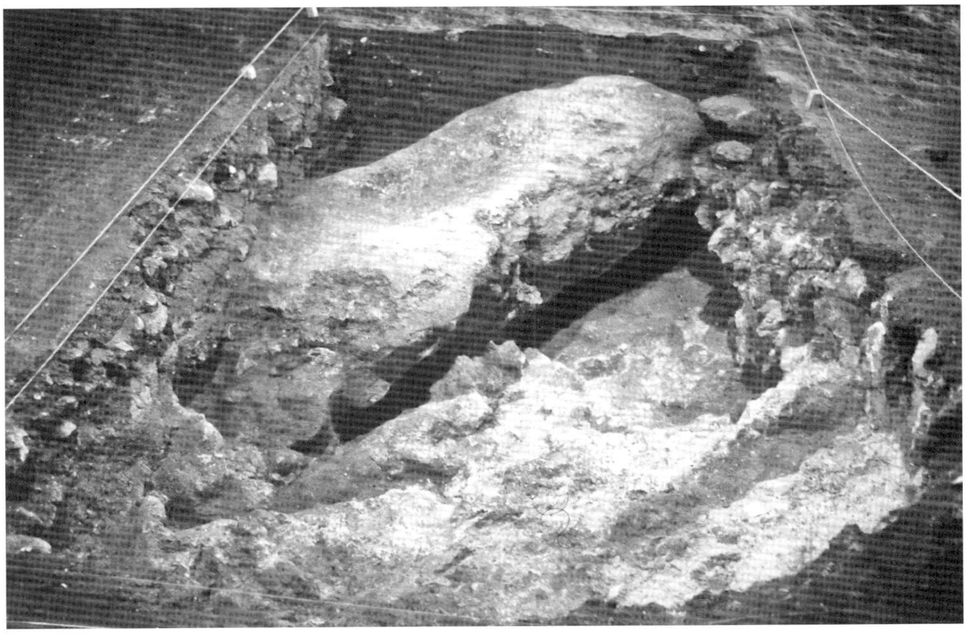

Fig. 1.12. Fallen block in Trench U (1989).

Çağatay, M.N., Görür, N., Flecker, R., Sakınç, M., Tünoğlu, C., Ellam, R., Krijgsman, W., Vincent, S., Dikbaş, A., 2006. Paratehyan-Mediterranean connectivity in the Sea of Marmara region (NW Turkey) during the Messinian. *Sedimentary Geology* 188/189, 171-187.

Degens, E.T., Ross, D.A., (Eds.) 1974. The Black Sea - Geology, Chemistry and Biology. *American Association of Petroleum Geologists*, Memoir N. 20, Tulsa, OK.

Degens, E.T., Paluska, A., 1979. Tectonic and climatic pulses recorded in Quaternary sediments of the Caspian-Black Sea Region. *Sedimentary Geology* 23, 149-163.

Degens, E.T., Stoffers, P., 1980. Environmental events recorded in Quaternary sediments of the Black Sea. *Journal of the Geological Society* (London) 137, 131-138.

Digis, A., 1986. Anbarlı-Şamlar (Istanbul) Arasındaki Bölgenin Jeolojisi. B. A. Thesis. Department of Geology, University of Istanbul.

Dinçer, B., Slimak, L., 2007. The Paleolithic of Turkish Thrace. Synthesis and recent results. TÜBA-AR 10, 49-61.

Ergin, M., Uluadam, E., Sarıkavak, K., Keskin, Ş., Gökaşan, E., Tur, H. 2007. Late Quaternary sedimentation and tectonics in the submarine Şarköy Canyon, western Marmara Sea (Turkey), in: Taymaz, T., Yılmaz, Y., Dilek, Y. (Eds.), *The Geodynamics of the Aegean and Anatolia*. The Geological Society of London, London, pp. 231-257.

Erinç, S., 1954. The Pleistocene history of the Black Sea and the adjacent countries with special reference to the climatic changes. *Review of the Geographical Institute*, University of Istanbul 1, 85-133.

Eriş , K.K., Ryan, W.B.F., Çağatay, M.N., Sancar, U., Lericolais, G., Ménot, G., Bard, E., 2007. The timing and evolution of the post-glacial transgression across the Sea of Marmara shelf south of Istanbul. *Marine Geology* 243, 57-76.

Erol, O., 1976. Quaternary shoreline changes on the Anatolian coasts of the Aegean Sea and related problems. *Bulletin Société Géologie de France*, Sér. 7, 18, 459-468.

Evans, G., Erten, H., Alavi, S. N., von Gunten, H.R., Ergin, M., 1989. Superficial deep-water sediments of the eastern Marmara basin. *Geo-Marine Letters* 9, 27-36.

Eyice, S., 1978. Tarihde Küçükçekmece. *İstanbul Üniversitesi, Edebiyat Fakültesi, Güney-Doğu Avrupa Araştırmaları Dergisi* 6/7, 57-120.

Farrand, W. R., McMahon, J.P., 1997. History of the sedimentary infilling of Yarımburgaz Cave, Turkey. *Geoarchaeology* 12, 537-565.

Giosan, L., Filip, F., Constatinescu, S., 2009. Was the Black Sea catastrophically flooded in the early Holocene? *Quaternary Science Reviews* 28, 1-6.

Gökaşan, E., Demirbağ, E., Oktay, F.Y., Ecevitoğlu, B., Şimşek, M., Yüce, H., 1997. On the origin of the Bosphorus. *Marine Geology* 140, 183-199.

Gökaşan, E., Gazioğlu, C., Alpar, B., Yücel, Z.Y., Ersoy, Ş., Gündoğu, O., Yaltırak, C., Tok, B., 2002. Evidence of NW extension of the North Anatolian Fault Zone in Marmara Sea; a new approach to the 17th August 1999 Marmara Sea Earthquake. *Geo-Marine Letters* 21, 183-199.

Görür, N., Çağatay, M.N., Sakınç, Sümengen, M., Şentürk, K., Yaltırak, C., Tchapalyga, A., 1997. Origin of the Sea of Marmara as deduced from Neogene to Quaternary paleogeographic evolution of its frame. *International Geology Review* 39, 342-352.

Grosswald, M.G., 1980. Late Weichelsian ice sheet of northern Eurasia. *Quaternary Research* 13, 1-32.

Gunnerson, C.G. and E. Özturgut, E., 1974. The Bosphorus, in: Degens, E.T., Ross, D.A. (Eds.), *The Black Sea - Geology, Chemistry and Biology*. American Association of Petroleum Geologists, Memoir N. 20, Tulsa, OK, pp. 99-114.

Güleç, E., Howell, F.C., White, T., 1999. Dursunlu—A new Lower Pleistocene artifact-bearing locality in southern Anatolia, in: Ullrich, H. (Ed.), *Hominid Evolution: Lifestyles and Survival Strategies*. Edition Archea, Berlin, pp. 349-364.

Güleç, E., White, T., Kuhn, S., Özer, I., Sağır, M., Yılmaz H., Howell, F.C., 2009. The Lower Pleistocene Lithic Assemblage from Dursunlu (Konya), central Anatolia, Turkey. *Antiquity* 83, 11-22.

Hovasse, R., 1927. Yarımburgaz Mağarası - La grotte de Yarim Bourgas. *Darülfünun Fen Fakültesi Mecmuası (Istanbul)* 5, 1-19 (French), 396-422 (Turkish).

Howell, F.C., 1988. Some observations on the cave of Yarımburgaz (Marmara, Turkey), its setting and infilling(s), and with particular reference to faunal residues recovered in a 1986 sounding. Report to Committee on Research and Exploration, National Geographic Society, Washington, D.C.

Howell, F.C., Arsebük, G., 1989. Yarımburgaz cave (1988). Report on investigations in the cave of Yarımburgaz (Marmara, Turkey). 1988 field season. Report to Committee on Research and Exploration, National Geographic Society, Washington, D.C.

Howell, F.C., Arsebük, G., 1990. Yarımburgaz Cave (1989). Report on investigations, and current status of researches in the cave of Yarımburgaz (Marmara, Turkey). 1989 field season. Report to Committee on Research and Exploration, National Geographic Society, Washington, D.C.

Hubbard, G.E., 1932. Turkish grottoes of Yarım Burgaz. *Pan-American Geologist* 57, 321-328.

Hughes, P.D., Woodward, J., Gibbard, P.L., 2006. Quaternary glacial history of the Mediterranean mountains. *Progress in Physical Geography* 30, 334-364.

Hughes, P.D., Woodward, J., Gibbard, P.L., 2007 Middle Pleistocene cold stage climates in the Mediterranean: New evidence from the glacial record. *Earth and Planetary Science Letters* 253, 50-56.

Kallel, N., Paterne, M., Labeyrie, L.D., Duplessy, J.C., Arnold, M., 1997 Temperature and salinity records of the Tyrrhenian sea during the last 18000 years. *Palaeogeography, Palaeoclimatology, Palaeoecology* 135, 97-108.

Kansu, Ş. A., 1963. Marmara bölgesi ve Trakya'da prehistorik iskan tarihi bakımından araştırmalar (1959-1962). Türk Tarih Kurumu, *Belleten* (Ankara) 27, 657-705. (French summary, pp. 672-676).

Kansu, Ş. A., 1967. İstanbul ve Türkiye Trakya'sının Tarihöncesi İskanına Ait Yeni Araştırma ve Buluşlar. *VI. Türk Tarih Kongresi Kongreye Sunulan Bildiriler*, Ankara, Seri IX, no. 6, 13-14.

Kansu, Ş. A., 1972. Yarımburgaz Mağarası'nda Türk Tarih Kurumu Adına Yapılan Prehistorya ve Araştırmalar, Tuzla Kalkolitiğinden Yeni Gözlemler. *VII. Türk Tarih Kongresi Kongreye Sunulan Bildiriler*, Ankara, Seri IX, 1 (7), 22-32.

Kappleman, J., Alçiçek, M., Kazancı, N., Schultz, M., Özkül, M., Şen, Ş., 2007. First *Homo erectus* from Turkey and implications for migrations into temperate Eurasia. *American Journal of Physical Anthropology* 135, 110-116.

Kerey, I.E., Meriç, E., Tünoğlu, M., Kelling, G., Brenner, R.L., Doğan, A.U., 2004. Black Sea-Marmara Sea Quaternary connections: new data from the Bosphorus, Istanbul, Turkey. *Palaeogeography, Palaeoclimatology, Palaeoecology* 204, 277-295.

Ketin, I., 1970. Türkiye'de önemli jeolojik aflörmanların korunması. *Türkiye Jeoloji Kurumu Bülteni* 13, 90-93.

Kocacan, H. R., 1921. Bir mağara nasıl tetkik olunur? *Tedrisat Mecmuası* 61, 12-18.

Kökten, K., 1963. Istanbul Batısında Eskitaş (Paleolitik) devrine ait yeni buluntular. *Ankara Üniversitesi Dil ve Tarih Coğrafya Fakültesi Dergisi* 20, 277-278.

Kroon, D., Alexander, I., Little, M., Lourens, L.J., Matthewson, A., Robertson, A.F.H., Sakamoto, T., 1998. Oxygen isotope and sappropel stratigraphy in the eastern Mediterranean during the last 3.2 million years, in: Robertson, A.H.F., Emeis, K.-C., Richter, C., and Camerlenghi, A. (Eds.), *Proceedings of the Ocean Drilling Program, Scientific Results*, Vol. 160, pp. 181-189.

Kuhn, S., 2002. Paleolithic archaeology in Turkey. *Evolutionary Anthropology* 11, 198-210.

Kuhn, S., 2004a. Upper Paleolithic raw material economies at Üçağızlı Cave, Turkey. *Journal of Anthropological Archaeology* 23, 431-448.

Kuhn, S., 2004b. From Initial Upper Paleolithic to Ahmarian at Üçağızlı Cave, Turkey. *Anthropologie* (Brno) 42, 275-288 (special issue).

Kuhn, S., Stiner, M.C., Güleç, E., 1999. Initial Upper Paleolithic in south-central Turkey and its regional context: a preliminary report. *Antiquity* 73, 505-517.

Kuhn, S., Stiner, M.C., Güleç, E., Özer, I., Yılmaz, H., Baykara, I., Açıkkol, A., Goldberg, P., Martínez, K., Ünay, E., Suata-Alpaslan, F., 2009. The Early Upper Paleolithic Occupations at Üçağızlı Cave (Hatay, Turkey). *Journal of Human Evolution* 56, 87-113.

Lindberg, K., 1952. Notes sur quelques grottes de la Turquie. *Annales de Speleologie (Spelunca)*, ser. 3, 7, 55-66.

Lourens, L.J., Hilgen, F.J., Raffi, I., Vergnaud-Grazzini, C., 1996. Early Pleistocene chronology of the Vrica section (Calabria, Italy). *Paleoceanography* 11, 797-812.

Magri, D., Kallel, N., Blancamaria, N., 2004. Paleoenvironmental changes in the Mediterranean Region 250-10 KYR BP, in: Batterbbe, R.W., Gasse, F., Stickley, C. (Eds.), *Past Climate Variability through Europe and Africa*. Springer, Dordrecht, The Netherlands, pp. 325-341.

Major, C., Ryan, W., Lericolais, G., Hajdas, I., 2002. Constraints on Black Sea outflow to the Sea of Marmara during the last glacial-interglacial transition. *Marine Geology* 190, 19-34.

Meriç, E., Algan, O., 2007. Paleoenvironments of the Marmara Sea (Turkey) Coasts from paleontological and sedimentological data. *Quaternary International* 167/168, 128-148.

Nafiz, H., Malik, A., 1933. Vertébrés fossiles de Küçükçekmece. *Bulletin*, Faculty of Science, Istanbul Universitesi. (in Turkish with French summary).

Okay, A.I., Kaşlılar-Özcan, A., Imren, C., Boztepe-Güney, A., Demirbağ, E., Kuş, I., 2000. Active faults and evolving strike-slip basins in the Marmara Sea, northwest Turkey: a multichannel seismic reflection study. *Tectonophysics* 321, 189-218.

Otte, M., Yalçınkaya, I., Taşkıran, H., Kozlowski, J.K., Bar-Yosef, O., Noiret, P., 1995a. The Anatolian Middle Paleolithic: new research at Karain cave. *Journal of Anthropological Research* 51, 287-299.

Otte, M., Yalçınkaya, I., Kozlowski, J., Bar-Yosef, O., Taşkıran, H., Noiret, P., 1995b. Evolution technique au Paleolithique ancien de Karain (Turquie). *L'Anthropologie* (Paris) 99, 529-561.

Otte M., Yalçınkaya, I., Leotard, J-M., Kartal, M., Bar-Yosef, O., Kozlowski, J.K., López Bayón, I., Marshack, A., 1995c. Epi-Palaeolithic of Öküzini cave (SW Anatolia) and its mobiliary art. *Antiquity* 69, 931-944.

Otte, M., Yalçınkaya, I., Kozlowski, J., Bar-Yosef, 0., Lopez Bayon, I., Taşkıran, H., 1998. Long-term technical evolution and human remains in the Anatolian Palaeolithic. *Journal of Human Evolution* 34, 413-431.

Özdoğan, M., 1984. 1983 Yılı doğu Marmara ve Trakya Araştırmaları. *II. Araştırma Sonuçları Toplantısı*, Ankara, pp. 221-232.

Özdoğan, M., 1985a. The Chalcolithic pottery of Yarımburgaz cave, in: *Studi di Paletnologia in Onore de Salvatore M. Pugliesi*. Università di Roma "La Sapienza", Rome, pp. 177-189.

Özdoğan, M., 1 985b. Marmara bölgesinde kültür tarihi ile ilgili bazı sorunlar ve buluntuların çözümüne jeomorfoloji araştırmalarının katkısı. *I. Arkeometri Sonuçları Toplantısı*, Ankara, pp. 139-162.

Özdoğan, M., 1987. Yarımburgaz Mağarası 1986 Yılı Kazı Çalışmaları. *V. Araştırma Sonuçları Toplantısı*, Ankara, pp. 323-346.

Özdoğan, M., 1989. Neolithic cultures of northwestern Turkey. A general appraisal of the evidence and some considerations, in: Bokonyi, S. (Ed.), *Neolithic of Southeastern Europe and its Near Eastern Connections. Varia Archaeologica Hungarica*, II. Instituti Archaeologici Academiae Scientiarum Hungaricae, Budapest, pp. 201-215.

Özdoğan, M., 1990. Yarımburgaz Mağarası. *X. Türk Tarih Kongresi*, pp. 373-388.

Özdoğan, M., Koyunlu, A., 1986. Yarımburgaz Mağarası. 1986 yılı çalışmalarının ilk sonuçları ve bazı gözlemler. *Arkeoloji ve Sanat* 32/33, 4-17. (Turkish, with English summary, pp. 15-17).

Paluska A., Degens, E.T., 1978. Climatic and tectonic events controlling the Quaternary in the Black Sea region. *Geologischen Rundschau* 68, 284-301.

Payne, S., 1972. Partial Recovery and Sample Bias: The results of some sieving experiments, in: Higgs, E.S., (Ed.), *Papers in Economic Prehistory*. Cambridge University Press, Cambridge, pp. 49-64.

Rink, W.J., Schwarcz, H. P., Grün, R.,Yalçınkaya, I., Taşkıran, H., Otte, M, Valladas, H., Mercier, N., Bar-Yosef, O., Kozlowski, J., 1994. ESR dating of the Last Interglacial Mousterian at Karain Cave, southern Turkey. *Journal of Archaeological Science* 21, 839-849.

Rohling, E.J., Thunell, R.C., 1999. Five decades of Mediterranean paleoclimate and sapropel studies. *Marine Geology* 153, 7-10.

Supko, P., Ross, D.A., Neprochnov, Y.P. 1978. *Initial Reports of the Deep Sea Drilling Project*. Vol. 42. Part 2. U.S. Government Printing Office, Washington, DC.

Rossignol-Strick, M., 1985. Mediterranean Quaternary sapropels, an immediate response of the African monsoon to variation of insolation. *Palaeogeography, Palaeoclimatology, Palaeoecology* 49, 237-263.

Rossignol-Strick, M., Nesteroff, W., Olive, P., Vergnaud-Grazzini, C., 1982. After the deluge: Mediterranean stagnation and sapropel formation. *Nature* 295, 105-110.

Rossignol-Strick, M., Paterne, M., 1999. A synthetic pollen record of the eastern Mediterranean sapropels of the last 1 Ma: implications for the time-scale and formation of sapropels. *Marine Geology* 153, 221-237.

Runnels, C., Özdoğan, M., 2001. The Palaeolithic of the Bosphorus region, NW Turkey. *Journal of Field Archaeology* 28, 69-92.

Ryan, W.F., 1972. Stratigraphy of Late Quaternary sediments in the Eastern Mediterranean, in: Stanley, D.J. (Ed.), *The Mediterranean Sea*. Dowden, Hutchinson and Ross, Stroudsburg, PA, pp. 149-169.

Ryan, W.B.F., Pitman, W. III., 1999. *Noah's Flood. The New Discoveries about the Event that Changed History*. New York, Simon and Schuster.

Ryan, W.B.F., Pitman III, W.C., Major, C.O., Shimkus, K., Moskalenko, V., Jones, G., Dimitrov, P., Görür, N., Sakınç, M., Yüce, H., 1997. An abrupt drowning of the Black Sea shelf. *Marine Geology* 138, 119-126.

Schrader, H.-J., 1979. Quaternary paleoclimatology of the Black Sea basin. *Sedimentary Geology* 23, 165-180.

Slimak, L., Roche H., Mouralis D., Buitenhuis H., Balkan-Atlı N., Binder D., Kuzucuoğlu C., Grenet M., 2004. Kaletepe Deresi 3 (Turquie), aspects archéologiques, chronologiques et paléontologiques d'une séquence pléistocène en Anatolie Centrale, *Palevol* 3, 411-420.

Slimak, L., Kuhn, S., Balkan-Atlı, N., Binder, D., and Dinçer, B., 2007. Kaletepe Deresi 3: de 'l'Acheuleem au Mousterien en Anatolie centrale. *Anatolia Antiqua/Eski Anadolu* 15, 257-273.

Slimak, L., Kuhn, S., Roche, H., Mouralis, D., Buitenhuis, H., Balkan-Atlı, N., Binder, D., Kuzucuoğlu, C., Guillou, H., 2008. Kaletepe Deresi 3 (Turkey): archaeological evidence for early human settlement in Central Anatolia. *Journal of Human Evolution* 54, 99-111.

Smith, A. D., Taymaz, T., Oktay, F., Yüce, H., Alpar, B., Başaran, H., Jackson, J.A., Kara, S., Şimşek, M., 1995. High-resolution seismic profiling in the Sea of Marmara (northwest Turkey): Late Quaternary sedimentation and sea-level changes. *Geological Society of America Bulletin* 107, 923-936.

Stanley, D.J., Blanpied, C., 1980. Late Quaternary water exchange between the eastern Mediterranean and the Black Sea. *Nature* 285, 537-541.

Şahin, N., Dinçer, B., Zimmerman, T. 2009. Neue Fundplätze des Älteren Paläolithikums bei Bursa in Nordwestanatolien (Türkei). *Archäologisches Korrespondenzblatt* 2/09, 1-10.

Şenyürek M., 1959. Tıkalı Mağaranın Paleolitik Endüstrisine dair bir not. *Belleten* 23, 9-58. (Turkish with English translation)

Şenyürek M., Bostancı E., 1958. Hatay Vilâyetinde Prehistorya Araştırmaları, *Belleten* 22, 147-166. (Turkish with English translation)

Şengör, A.M.C., 1979. The North Anatolian transform fault: its age, offset and tectonic significance. *Journal of the Geological Society* (London) 136, 269-282.

Şengör, A.M.C., Tüysüz, O., İmren, C., Sakınç, M., Eyidoğan, H., Görür, N., Le Pichon, X., Rangin, C., 2005. The North Anatolian Fault: a new look. *Annual Review of Earth and Planetary Sciences* 33, 37-112.

Taşkıran, H., 2008. Réflexions sur l'Acheuléen d'Anatolie. *l'Anthropologie* 112, 140-152.

Thunell, R. C., 1979. Climatic evolution of the Mediterranean Sea during the last 5.0 million years. *Sedimentary Geology* 23, 67-79.

Tuenter, E., Weber, S.L., Hilgen, F.J., Lourens, L.J., Ganopolski, A., 2005. Simulation of climate phase lags in response to precession and obliquity forcing and the role of vegetation. *Climate Dynamics* 24, 279-295.

Tzedakis, P.C., 2005. Towards an understanding of the response of southern European vegetation to orbital and suborbital climate variability. *Quaternary Science Reviews* 25, 1585-1599.

Tzedakis, P.C., 2007. Seven ambiguities in the Mediterranean paleoenvironmental narrative. *Quaternary Science Reviews* 26, 2042-2066.

Tzedakis, P.C., Andrieu V., de Beaulieu, J.-L., Crowhurst, S., Follieri, M., Hooghiemstra, H., Magri, D., Reille, M., Sadori, L., Shackleton, N.J., Wijmstra, T., 1997. Comparison of terrestrial and marine records of changing climate of the last 500,000 years. *Earth and Planetary Science Letters* 150, 171-176.

Tzedakis, P.C., Andrieu, V., Birks, H.J.B., de Beaulieu, J.-L., Crowhurst, S., Follieri, M., Hooghiemstra, H., Magri, D., Reille, M., Sadori, L., Shackleton, N.J., Wijmstra, T.A., 2001. Establishing a terrestrial chronological framework as a basis for biostratigraphical comparisons. *Quaternary Science Reviews* 20, 1583-1592.

Tzedakis, P.C., Hooghiemstrab, H., Pälikec, H., 2006. The last 1.35 million years at Tenaghi Philippon: revised chronostratigraphy and long-term vegetation trends *Quaternary Science Reviews* 25, 3416-3430.

van der Wiel, A.M, Wijmstra, T.A., 1987a. Palynology of the lower part (78-120 m) of the core Tenaghi Philippon II, Middle Pleistocene of Macedonia, Greece. *Review of Palaeobotany and Palynology* 52, 73-88.

van der Wiel, A.M, Wijmstra, T.A., 1987b. Palynology of 112.8-197.8 m interval of the core Tenaghi Philippon III, Middle Pleistocene of Macedonia. *Review of Palaeobotany and Palynology* 52, 89-117.

Vergnaud-Grazzini, C., Ryan, W.B.F., Cita, MG., 1977. Stable isotopic fractionation, climate change and episodic stagnation in the eastern Mediterranean during the late Quaternary. *Marine Micropaleontology* 2, 353-370.

Viquesnel, A., 1868. *Voyage dans la Turquie d' Europe. Description physique et geologique de la Thrace, 1855-1868*. 2 vols. Paris.

Watts, W.A., 1988. Europe, in: Huntley, B., Webb III, T. (Eds.), *Handbook of Vegetation Science* 7. Vegetation History. Kluwer, Dordrecht, pp. 155-192.

Yalçınkaya, I., 1981. Le Paléolithique inférieur de Turquie, in: Cauvin, J., Sanlaville, P. (Eds.) *Préhistoire du Levant*. Colloques Internationaux du C.N.R.S., No. 598. C.N.R.S., Paris, pp. 207-218.

Yalçınkaya, I. 1989. *Alt ve Orta Paleolitik Yontmataş Endüstrileri Biçimsel Tipolojisi ve Karain Mağarası*. Atatürk Kültür, Dil ve Tarih Yüksek Kurumu, Türk Tarih Kurumu Yayınları, VI. Dizi, Ankara.

Yalçınkaya, I., Otte, M., Bar-Yosef, 0., Kozlowski, J., Leotard, J.-M., Taşkıran, H., 1992. Karain 1991. Recherches Paléolithiques en Turquie du Sud. Rapport provisoire. *Paléorient* 18, 109-122.

Yalçınkaya, I., Leotard, J.-M., Kartal, M., Otte, M., Bar-Yosef, O., Carmi, I., Gautier, A., Gilot, E., Goldberg, P., Kozlowski, J.K., Lieberman, D., Lopez-Bayon, I., Pawlikowski, M., St. Thiebault-Ancion, V., Patou, M., Emery-Barbier, A., Bonjean, D., 1995. Occupations tardiglaciaires du site d'Öküzini (Sud-Ouest de la Turquie): résultats préliminaires, *l'Anthropologie* 99, 562-583.

Yalçınlar, I., 1971. Sur quelques depots Quaternaires fossilifères en Turquie. *Quaternaria* 15, 211-215.

Yalçınlar, I., 1983. *Terrains a vertébrés Néogènes et Quaternaires en Turquie et leurs caractères géomorphologiques*. İstanbul Üniversitesi Edebiyat Fakültesi Yayınları, no. 2741. (Turkish, with French summary).

Yaltırak, C., 2002. Tectonic evolution of the Marmara Sea and its surroundings. *Marine Geology* 190, 493-529.

Yaltırak, C., Sakınç, M., Akso, A.E., Hiscott, R.N., Galleb, B., Ülgen, U.B. 2002. Late Pleistocene uplift history along the southwestern Marmara Sea determined from raised coastal deposits and global sea-level variations. *Marine Geology* 190, 283-305.

Yanko-Hombach, V., Gilbert, A.S., Panin, N., Dolukhanov, P.M. (Eds.). 2007. *The Black Sea Flood Question: Changes in Coastline, Climate, and Human Settlement*. Springer, New York.

Zubakov, V. A., 1988. Climatostratigraphic scheme of the Black Sea Pleistocene and its correlation with the oxygen-isotope scale and glacial events. *Quaternary Research* 29, 1-24.

CHAPTER 2

THE SEDIMENTARY INFILLING OF YARIMBURGAZ CAVE, TURKEY

William R. Farrand – Jill P. McMahon

DESCRIPTION OF THE CAVE

The Yarımburgaz cave system (as described in Chapter 1) consists of two passages; the longer, lower passage (or LC) penetrates about 500 m into the limestone ridge, but the shorter, higher one (or UC) reaches a dead end only 50 m from the present entrance. A short transverse corridor connects the two longer passages, sloping down from the UC to the LC about 25 m inside the cave entrance. The mouths of the two passages are, respectively, 11.5 and 18.6 m above sea level and face southward (Figure 2.1), overlooking the Sazlıdere valley, the marshy floodplain which is only about 1 m above sea level.

The lower passage of Yarımburgaz is still an active karstic cave, with growing speleothems and pools of standing water in its dark, inner recesses. Evidence of running water exists in the form of rounded quartz and chert pebbles and stratified sand on the existing cave floor. These exotic materials indicate that an insurgence exists (or existed in the past) in the inner reaches of the cave. We excavated similar deposits in Lower Paleolithic levels in the outer part of the LC, as described below. Also, deposits of very fine silt and clay occur in the active portion of the cave, especially where lateral karstic conduits join the main passage. These fine deposits appear to be reworked surface soil or decalcification clays washed in from other parts of the karstic system. Again, they have counterparts in the Lower Paleolithic deposits.

The UC was strongly modified in Byzantine and classical times both inside and outside the cave mouth (as detailed in Chapter 1). These modifications removed prehistoric deposits to an undetermined extent, but some Chalcolithic and Paleolithic deposits remain. The LC was also disturbed by illicit excavators or treasure seekers. Excavation trenches revealed modern pits as deep as 2 or 3 m, and any evidence of prehistoric deposits intermediate in age between the Lower Paleolithic and the modern, mixed debris, i.e., Middle/Upper Paleolithic and Chalcolithic, which are present in the UC, are missing from the LC.

We believe that both the upper and lower passages formerly continued out beyond their present mouths at the times of the Paleolithic occupations, when the Sazlıdere valley had not been eroded as deeply as now. We found no prehistoric remains in the vicinity of the modern drip line, in the area where one expects rather intense human utilization of a cave. On the other hand, erosion of preexisting cave sediments was very clearly in evidence at the mouth of the lower passage (our Trench Z-88).

The deepest trench (V-88 in the LC, adjacent to Özdoğan's A.71 trench) reached 5 m below the present floor of the LC, but the lowest 3 m appear to be barren of artifacts or bones. None of the excavation trenches reached bedrock at any point along the central axis of the lower cave.

Figure 2.1.
Photograph of upper and lower entrances of Yarımburgaz cave viewed from the road below the site.

Figure 2.2.
Cross-section of the upper cave along the north faces of Trenches A.40, A.29, and A.18, with a lateral section of the east face of Trench A.29, as extended in 1990. Except for the sands of Level 11 in A.29 ext., the stratigraphy follows that of Özdoğan. The "Slump" area in A.29 ext. marks a portion of the face that collapsed since 1986. Redrawn from Özdoğan and Koyunlu (1986).

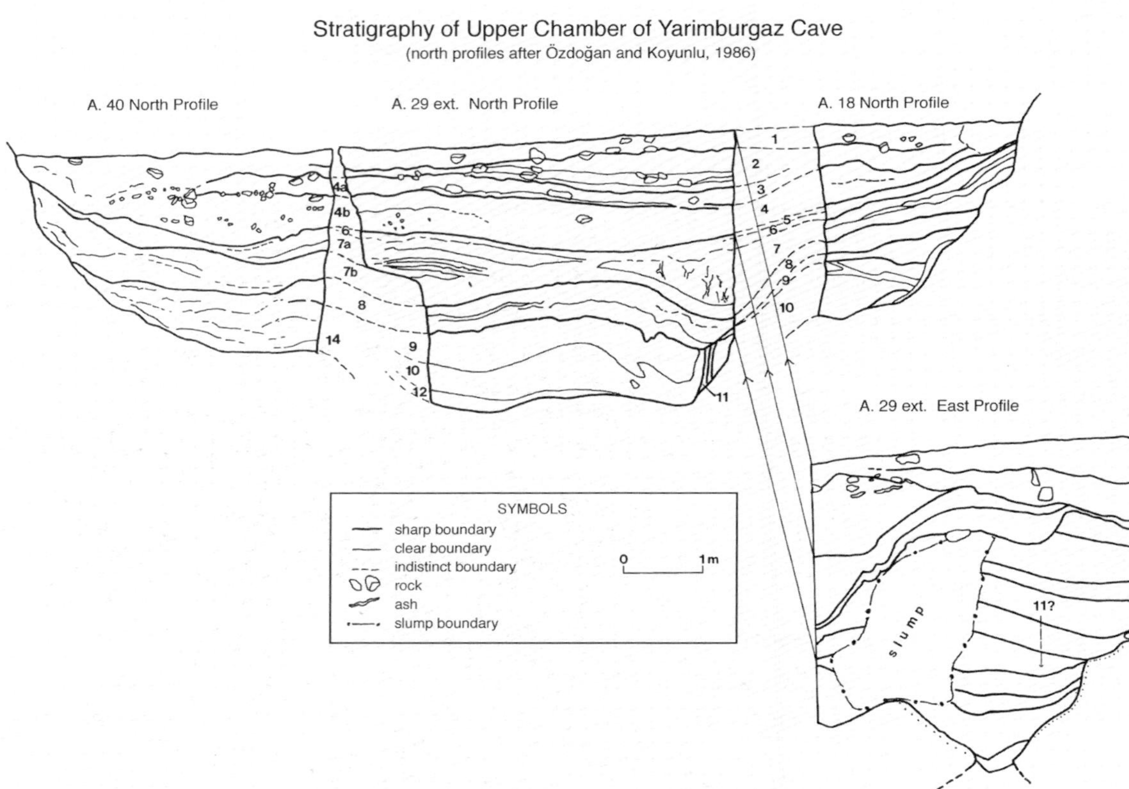

GEOARCHAEOLOGICAL OBSERVATIONS

In 1989 and 1990 the authors examined the sections exposed by the excavators, collected sediment and micromorphology samples, and began a stratigraphic synthesis of the deposits in both the lower and upper caves. The sediment study was by McMahon, and the essential results are reported here. Charcoal for radiocarbon dating was collected from the upper cave. ESR dating of cave bear teeth and uranium-series dating of travertine was conducted by B. Blackwell and H. P. Schwarcz (Blackwell et al. 1990; Blackwell et al. this volume). Geomorphological observations of the terrain surrounding Yarımburgaz Cave were made, and preliminary conclusions are presented in Farrand (1995) and Farrand and McMahon (1997). This paper will deal only with the character and history of the cave infilling.

Upper Chamber

The stratigraphy of the upper cave deposits is illustrated in section drawings supplied by M. Özdoğan, several of which were published by Özdoğan and Koyunlu (1986). One section is redrawn here as Figure 2.2, which represents a transverse section along the north faces of trenches A.18, A.29, and A.40 (Figure 2.2) as well as the east face of trench A.29, which parallels the long axis of the upper cave.

A typical view of sediments of Trench A.18 is shown in Figure 2.3. The section drawing in Figure 2.2 depicts almost all of the stratigraphic units ("Levels") designated by Özdoğan, except the breccia in Level 13, which is not present in those sections. Table 2.1 is a summary of Özdoğan's unpublished descriptions of his levels as we are able to interpret them from translations graciously prepared for us by M. Özbaşaran. We have modified and amplified the descriptions in places where we have made additional observations. Furthermore, it is now clear that the earlier excavators had some misconceptions about certain features. In particular, "calcareous" veins, concretions, or particles are in most all cases actually composed largely of secondary phosphate minerals (see below). All in all, Özdoğan's section drawings are reasonably faithful to the existing sections that we were able to revisit.

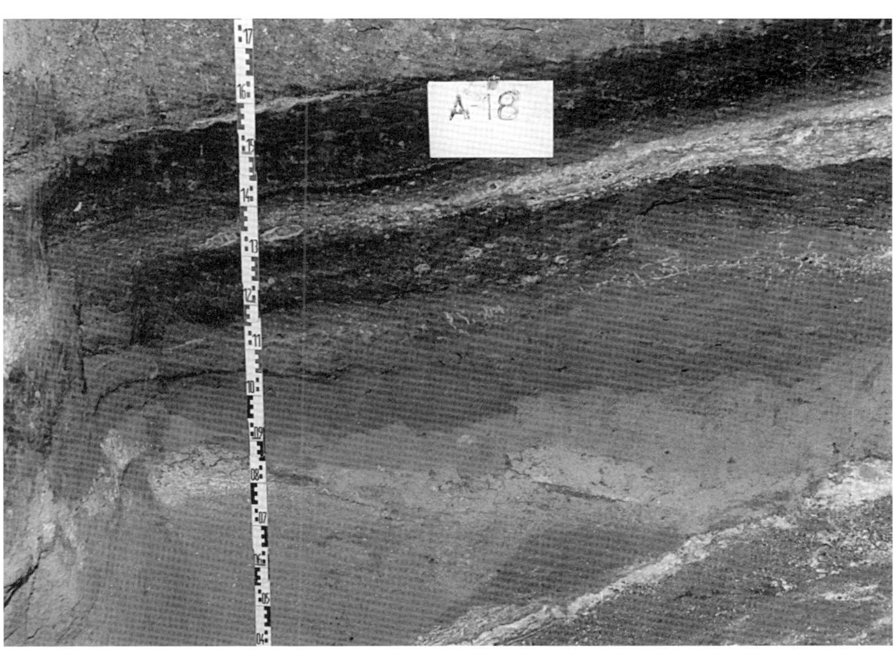

Figure 2.3. Photograph of the north face of Trench A.18 in the Upper Cave. The upper black horizon is Level 5 and 6. The lower black horizon is the base of level 7, and the white streak in the lower right is the bedrock surface. Compare the drawing in Figure 2.2.

Figure 2.4. Close-up photograph of cross-bedded sand in Level 11, Trench A.29, upper cave. The cave entrance is to the right.

There has been much disturbance, both human and natural, of the UC deposits. The most recent disturbance was by Italian film makers who excavated a large pit in the center of the upper cave and filled it with water (Özdoğan and Koyunlu 1986; Özdoğan, pers. comm.). Some 1600 years earlier a Byzantine church occupied the upper cave. The remaining evidence is largely in the form of niches, steps, arches and apses carved into the bedrock walls and ceiling (Figs. 7, 9, 10 in Kansu 1970), as well as stone walls constructed in the transverse corridor leading to the LC. Undoubtedly the cave floor was also modified at that time, but any evidence is obscured by later disturbance. Still much earlier collapse and/or subsidence occurred naturally as seen in the steeply dipping strata overlying and surrounding the prominent deposit of beach sand (Level 11) that was found in Trenches A.18 and A.29 (Figure 2.2), discussed below. The dip angles are much too steep, near vertical in places, to be explained as original depositional slopes. The cause of this disturbance is not clear, but it may be related to underlying karstic cavities, the presence of which can only be surmised.

One striking characteristic of the UC deposits is the presence of a thick unit of well-sorted quartz sand and fine gravel, Level 11 in Özdoğan's Trench A.29, that underlies Upper and Middle (?) Paleolithic levels. (The presence of Middle Paleolithic industries is not certain.) We enlarged Trench A.29 in 1990 revealing a deposit of sand at least 3 m thick, much thicker than previously known. The sand is cross-stratified, with bedding dipping in several different directions, typical of a beach deposit (Figure 2.4). There are alternations in grain size throughout the beach deposit, from well-sorted fine sand to somewhat poorly sorted sand with small gravel (up to 5-6 mm maximum), with variable cementation throughout the deposit. There are also intercalations of poorly sorted, loamy lenses, perhaps mud flows or slumps. Contrary to Özdoğan's interpretation, the sand rests on large angular limestone blocks, not bedrock. The blocks, as much as a meter in maximum dimension, may be part of the rockfall deposit exposed in Trench A.40 (Level 13, "Breccia"). The surfaces of the blocks have thick weathering rinds.

No fossils at all were found in the sand and gravel to confirm its marine origin. However, its altitude, about 12 m above sea level, and the fact that it is overlain and truncated by deposits of relatively young age, ca 10,000 to 24,500 B.P. by radiocarbon (see Table 2.1), makes it a good candidate for the Last Interglacial beach. This would be Tyrrhenian in the broad sense or MIS 5 in the deep-sea chronology. The age of the beach would, therefore, be somewhere between 125,000 and 85,000 B.P., if this correlation is valid.

The strata overlying the beach deposit enclose Chalcolithic occupational debris (ceramics, mollusk shell, charcoal in Levels 2 through 4) and rather non-diagnostic Paleolithic chert flakes and cores (Levels 6, 7, and probably 8), according to Özdoğan and Koyunlu (1986). Mollusks identified in Levels 3 and 4 include marine species *Mytilus galloprovincialis* Lamarck, *Ostrea edulis* (Linne), *Cerastoderma edule* (Linne), and the land snail *Helix pomatia* Linne (Meriç et al. 1988). Levels 6, 7, and 8 to some extent, are characterized by dark brown and black, greasy organic sediments (up to 24 per cent organic matter—see Figure 2.3), perhaps bat guano, in contrast to the distinctly lighter grayish to yellowish brown Chalcolithic strata. Levels 6 through 10 also contrast with the overlying levels in the degree of disturbance. Steep dips and undulating contacts suggest that the cause of the natural disturbance was largely or completely dissipated before the Chalcolithic occupation began sometime before ca. 7330 years B.P.

Stratigraphically below the beach deposit, Levels 12 through 14 are quite different in character from the overlying levels. This is best seen in Trench A.40 where Level 12 is a thin, laminated, yellowish brown crust that covers a heavy breccia (Level 13) of angular limestone blocks, which in turn overlies a dark reddish brown loam (Level 14) prominently marked by white veins and flecks of secondary phosphate mineralization. The breccia appears to be limited in extent to the area of Trenches A.40 and A.39, although it may lie below the bottom of Trench A.29. In any case, its occurrence in Trenches A.39 and A.40 is in close proximity to the opening of the transverse corridor that leads down to the LC. This may be fortuitous but it conceivably is related to the opening or enlarging of that corridor.

Levels 12 through 14 are reminiscent of strata in the LC, described below, namely strata X (very rocky) through T (reddish brown loam). Archaeologically, all these levels contain similar pebble tools, manuports, and bones. Level 13 encloses several hearths and charcoal fragments, according to Özdoğan's notes, but no hearths or charcoal were detected in any level of the lower chamber.

Lower Chamber

A summary section has been prepared (Figure 2.5) that follows more or less the midline of the LC. The trenches are designated by capital letters followed by the year in which in trench was excavated. (The excavators did not use the letters Q, W, and X for the trenches because they do not occur in Turkish. They are used, however, in designating lithostratigraphic units; see next paragraph.) Trench U is not included here because it was very shallow over bedrock and its stratigraphy was not informative. Trench A.71, excavated by Özdoğan, in 1986 is contiguous to our Trench V-88. The level numbers shown in each trench were assigned by the excavators independently of those in the other trenches, excepting those of trenches S-89 and T-89, which were dug simultaneously so that the correlations were immediately apparent during excavation.

We have grouped the excavation levels into lithostratigraphic units designated (from the top down) stratum Z through stratum R, as can be seen in Figure 2.5. A summary of the stratification is given in Table 2.2 which shows the presence or absence of a given stratum in the various trenches, the archaeological levels occurring in that stratum, as well as the abundance of artifacts recovered from each stratum. An enlarged drawing of the east and south faces of Trench A.71/V-88 in Figure 2.6 shows the stratigraphy described in the following paragraphs in greater detail. This trench exposes the most complete stratigraphic section of the LC.

Table 2.1. Descriptions of excavated levels, upper cave, Yarımburgaz , prepared from unpublished descriptions by M. Özdoğan as modified by our field observations. Key radiocarbon dates for Levels 2, 3, 4, 6, 7, and 8 are included (from Farrand 1995).

STRATIGRAPHY

TRENCH A.29 and A.18

Descriptions are based on Özdoğan's unpublished notes with additional information from McMahon, as noted in brackets.

Level	Description
0	Dark gray, mixed layer
1	Yellowish brown, fine-grained, loose, stony layer; 50 - 75 cm thick; Byzantine.
2	Grayish brown, fine-grained, clayey to sandy, with small stones and abundant charcoal and mollusks; some larger stones in middle; 5 - 35 cm thick; basal 5 cm much richer in mollusk shell and appears to lie on a constructed floor; color changes to light orange brown in lower part; this layer is truncated by Level 1; Chalcolithic. $^{14}C = 5930 \pm 110$ (GrN-15534).
3	Light yellowish brown, clayey to silty layer, with a little sand and few small stones; a few small charcoal fragments in general, but common on east side; some larger limestone blocks to the west, with interstitial sandy mud; 25 - 35 cm thick; artifacts rare in lower part. $^{14}C = 6880 \pm 90$ (GrN-15528).
4	Grayish brown, fine-grained, sandy not clayey, with medium-size rock fragments; 65 - 70 cm thick; to the west, a lens of small and medium-size rocks; contact with Level 3 marked by dark-colored laminae; some small charcoal fragments; Level 4 slopes distinctly to the north (into the cave); Chalcolithic. $^{14}C = 7330 \pm 60$ (GrN-15529).
	[McMahon: Sand fraction predominantly rounded to angular quartz and phosphatic concretions, bone fragments, occasional chert, feldspar and sandstone.]
5	Homogeneous mass of variable thickness, 0 - 5 cm; below the top 2 cm the deposit is dark brown, hard and sandy [loam]; in the lower part cemented masses of iron oxide and small "calcareous" crystals, "like iron slag;" Level 5 pinches out to the west; a few potsherds.
	[McMahon: Sand fraction predominantly rounded to angular quartz, phosphatic concretions, and occasional chert, feldspar, and sandstone.]
6	Black with reddish brown spots and lenses, soft with very little sand; some "calcareous" concretions; 3 - 10 cm thick; slopes down and thickens to north; truncated by a pit on the south; undiagnostic flint flakes. $^{14}C = 8110 \pm 320$ B.P. (GrN-15532).
7	Complex stratum, 25 - 70 cm thick; black to reddish brown, soft and clayey loam, with local lenses and laminae variable in color; in NE prominent vertical cracks filled with white "calcareous" residue; Level slopes sharply to the north (into the cave); truncated by Level 1 in SE; some flint cores and "Levallois" flakes. $^{14}C = 10{,}000 \pm 80$ (Beta-42734) and $11{,}840 \pm 100$ (Beta-42733)
	[McMahon: Upper Level 7: pale brown silty loam with small pebbles, rare charcoal; collophane-lined vugs; sand fraction is predominantly rounded to angular quartz and cemented silt or clay-size grains with embedded quartz grains.
	Lower Level 7: dark gray, greasy black loam, 24% organic matter.]
8	Transitional level more like Level 9 than Level 7, but with distinct contacts; 10 - 35 cm thick; harder than Level 7, but minor cementation only in lower lenses/laminae; small pebbles, some sand, and reddish fragments(?); sand increases in lower part; colors vary from light orange gray to black and dark brown; no stones or bones. $^{14}C = 24{,}500 \pm 240$ B.P. (Beta-42732).
9	Pale brown [loam], 12 to 55 cm thick, soft, massive level with sand, rare [quartz and chert] pebbles, and clay clods; also irregular "calcareous particles" which increase to E; truncated on the E.
	[McMahon: Some rounded and weakly to firmly cemented silt-clay lumps; also granule-size pieces of collophane; sand fraction predominantly quartz and cemented aggregates with embedded quartz grains.]

Level	Description
10	Fine-grained deposit with a little sand and gravel; variably (red) brown in color; 30 - 65 cm thick; sand increases upward and deposit is harder; more clayey, darker and softer in lower parts; some [collophane] fragments and wavy lenses; contact with Level 11 not distinct.
	[McMahon: Medium brown to pinkish gray silt loam with collophane crusts and white lumps, probably limestone fragments with phosphatic coatings; sand fraction predominantly quartz and cemented silt-clay aggregates with embedded quartz grains, occasional chert and feldspar.]
11	Sandy deposit, over 3 m thick, largely in SE quarter of Trench A.29, but also seen in Trench A.18 to the E; stratification throughout the Level dips southward (toward cave entrance)
	--top 50 cm, orange sand with some rocks, [gravel fraction almost entirely phosphatic concretions];partly cemented; color changes to light grayish yellow with depth;
	--next 50 cm, uncemented reddish, sandy loam, becoming lighter in color downwards;
	[McMahon: at -2.80 to -2.90 cm below datum, thick unit of very clean, well-sorted fine sand; at least 4 fining-upward cycles; occasional pebbles to 5 mm diam.; sand is subangular to rounded, white, yellow, and pink quartz, chert, feldspar, and rare sandstone grains with rust-colored coatings.]
	--next 20 cm, orange, lightly cemented layer;
	--next 20 cm, rather hard, cemented layer with [phosphatic] lenses;
	[additional observations in 1990 : much of the sand and gravel is cross-laminated; the lowest 75-80 cm of this Level is cemented gray gravelly sand overlying gray clay with white flecks.]
	[McMahon: at -3.30 to -3.35 cm below datum, Red silty clay with flecks of white fine-grained material and MnO_2 streaks and fragments; sand fraction predominantly cemented white and orange silt-clay aggregates, some quartz.]
	--this deposit overlies and fills gaps between large fallen limestone blocks with thick weathering rinds.
	[The lower part of the sequence is not exposed in Trench A.29.]
	TRENCH A.40
12	Light yellow, "calcareous" laminations, 2 - 5 cm total thickness, hard in part, draped over the breccia of Level 13;
13	Breccia; reddish brown to gray, black, and white; 50 - 60 cm thick, cemented mass of larger and smaller limestone rubble, with some sand in places; includes small pebbles, some manuports, and bones; two hearths observed, with pieces of charcoal(?) and burnt bone and stones; deposit concentrated in S half of A.40 and in A.39, and pinches out to W against rock wall and to N in middle of trench; slopes down to E (toward cave center) but must lie below bottom of Trench A.29.
14	Dark reddish brown, (sandy) loam, soft but more compact with depth; 105 - 110 cm thick; includes numerous white and orange flecks (decaying limestone fragments?) and conspicuously veined with vertical and oblique white [phosphatic] streaks, which become more abundant with depth; encloses a 25-cm, angular "boulder" of stratified fine sandstone, rusty-colored above grading to light brown below; a few flint artifacts, especially pebble tools, throughout this level, increasing in number to S.
	[McMahon: Upper Level 14: Reddish brown loam matrix with pinkish white flecks; small quartz pebbles; collophane in some vugs; veins filled with yellowish white fine-grained material; rare pieces of weathered sandstone; sand fraction predominantly quartz and cemented aggregates with embedded quartz grains, occasional chert and feldspar.
	Lowest Level 14: Dark reddish brown silt loam, densely veined; has sticky clayey feel; black organic(?) pieces, and white flecks and white, fine-grained vein fillings; sand fraction predominantlly quartz and cemented silt-clay aggregates, nodular yellow phosphatic concretions, some chert, feldspar and weathered sandstone pieces.
15	Weathered rind on bedrock surface.

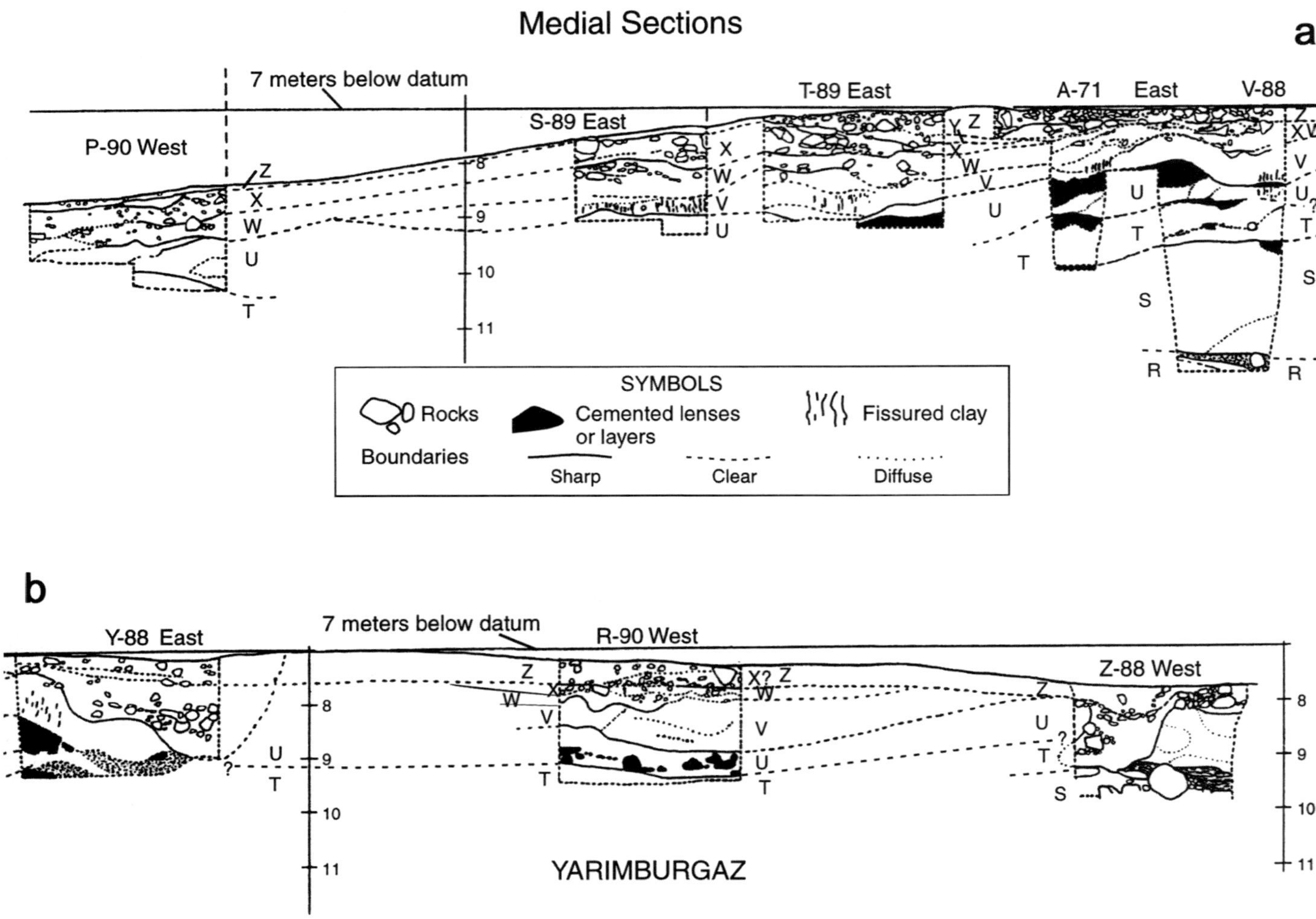

Figure 2.5. Longitudinal section, approximately 43 m long, through the lower cave, Yarımburgaz. Trench Z-88 is under the drip line at the cave entrance. P-90 is the innermost trench excavated. Excavation levels given in arabic numerals. Geologic strata R through Z are labeled in the area between the trenches.

Stratum Z consists of modern disturbance and fill. This heterogeneous stratum includes Byzantine to modern debris in a rather rocky sediment. Color ranges from light grayish brown to dark brown, even black, as a function of the amount of organic matter present. Rock fragments are rather angular and range up to 20 or 30 cm in length. This stratum is about 30 to 40 cm thick in the central and outer part of the excavated area, but it thins to 10 to 15 cm towards the cave interior (in trenches P-90 and S-89). In Trench Y-88, however, a pit at least two meters deep has been refilled with debris (Levels 0a and 0b) in very recent times, as seen by the presence of plastic and barbed wire. Presumably these deposits and the pit itself are the results of illicit excavations, and they are not related to the prehistoric site-formation processes. A major unconformity separates Stratum Z from all the underlying beds.

Stratum Y is a thin-to-discontinuous layer of soft reddish to yellowish brown loam with few rocks. It was distinguished only in the central part of the LC excavated area, namely in Trenches A.71, V-88, and T-89. In the latter it reaches a maximum thickness of ca 25 cm.

Table 2.2. Summary stratigraphic chart showing the general character of each geological stratum, the archaeological levels assigned to each stratum in each LC trench. Bold numbers are archaeological levels. The symbol |||||||| indicates that that stratum was not present in a given trench. Artifact abundance codes: 0= absent; * = rare (<20); ++ = common (200-300); +++++ = very common (> 1000).

<table>
<tr><th>SEDIM. CYCLE</th><th>STRAT.</th><th>DESCRIPTION</th><th>P-90</th><th>S-89</th><th>T-89</th><th>A.71</th><th>V-88</th><th>Y-88</th><th>R-90</th><th>Z-88</th><th>N-90</th><th>ARTIFACT ABUND.</th></tr>
<tr><td>MODERN</td><td>Z</td><td>Modern debris, rocky, angular, some large boulders</td><td>0</td><td>0</td><td>0</td><td>0</td><td>00</td><td>0a pit</td><td>0</td><td>0,1,2,3,5,6</td><td>0,1,2,?</td><td>*</td></tr>
<tr><td></td><td>Y</td><td>Fine-grained, light brown w/few rocks; discontinuous from trench to trench</td><td>||||||||
							</td><td>							
							</td><td>1</td><td>1</td><td>1</td><td>0b(west)</td><td>0(part)</td><td>							
							</td><td>							
							</td><td>*</td></tr>
<tr><td>3d</td><td>X</td><td>Submodern(?); very rocky w some meter-size boulders, usually angular, but no crusts</td><td>1</td><td>1,2,3</td><td>2</td><td>2,3</td><td>2,3</td><td>1a</td><td>1</td><td>							
							</td><td>							
							</td><td>++</td></tr>
<tr><td></td><td>W</td><td>Light to dark grayish brown silt loam, firm, w/large rocks, usually chalky</td><td>2</td><td>4</td><td>3,4</td><td>4,5,6,7</td><td>4,5,6,7</td><td>1b</td><td>2</td><td>							
							</td><td>							
							</td><td>+++++</td></tr>
<tr><td></td><td>V</td><td>Reddish brown clay loam w/ scattered pebbles; fissured</td><td>							
							</td><td>5</td><td>5</td><td>8,9</td><td>8,9</td><td>1c,1d,2a, 3a,3b</td><td>3.4</td><td>							
							</td><td>							
							</td><td>++</td></tr>
<tr><td>2d</td><td>U</td><td>Light grayish to yellowish brown, firm to friable loam w/few rocks, but abundant cemented masses</td><td>3</td><td>6</td><td>6</td><td>10-13</td><td>10-13?</td><td>4 upper</td><td>5</td><td>4 upper</td><td>							
							</td><td>*</td></tr>
<tr><td></td><td>T</td><td>Yellowish brown, firm loam, redder and clayier in places, scattered pebbles common, collophane vugs and masses</td><td>4</td><td>n.d.</td><td>Crust</td><td>14,15</td><td>a,b</td><td>4 lower</td><td>5</td><td>4 lower</td><td>							
							</td><td>*</td></tr>
</table>

1st	S	Rockfall of meter-size boulders; matrix varies from smooth clay to gritty sand with pebbles; laminated crust at top	n.d.	n.d.	n.d.	16 + traver-tine	c,d	4 lower (on east side), 5	n.d.	9 – 13	3 – 6	0
	R	Rounded pebbles and cobbles					d′	n.d.		n.d.	n.d.	0

<u>Stratum X</u> is a very rocky stratum with relatively sparse, pale brown matrix, and a moderate number of bones and artifacts, especially in Trench R-90. Near the cave walls the rocks in this layer tend to be rather angular, but toward the center of the cave they are more rounded (corroded rather than rolled). These deposits reflect a period of rather intense breakdown of the cave roof, but whether the cause was climatic or due to earthquakes or something else is yet to be determined. As mentioned above, this rocky deposit resembles Level 13 in the upper cave, but the contemporaneity of the two deposits has not been demonstrated.

<u>Stratum W</u> is light to dark grayish brown, redder in some areas, firm loam to clay loam, containing scattered rocks, some with heavy weathering rinds. Stratum W is 40 to 75 cm thick, but it is thin-to-discontinuous in the front part of the cave (Trenches Y-88 to Z-88). About two-thirds of all artifacts recovered from the LC occur in this stratum. An apparent unconformity separates Stratum W from Stratum V, as inferred from the change in sediment type and the presence of fissuring in Stratum V.

<u>Stratum V</u> is dark reddish brown clay to clay loam with only a few small rocks. Patches (or pebbles?) of yellow ocherous sand, presumably ripped up from underlying sediments, and isolated, well-rounded chert pebbles, 1-2 cm diameter, occur throughout. This stratum is up to one meter thick, especially in Trenches Y-88 and R-90, but in trench P-90 it is either missing or it merges with Stratum W. It seems to be eroded from the cave entrance (Trench Z-88). In all trenches this clay loam is conspicuously fissured, forming vertically oriented prismatic blocks resembling a pedogenic structure, but probably to be attributed to desiccation during a period of non-deposition that followed its accumulation. Artifact abundance drops off conspicuously except in Trench Y-88.

<u>Stratum U</u> is light grayish brown, reddish yellow, or yellowish brown hard loam with numerous cemented lenses and nodules, few large rocks, but rather frequent isolated chert pebbles. Stratum U is up to 75 cm thick. Strong cementation in the top of this stratum appears to be by collophane, presumably derived from the decay of bone apatite. Vugs of pinkish orange, mammillary collophane occur in many places. Only a small number of artifacts can be attributed to this stratum. The cemented, undulating upper surface is the basis for postulating a probable unconformity at the top of Stratum U.

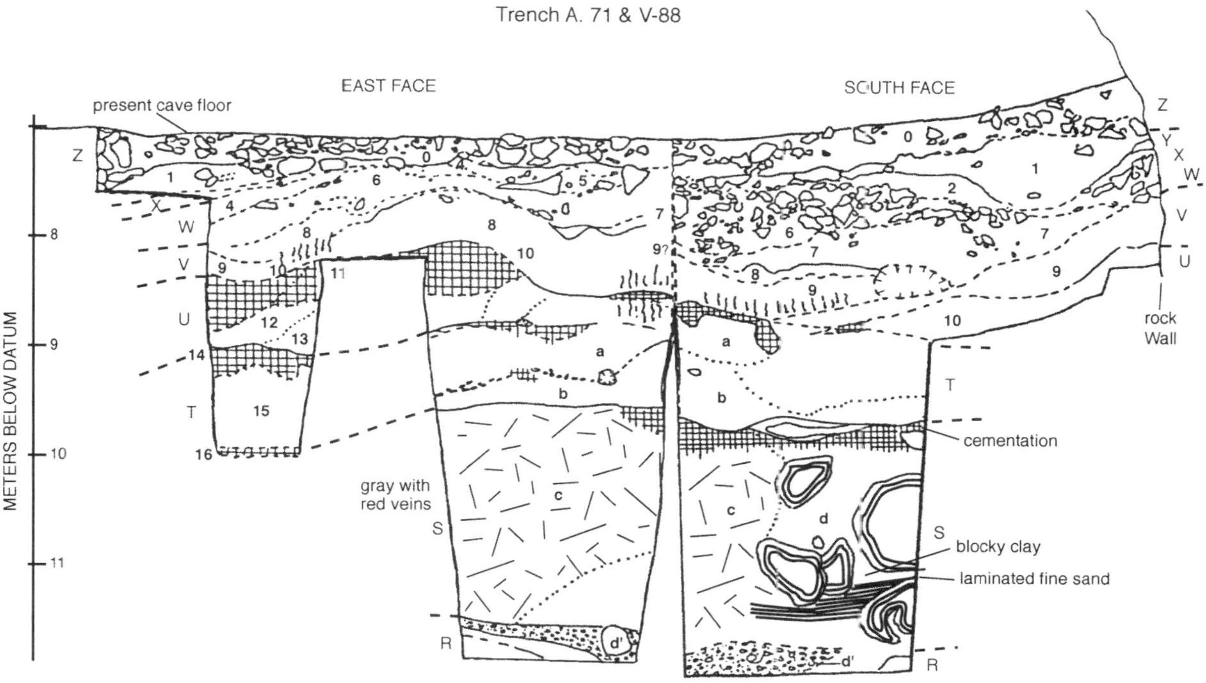

Figure 2.6. Detailed section drawing of east and south faces of Trench A.71/V-88. See text for explanation.

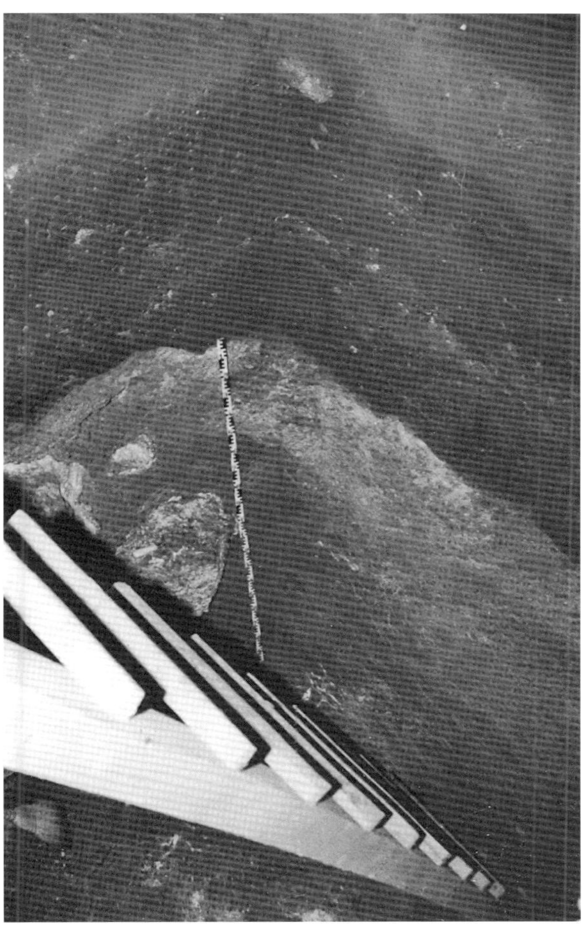

Stratum T is yellowish brown to dark reddish brown clay loam with some cemented zones, collophane vugs, and lenses of sand and gravel. The reddish brown clay loam in the lower part is relatively soft with many scattered, well-rounded small pebbles and collophane concretions (Figure 2.7). The upper surface is marked by a cemented zone, similar to the top of Stratum U, and likewise seems to represent an unconformity. This layer is about 70 cm thick in Trench V-88, but was not reached in Trench S-89. In Trench T-89 the excavation stopped on a hard crust, probably the top of Stratum T. It is questionably present at the bottom of Trenches P-90 and R-90. Only three artifacts came from this layer.

Figure 2.7.
Photo view looking down into Trench V-88. The top of the stadia rod indicates the position of the unconformity between Stratum T (dark brown) and Stratum S (pale brown), the latter being strongly cemented with phosphate at the top. A large limestone block within Stratum S can be seen to the left of the stadia rod

Figure 2.8.
Photograph of the north face of the lowermost sediments in Trench V-88. To the left is very smooth clay which is interrupted by a pebbly lens (Level d') coming from the right. Prominent white veins of phosphate (collophane) penetrate the sediments on the right. Compare Figure 2.7.

Stratum S is a thick, complex deposit of large, heavily weathered limestone blocks (up to 1 m diameter) scattered throughout a reddish brown, smooth clay loam that interfingers with a gritty sand facies (Figure 2.8). It is known completely only in Trench V-88, where it is just over 2 m thick. It also seems to occur at the bottom of Trench Y-88 and in the lower half of Trench Z-88 at the cave entrance. The limestone blocks commonly measure 50 to 100 cm across and are matrix-supported, i.e., they usually are not touching each other. Moreover, they are heavily corroded and have developed laminated weathering rinds, as much as 9 to 10 cm thick, apparently resulting from reaction of phosphate-bearing solutions with the limestone. In addition, the upper contact of this deposit is more or less continuously cemented (with collophane?), in places resembling travertine (which it is not), once again suggesting an unconformity at the top of Stratum S.

In Trench V-88 the sediments on the medial (east) and lateral (west) sides of the trench are quite dissimilar. The west wall of the trench, like that of Trench Y-88, reveals large blocks of limestone scattered throughout a very smooth reddish brown to reddish yellow clay (Level d), but the east wall (Figures 2.6 and 2.7), which is situated more or less along the mid-line of the cave, has no blocks and exposes a very firm, very pale gray, gritty sediment (Level c) crisscrossed with prominent bright red veins (or fissures with a bright red filling). The gray sediment appears to be a gleyed deposit, that is, deposited under reducing (subaqueous) conditions, and its grittiness is to be attributed primarily to incipient cementation rather than a sandy component (see below). Small chert pebbles are common in the top 15 to 20 cm of this layer. The bright red coloration appears to be a hematite-stained material that has filtered in along a system of pervasive cracks. The north and south walls of Trench V-88 show clearly the abrupt interfingering of this gray-and-red fissured facies and the smooth reddish brown clay of the west wall. Where the two facies meet, the cracks in the matrix are filled with a chalky white phosphatic material. There are also lenses of pebbles (Level d′) that project from the gray-and-red sediment laterally into the brown clayey facies (Figure 2.8).

The origin of this deposit and of the unusual gray-and-red alteration is problematic. The smooth brown clay shows some faint laminations in places, suggesting deposition in standing water; in other places, especially deep in Trench V-88 and Levels 9 through 11 in Trench Z-88, it gives way to a laminated, yellowish brown sandy clay loam, the laminations being highlighted by black staining, probably manganese oxide. The pebbly facies suggest that running water persisted along the midline of the cave while the fine clay accumulated nearer the cave wall.

Stratum R consists of stratified sand, small gravel, and well-rounded chert and quartz pebbles constituting Level d' at the bottom of Trench V-88 (Figure 2.8). This sediment continues below the present floor of V-88 where a small sondage exposed a bed of chert pebbles, commonly 3 to 4 cm diameter. In this very limited exposure these stratified deposits seem to dip toward the cave entrance.

SEDIMENT ANALYSIS

Results of grain-size analysis are shown in Figures 2.9, 2.10, and 2.11, for the upper cave, the beach deposit, and the lower cave, respectively. Table 2.3 lists the percentages of organic matter and of acid-soluble residue for selected samples from the upper and lower caves.

The granulometric histograms of the raw (non-acid-treated) sediments show that the cave sediments have polymodal distributions (as commonly found by Farrand 2001), except the beach sands, which are mostly unimodal. Modes in gravel/granule, sand, silt, and clay sizes are commonly present, but the relative weighting of the different modes varies from level to level. Clay-sized modes (smaller than 9 phi, or <.002 mm) are generally prominent, and the silt mode tends to be larger than the sand mode, especially in the upper cave. For the most part, however, there are not striking differences in granulometry between the sediments from the LC and the UC. The <2 mm fractions are largely loams or clay loams.

Since much of the sediment is slightly to strongly cemented, it is of interest to examine the original, non-cemented sediment. After removal of organic matter with H_2O_2, the samples were reacted with 10 per cent HCl with mild heating. This reaction removed any $CaCO_3$ as well as phosphate cements. As seen in Table 2.3, the total acid-soluble fraction varied considerably, from a few percent in the sands to well over 50 percent in many levels. Figure 2.12 compares histograms of several samples before and after acid treatment. Two patterns are seen. Samples YBZ-62 and 63 from Strata U and V in the mid-depths show an increase in the sand fraction, centered on 2 and 3 phi (0.25 to 0.125 mm). Since it is not possible that the sand actually increased, the decrease in clay implies that much of the original "clay" was in fact finely divided cement, independently known to be mainly phosphate, that was dissolved in the acid. Another pattern is shown by sample YBZ-73, top right in Figure 2.11 (the gritty gray-and-red deposits of Level c of Stratum S), from lower depths in Trench V-88. Here, after acid treatment, the clay percentage increases conspicuously, and the sand fraction almost disappears, while the silt fraction remains largely unchanged. In this case it appears that the sand fraction was composed mostly of sand-size aggregates of silt and clay-sized mineral grains cemented by phosphate. Upon acid treatment the cement was removed and the embedded mineral grains are added to the silt and clay bars of the histogram.

It is of interest to note (also in Figure 2.12) that sample YRM-14, the very smooth clay in the lower part of Trench V-88, has a grain-size distribution very similar to the acid-treated YBZ-73. Although its acid-treated histogram is not shown, YRM-14 had much less secondary phosphate than YBZ-73, 18 vs 48 percent (Table 2.3). One can infer, therefore, that the gray-and-red gritty sediments of Level c in this trench were originally similar to the smooth clay of Level d before diagenetic cementation occurred.

Also in Figure 2.12 is a histogram of decalcified limestone bedrock from the cave, the insoluble residue of which is >90 per cent clay-sized material with just a trace of silt. If this limestone sample is typical, one can conclude that the smooth clay of YRM-14 cannot be simply the residue of limestone weathering, and that its silt component must be derived from elsewhere.

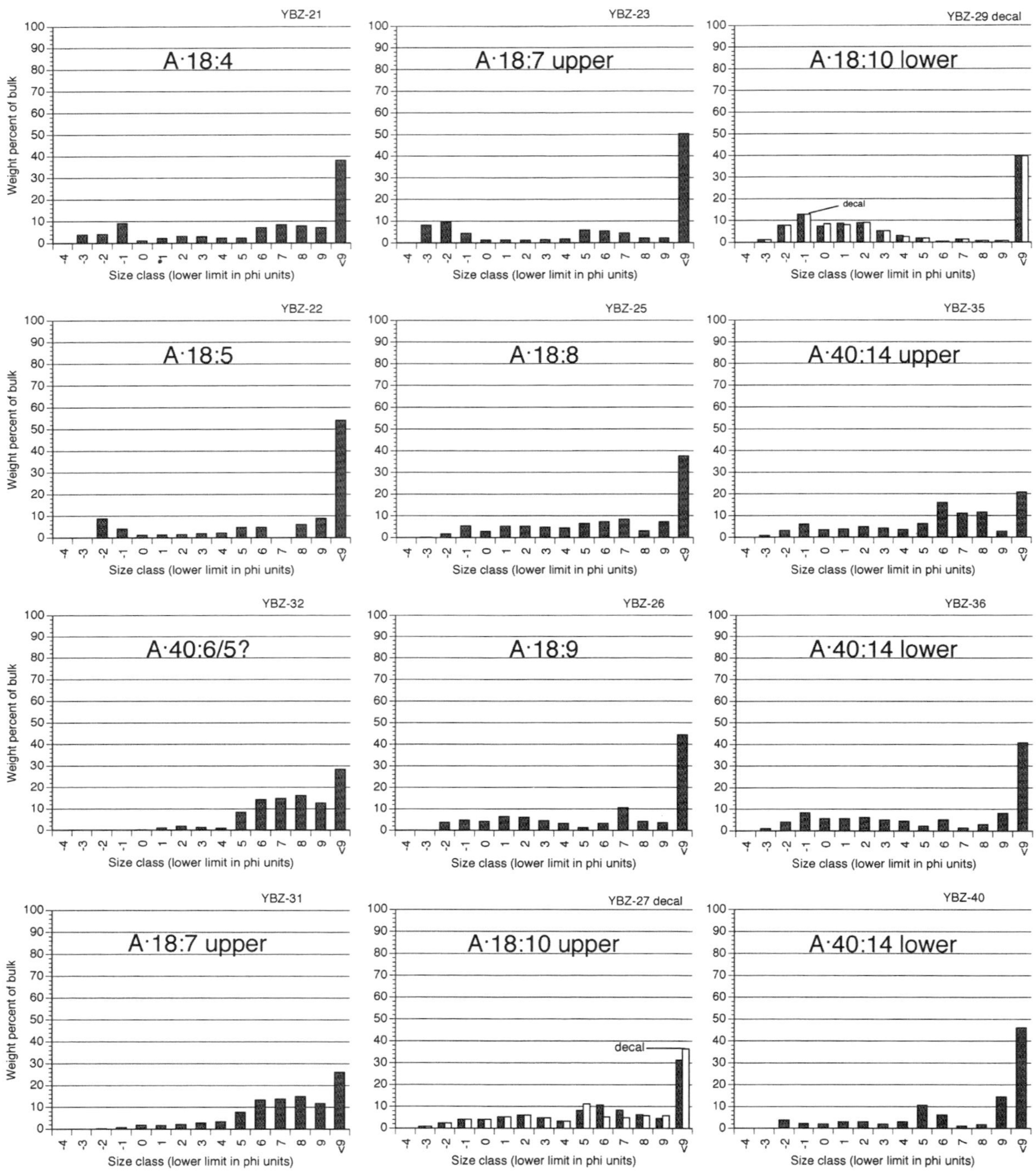

Figure 2.9. Granulometric histograms of sediment samples, upper cave, Yarımburgaz , plotted at one phi intervals from -5 to -4 phi (32-16 mm) at the far left and >9 phi (<.002 mm) at the far right. The stratigraphic sequence of samples moves from uppermost at the top of the left column down to the lowest at the bottom of the right column. Open histograms are acid-treated samples.

As mentioned at the outset, exotic sediments occur widely in the cave, both in the dark interior and in the outer, habitable chambers. In addition to the obvious deposits of quartz-rich sand and quartz and chert pebbles at the very bottom of Trench V-88 (Level d'), finely laminated fine sand occurs at intermediate levels in Stratum S (Tr. V-88) and Stratum S (Level 9) in Trench Z-88, just to mention a few occurrences.

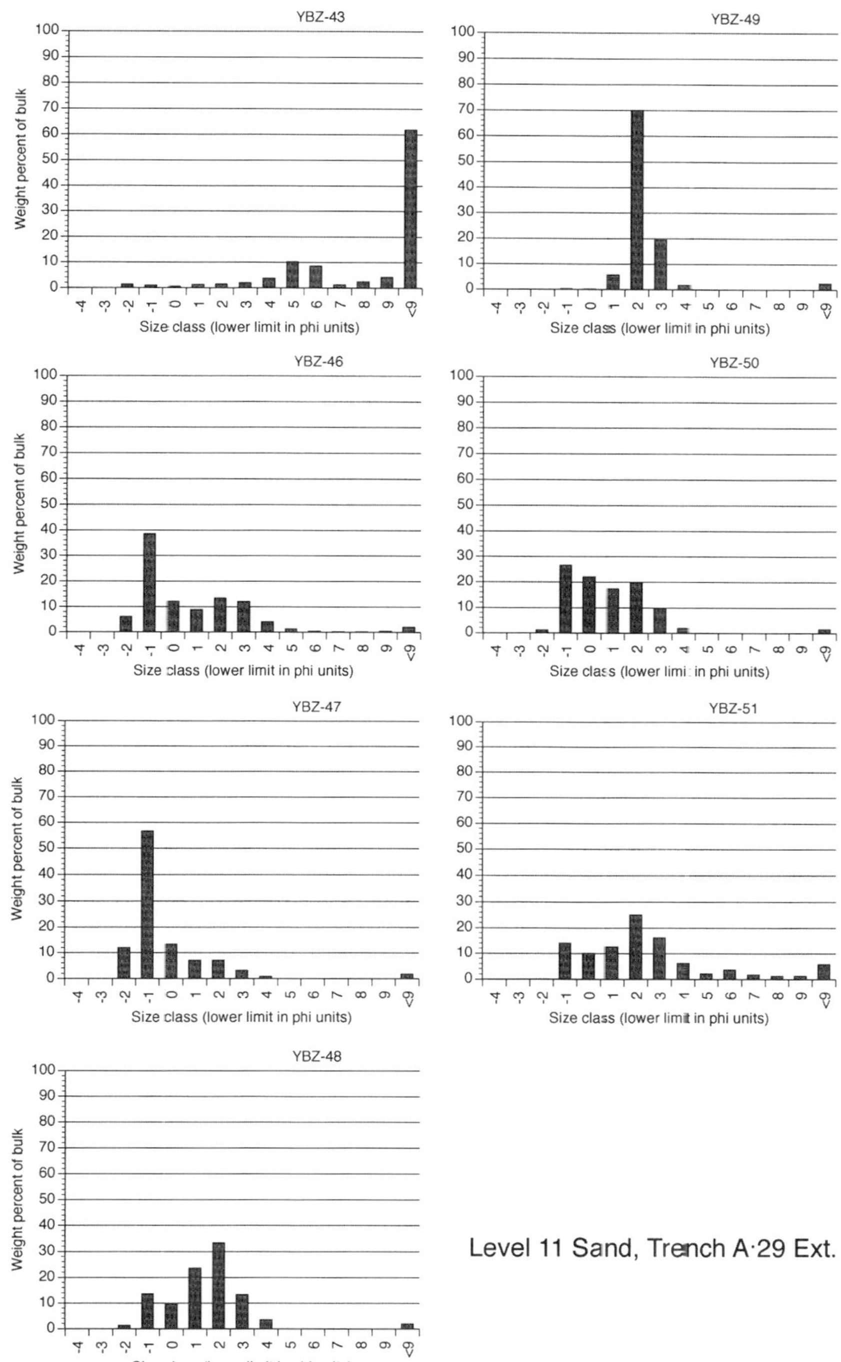

Figure 2.10. Granulometric histograms of Level 11, upper cave, Yarımburgaz. Conventions as in Figure 9. Except for sample YBZ-43, all these histograms show unimodal distributions of moderately to well-sorted sand.

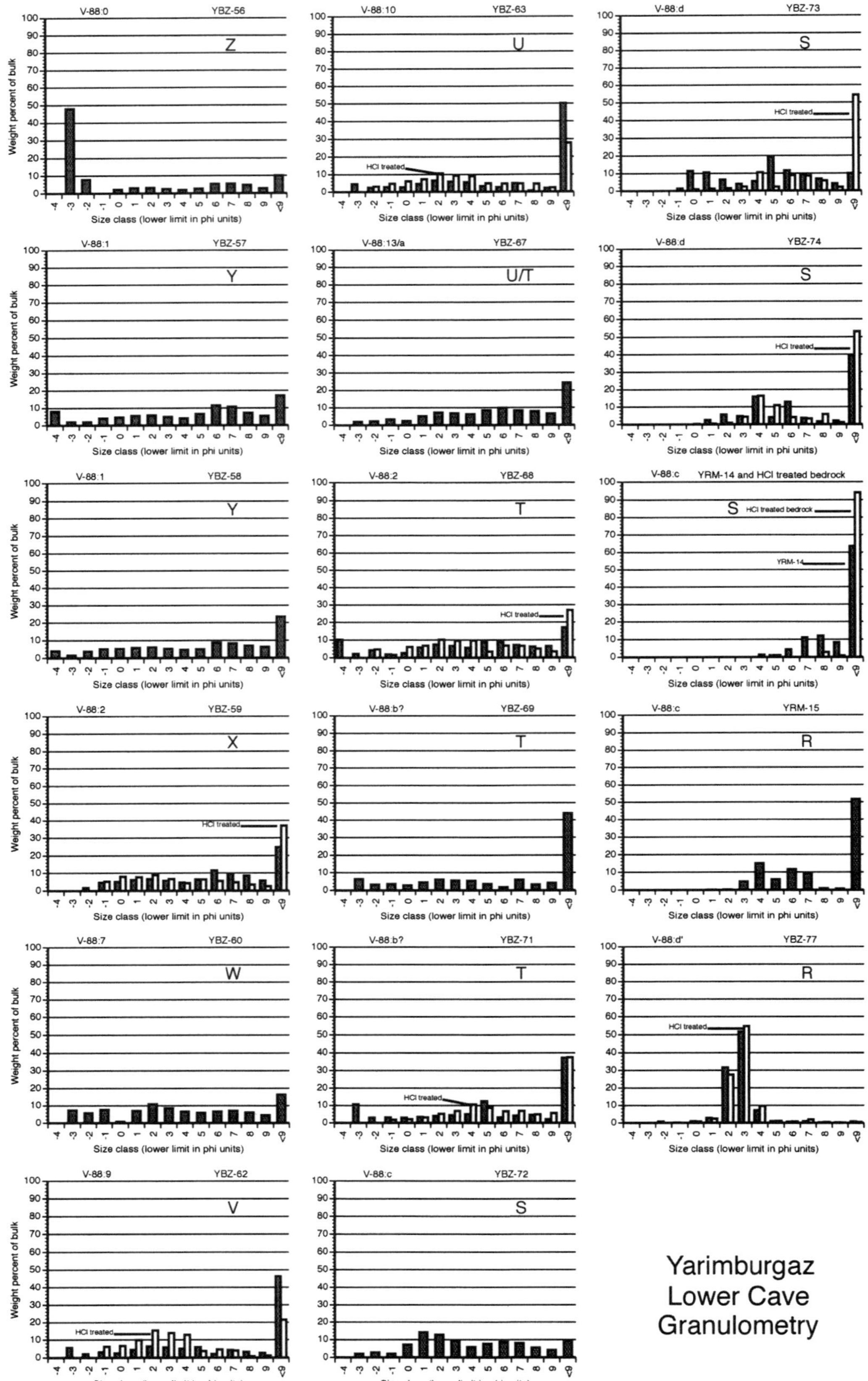

Figure 2.11. Granulometric histograms of sediments in the lower cave, Yarımburgaz. Conventions as in Figure 9. Both the excavation levels and lithostratigraphic units are indicated.

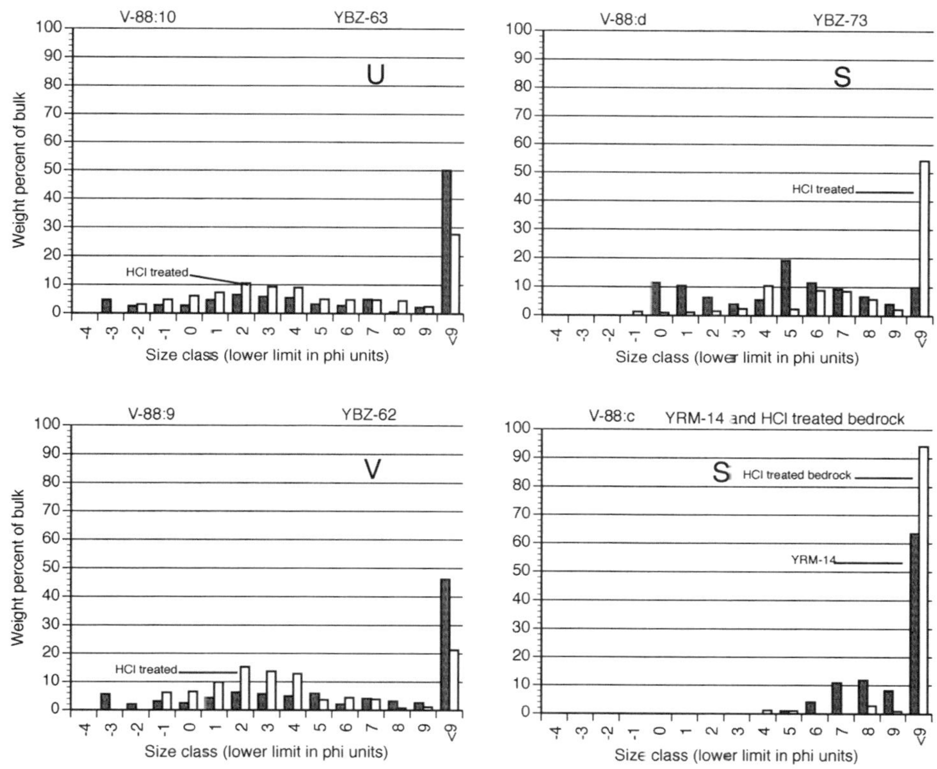

Figure 2.12. Comparison of histograms of raw (solid bars) and acid-treated (open bars) sediments and bedrock from Trench V-88, lower cave, Yarımburgaz. Note that YRM-14 has not been acid-treated.

We have observed grains of quartz, feldspar, chert, and sandstone in the sand fractions of most of the loamy deposits in the upper cave as well. Heavy minerals identified include goethite, rutile and anatase, both in sandy horizons in the LC and in the beach sands in the UC. None of these sand-size minerals comes from the parent bedrock. We presume that they were washed out from the cave interior, having entered there through an as-yet-unidentified insurgence. The beach sand deposit in the upper cave, in contrast, appears clearly to have been washed into the cave through the entrance, but the mineralogy of the sands suggests that they all were derived from more or less similar sources outside the cave.

The phosphate component in the cemented facies, vugs, and disseminated through the clayey sediments was identified by X-ray diffraction analysis (XRD). This material consists commonly of hydroxyapatite and carbonate-hydroxyapatite. Crandallite and its hydrated form, montgomeryite, occur as disseminated cement in the more clay-rich strata such as Level 14 in the upper cave and Stratum S in the LC, although they are not limited to those strata. Leucophosphite also occurs and appears to be associated with dark, organic-rich layers. The primary component of the white flecks in the beach sand of Level 11 of the upper cave and in the white veins of Level 12a of the LC, initially thought to be calcite or aragonite from marine shells, is taranakite, $(K,NH_4)AL_3(PO_4)_3(OH).9H_20$. All of these phosphate minerals are thought to have originated from phosphate leachate from guano and bone decay reacting with cave clay.

Table 2.3. Organic Matter and Acid-Soluble Content (McMahon) (determined by weight loss in H_2O_2 and warm HCl, respectively)

SAMPLE	TRENCH	LEVEL	ORG. MAT. %	ACID-SOL. %
UPPER CAVE				
YBZ-21	A.18	4	8.0	38.9
YBZ-22	A.18	5?	11.3	37.5
YBZ-32	A.40	6?	14.1	16.4
YBZ-23	A.18	7 upper	13.1	40.7
YBZ-31	A.18	7 upper	2.1	28.3
YBZ-24	A.18	7 lower	23.9	34.4
YBZ-34	A.40	7 lower	18.6	17.7
YBZ-25	A.18	8	5.0	43.4
YBZ-26	A.18	9	3.0	21.0
YBZ-27	A.18	10 upper	2.0	21.3
YBZ-29	A.18	10 lower	1.5	16.5
YBZ-43	A.29	11(10?)	1.3	24.6
YBZ-46	A.29	11(-2.50)	<1	2.9
YBZ-47	A.29	11(-2.63)	<1	4.0
YBZ-48	A.29	11(-2.75)	<1	3.8
YBZ-49	A.29	11(-2.90)	<1	2.9
YBZ-50	A.29	11(-3.00)	<1	3.5
YBZ-51	A.29	11(-3.15)	<1	3.6
YBZ-52	A.29	11(-3.35)	3.0	50.9
YBZ-54	A.29	11(-4.05)	1.9	57.8
YBZ-35	A.40	14 upper	4.1	43.0
YBZ-36	A.40	14 lower	4.0	38.6
YBZ-37	A.40	14 lower	6.6	57.2
YBZ-40	A.40	14 lower	6.8	54.1
LOWER CAVE				
	TR.V-88 STRATA			
YBZ-56	Z	0	3.0	52.1
YBZ-57	Y	1	2.7	42.3
YBZ-58	Y	1	2.4	36.9
YBZ-59	X	2	2.3	28.8
YBZ-60	W	7	0.7	32.5
YBZ-62	V	9	2.8	65.8
YBZ-63	U	10	2.7	59.8
YBZ-67	U/T	13/a?	1.3	26.6
YBZ-68	T	a	1.9	51.2
YBZ-69	T	b?	2.2	36.9
YBZ-71	T	b?	1.6	25.9
YBZ-72	S	c	2.0	37.0
YRM-14	S	c	13.2??	18.1
YRM-15	S	c	6.8	17.8
YBZ-73	S	d	1.9	47.8
YBZ-74	S	d	1.6	33.8
YBZ-77	R	d'	0.6	6.3

CHRONOLOGY OF THE LOWER CHAMBER

The age of the artifact assemblages in the LC is still largely speculative. Electron-spin resonance (ESR) results on ten *Ursus deningeri* teeth suggest an age between 200-220 ka (linear uptake model) and 270-390 ka (recent uptake model) (see Blackwell et al. this volume). A uranium-series date on a collophane crust below the major *Ursus*-bearing horizon (i.e., Stratum W) of 160 ka is consistent with the ESR estimate since the crust must be younger than the original deposition (Blackwell et al. 1990; Blackwell et al., this volume). On the other hand, Howell (personal communication) saw a "substantial, even close resemblance" between the Yarımburgaz faunal assemblage and other sites in central and western Europe the ages of which are much older than the Yarımburgaz radiometric dates, ranging from 350,000 to as much as 780,000 years ago. Santel and von Koenigswald (1998, this volume) find microfaunal taxa in the LC sediments characteristic of those in the Russian Oka Glaciation and the following Likhvin Interglaciation, generally correlated with the Elster Glaciation and Holsteinian Interglaciation of western Europe, in turn presumably equivalent to MIS 8 and 7, respectively.

The cross-stratified beach deposits in the upper cave constrain the upper age of the Lower Paleolithic deposits in the LC as well as Levels 12 through 14 in the UC, if our assessment of the Last Interglacial age of the beach (given above) is correct. In that case, the LC deposits should be at least older than the Last Interglacial. There is a problem, however, in understanding the nature of deposition in the lower cave during the Last Interglaciation because the lower passage is some 7 m lower than the beach sands in the upper cave. Why, then, do these sands not occur also in the lower cave? Perhaps the mouth of the LC was blocked by deposits that have since been eroded away. In that case, the transverse passage between the upper and lower caves must have been blocked as well, or the sands would have washed down. Another possible scenario calls on active karstic streams that eroded away all traces of the beach sand and associated sediments in the LC, as well as any Middle or Upper Paleolithic artifacts that might have been present. The UC does not have an interior karstic passage so these sediments may have been preserved there.

SEDIMENTARY HISTORY OF THE YARIMBURGAZ CAVES

The sedimentary infilling of Yarımburgaz can be subdivided into six sequences. A "sequence" is used here in the sense of sequence stratigraphy (Salvador 1994: 50-51), being a group of sedimentary beds bounded by unconformities—erosional or non-depositional surfaces—indicative of a temporal hiatus in the depositional history. The sequence stratigraphy is summarized in Table 2.4.

The earliest sedimentary sequence, Sequence 1, comprising Strata R and S of the lowermost 2 to 2.5 meters in Trench V-88, begins with sorted and stratified quartz sand and pebbles. Above this level are very fine smooth clays that interfinger with gritty, cemented sediments, indicating quiet water deposits, into which large limestone blocks fell from the cave ceiling from time to time. Sediments of this sequence were also reached in the northwest corner of Trench Y-88 (Level 5) and apparently in Trench Z-88 (Level 12) at the present LC entrance, but not in the other trenches. The top of the first sequence is strongly cemented by phosphates and irregular in configuration, suggestive of an erosional unconformity and weathering.

Sequence 2 (Strata T, U, and V) consists of dark reddish brown clay loams with widely scattered quartz and quartzite pebbles. A light yellow or orange loamy horizon (Stratum U) occurs

Table 2.4. Sequence stratigraphy of Yarımburgaz Cave deposits

SEQUENCE	STRATA	DESCRIPTION
Post-6	1 in UC	Byzantine overburden and modern disturbance
	Z in LC	
6	2, 3, 4, 5, 6 in UC	Grayish brown silty clay loam with small stones; Chalcolithic
5	7, 8, 9, 10 in UC	Angular unconformity, brown, organic-rich loam with some Upper Paleolithic artifacts
4	11 in UC	Angular unconformity, beach sand
3	12, 13 in UC	Rocky with little matrix, abundant bones and artifacts
	Y, X, W in LC	
2	V, U, T in LC	Fissured dark reddish brown loam (V), orange loam (U) with collo-phane, reddish brown clay loam with scattered pebbles (T)
1	S, R in LC	Strong phosphate concretion (top), inter-fingering smooth clay and gritty sediment, overlying waterlaid sand with pebbles.

consistently in the middle of this sequence. Numerous vugs are lined with secondary phosphate (collophane) and scattered masses of sediments are firmly cemented with collophane. A firm, dark reddish brown loam with conspicuous vertical fissures (Stratum V) marks the top of this sequence. The fissures suggest desiccation of sediments on the cave floor during an extended period of time when little or no deposition was taking place, and they imply that the cave was open to the external atmosphere at the time. Level 14 in the upper cave, the dark reddish brown loam with abundant secondary phosphate, seems to fit into this sequence, as does Level 4 in Trench Z-88 at the entrance of the LC. Abundant animal bones and some artifacts imply the earliest human and animal utilization of the lower chamber took place during the final phase of the second sequence (layer V).

The Sequence 3 sediments (Strata W, X and Y) are quite different, being very rocky with relatively little matrix between the rock fragments. Levels 12 and 13 in the upper cave may be part of this sequence. Bones and artifacts are very abundant in these sediments. Whereas the sediments of the first and second sequences are logically products of slow deposition by running water or mudflows issuing from the cave interior, these third-sequence sediments may be the result of earthquakes that detached rocks from the ceiling of the cave during a relatively short time span.

Sequence 4 is exceptional, namely in recording the influx of beach sand into the upper cave mouth. As outlined above, it is not clear why the beach deposit is not found in the LC. In any case, if the beach was deposited during the Last Interglaciation (MIS stage 5) and the Strata W, X, and Y in the LC and Level 13 in the upper cave (third sequence) accumulated during MIS 7 or earlier (based on ESR dates), there must be a sizeable temporal hiatus of nearly100 kyr, spanning MIS 6, between sequences 3 and 4.

Sequence 5 encompasses Levels 7 through 10 in the upper cave, with a scattering of Paleolithic artifacts. These brown loams with scattered pebbles and organic-rich lenses overlie the beach deposits unconformably and have slumped into one or more cavities in the central part of the upper cave.

Finally Sequence 6 (Levels 2 though 6 in the Upper Cave) is represented by pale grayish brown silty to clayey loams with some small stones and Chalcolithic potsherds, dated between 6 ky and 8 ky BP. These beds are essentially horizontal, and they truncate Levels 7 through10 in an angular unconformity.

A covering of sediments and detritus, Byzantine and younger, of variable thickness, caps Sequence 6 in the UC and lies over the third sequence in the LC. It is usually thin (10 to 20 cm), but deep pits with 20th-century debris (plastics, etc.) occur in places in both the upper and lower caves. The absence of Paleolithic and Neolithic artifacts from the LC implies a very long time gap between the third sequence accumulation and modern debris that caps it. Perhaps post-Lower Paleolithic sediments were removed by erosion, but there is no clear evidence for erosion at the top of Sequence 3 sediments, except the erosion in Trench Z-88. In Trench Z-88, just inside the present drip line of the LC, the uppermost sediments are Byzantine and modern. They overlie a very irregular surface that appears more likely to be an erosional unconformity than a modern pit (see Trench Z-88 West in Figure 2.5). The absence of any artifacts in Trench Z-88 at the cave mouth, which would certainly have been an important activity area or at least the site of heavy traffic, seems to confirm the idea that erosion has removed sediments equivalent to Sequence 3 and the upper part of Sequence 2 of the cave interior, as well as any younger sediments that may have been present there.

SUMMARY

Geologic study demonstrate that the Yarımburgaz caves had a complex sedimentary history, beginning well before there was any evidence of human presence in the caves. The lithostratigraphic units are divided into six unconformity-bounded sequences (Table 2.4). Bedrock was not reached in any excavation trench in either the upper or the lower cave, except in Trench U-89 where the deposits are very thin and close to the cave wall in Trenches N-90, R-90, and T-89.

The lowest deposits encountered, pebbles and sand, were deposited by high velocity water, which then gave way to quiet-water deposition of fine clays, presumably at a time when the cave was not open to the exterior, or when the cave entrance was much farther out than it is now. The middle levels in the LC deposits are dominated by reddish brown clayey loams with scattered grit and isolated small chert pebbles (1-2 cm), along with an increasing number of animal bones, largely cave bear, and stone implements imported by humans. The final evidence of Lower Paleolithic occupation of the LC, such as has escaped erosion, occurs in very rocky deposits, suggesting that the cave was rather wide open, perhaps similar to today. However, the exact position or configuration of the cave mouth at that time is not clear. There was considerable fragmentation and breakdown of the cave roof, perhaps induced by earthquakes, dropping numerous slabby limestone fragments to the cave floor.

The next identifiable event was the influx, into the upper cave, of a large quantity of beach sand, which is tentatively correlated with a high sea level of the Last Interglaciation, or MIS 5. However, it is not clear why this sand did not also flood into the lower chamber of the cave. Either its entrances were blocked, or the sand has been completely removed from the lower cave by subsequent erosion. During Late Pleistocene and Holocene time the UC was frequented by Middle/Upper Paleolithic, Chalcolithic, and finally Byzantine peoples. Presumably the lower cave was also occupied, but any evidence was removed or obscured by very recent human disturbances and/or by erosion within the active karstic system.

ACKNOWLEDGEMENTS

The authors gratefully acknowledge the encouragement and support of the late F. Clark Howell in all stages of this work. Financial support was largely from National Geographic Society grants to Howell. We are pleased also to recognize the excellent field work by the students from Istanbul University, under the direction of Professor Güven Arsebük who were very professional in carrying out the daily routine of excavation and very helpful in conveying to us the stratigraphic subtleties that they were uncovering.

REFERENCES CITED

Blackwell, B., Schwarcz, H.P., Porat, N., Howell, F.C., and Arsebük, G., 1990. Electron spin resonance (ESR) dating of *Ursus* teeth from Yarımburgaz, Turkey. *Geological Society of America*, Abstracts with Programs 22, A120-A121.

Farrand, W.R., 2001. Archaeological sediments in rockshelters and caves, in: Stein, J.K., Farrand, W.R., (Eds.), *Sediments in Archaeological Context*. University of Utah Press, Salt Lake City, pp. 29-66.

Farrand, W.R., 1995. Geoarchaeology of Yarımburgaz cave, Turkey, in: Bermúdez De Castro, J.M., Arsuaga, J.L., Carbonell, E. (Eds.), *Evolución Humana en Europa y los Yacimientos de la Sierra de Atapuerca*. Junta de Castilla y León, Spain, pp. 19-36.

Farrand, W.R., McMahon, J.P., 1997. History of the sedimentary infilling of Yarımburgaz Cave, Turkey. *Geoarchaeology* 12, 537-565.

Kansu, S.A., 1970. Yarımburgaz (Küçükçekmece-İstanbul) Mağarası'nda Türk Tarih Kurumu adına yapılan prehistorya araştırmaları ve Tuzla Kalkolitiğinde yeni gözlemler. *VII. Türk Tarih Kongresi, Ankara 25-29 Eylül 1970, Kongreye Sunulan Bildiriler*, I. Cilt, pp. 22-32.

Meriç, E., Özdoğan, M., Sakınç, M., Açkurt, F., 1988. Yarımburgaz Mağarası kazılarında gözlenen mollusk kavkıları. *Arkeoloji ve Sanat* 40/41, 28-32.

Özdoğan, M., Koyunlu, A., 1986. Yarımburgaz Mağarası. 1986 Yılı Çalışmalarının ilk sonuçları ve bazı gözlemler. *Arkeoloji ve Sanat* 32/33, 4-17.

Salvador, A. (Ed.), 1994. *International Stratigraphic Guide*, 2d Ed. International Union of Geological Sciences and Geological Society of America, Boulder, CO.

Santel, W.T., von Koenigswald, W., 1998. Preliminary report on the middle Pleistocene small mammal fauna from Yarımburgaz Cave in Turkish Thrace. *Eiszeitalter und Gegenwart* 48, 162-169.

CHAPTER 3

ELECTRON SPIN RESONANCE (ESR) AND $230^{TH}/234^{U}$ DATING IN THE LOWER CHAMBER AT YARIMBURGAZ, TURKEY

Bonnie A.B. Blackwell – Henry P. Schwarcz – William R. Farrand
Joyce Lundburg – Anne R. Skinner – Mimi N. Divjak – Joel I.B. Blickstein

INTRODUCTION

At Yarımburgaz Cave, the Lower Chamber contains one of the few samples of Middle Pleistocene fauna co-occurring with Paleolithic artefacts in Turkey. This cave also sits near the major faunal migration route between Europe and the Levant, making it critical for understanding the pattern and timing for migrations by various species, including hominins, between the two continents. Absolute dating is crucial to correlating this site with the general European and Levantine Pleistocene stratigraphy.

To ascertain that no systematic errors arise from the methods used in dating a site, one should apply at least two different methods, when more than one method is applicable (Blackwell and Schwarcz 1993b). Therefore, standard and isochron $^{230}Th/^{234}U$, standard ESR, and coupled ESR-$^{230}Th/^{234}U$ were used to date travertines and teeth from Yarımburgaz Cave. $^{230}Th/^{234}U$, an U-series method, has long been used to date travertine layers in archaeological deposits (e.g., Blackwell *et al.* 1983; Schwarcz and Blackwell 1992). Blackwell and Schwarcz (1995) describe the method in detail. Several fragmentary travertine and collophane deposits within the cave were collected for $^{230}Th/^{234}U$ dating. Electron spin resonance (ESR) analyses date the actual fossils, rather than other stratigraphically associated materials. For tooth enamel, ESR dates have compared well with ^{14}C dates for samples up to 35-45 ka in age (see references in Blackwell, 2001, 2006). ESR dates for fossils from older sites agree with the expected geological or faunal estimates or with $^{230}Th/^{234}U$ or TL dates on associated materials (see references in Blackwell 2001, 2006). In order to improve the chronology for Yarımburgaz, 17 teeth from seven layers within four strata in the cave were dated with standard ESR and two with coupled ESR-$^{230}Th/^{234}U$.

STRATIGRAPHY AND GEOLOGY

The Yarımburgaz Cave complex developed in Eocene marly limestone 11-18 m above modern sea level. It occurs in the Sazlıdere valley on the northern tip of the Küçükçekmece lagoon on the northern shore along the Sea of Marmara, Thrace, Turkey (Figure 1.2). The site contains two chambers, the Upper Chamber (UC), averaging 50 x 15 m with more than 3 m of fill (Figure 2.2), and the Lower Chamber (LC), 6-10 m wide, but connecting to a much more extensive fissure cave system about 500 m long. The LC contains sediment near the opening averaging more than 5 m in thickness (Figure 2.5). Chalcolithic, Neolithic and Paleolithic occupations have been recognized in a complex sequence of thinly-bedded layers (Farrand and McMahon, this volume;

Ozdoğan and Koyunlu 1986). In the Lower Chamber, the bulk of the artefacts were recovered from Strata V, W and X (Farrand and McMahon, this volume).

The sedimentary units show successive episodes of fluvial and gravity movement of karst sediment, as well as interglacial marine deposition (Farrand and McMahon, this volume; Figure 2.5). Important unconformities occur in each chamber, bracketing Layer 11 in the UC, and strata below W and T in the LC. In the UC, Layer 11, a thick, crossbedded sand and gravel unit, represents a beach deposit, likely correlated with the Tyrrhenian or Marine Isotope Stage (MIS) 5e high seastand. Layers 12 and 13 in the UC probably correlate with Stratum Y, X, and W in the LC. Weathering and desiccation have produced some sediment alteration (Farrand and McMahon 1997, this volume), including several deposits that appear to be secondary collophane deposits. The collophane probably derived from partial dissolution of phosphates in the cave sediment, including bone and dental tissues, followed by subsequent precipitation within interstitial void spaces lower in the sequence.

Vertebrate remains from the LC include rodents, birds, lagomorphs, ursids, canids, felids, equids, cervids, and bovids, suids, hyenas, and one human incisor (Stiner, Howell, this volume). Numerous teeth and post-cranial bones from *Ursus* cf. *deningeri* (the Middle Pleistocene cave bear) and some from *Ursus* cf. *spelaeus* (the Late Pleistocene cave bear) occur in Strata X-T, especially in W (Stiner, Tsoukala, this volume). Although the fauna generally suggest a Middle Pleistocene date some faunal elements provide divergent indications of age. To determine more precise ages, carbonate and phosphate deposits were dated with ^{230}Th/^{234}U and cave bear molars were dated by ESR and coupled ESR-^{230}Th/^{234}U.

^{230}TH/^{234}U DATING OF SEDIMENTARY CARBONATES

Of the 15 samples originally collected for ^{230}Th/^{234}U dating, most proved unsuitable for analysis due to some combination of their extreme porosity, high detrital contents, and high concentrations of organic materials. Ultimately, five samples with low porosity and little obvious detritus or organic matter were analyzed from three strata (Table 3.1). These samples, however, also proved to be porous, mainly due to calcite dissolution, collophane cementation, or reactions rims formed on limestone *éboulis* clasts. Although 89YAR4 was a planar laminated tufa from Stratum Z, it contained limestone *éboulis* fragments, which made it unsuitable for dating. From

Table 3.1. ^{230}Th/^{234}U dating samples from Yarımburgaz

Sample Number		Type	Trench	Layer	Stratum	Depth	Find
Catalogue	Field					(m)	Spot
89YAR4		calcareous tufa	Z	6	Z		
89YAR9		travertine drapery fragment	T	2	X (top)		
89YAR37	T11	travertine drapery fragment	T	2	X (top)	6.80	a/2
90107	P1	stalactite	P	1	X	8.85	c-d/1-2
89YAR7		collophane[1]	Y	5 (top)	S		
T57d1[2]		*Ursus* cf. *deningeri* dentine	V	0	Z		
QT47d5[2]		*Ursus* cf. *deningeri* dentine	S	4	W	7.98	d/1

1 Probably secondary
2 ESR analyses also performed on enamel.

Stratum S, 89YAR7 was a dense, yellow collophane crust that formed post-depositionally in the sediment, possibly as a reaction rim. Both 89YAR9 and 89YAR37 were travertine drapery fragments from the top of Stratum X. Sample 90107 was a small stalactite fragment found embedded in Stratum X.

Analysis

Samples were analyzed initially according to the method described by Gascoyne *et al.* (1978) using ion exchange resins to separate U and Th from the matrix, followed by α spectrometry, as was the protocol in 1985. When this method proved unsatisfactory, several samples were analyzed according to the bone protocol (e.g., Rae and Ivanovich 1986), which, given the methods then available, usually produced better U and Th yields for samples with high organic concentrations. For all but 89YAR7, however, the bone protocol did not significantly improve the results. Isotope ratios, concentrations, dates, and their errors were calculated following Gascoyne's program (see Ivanovich and Harmon 1982).

^{230}Th/^{234}U Dating Results

Although 12 analyses were attempted, only sample 89YAR7 provided a reliable age determination based on its geochemical criteria. For most samples, either U or Th was lost during the preparation. In most cases, the extremely low U or Th count rates did not make it worthwhile measuring the concentrations for more than a few minutes. Of those samples for which sufficient U or Th was separated from the samples to justify counting the U and Th, only 89YAR7 yielded enough U and Th (*i.e.*, more than 5% chemical yield) to produce a reliable age (Table 3.2). Furthermore, for those samples with very low U yields, the U concentrations calculated did not agree well with those measured in the same samples by NAA, suggesting either that the U data from such low yields are unreliable, or that these samples had strongly heterogeneous U concentrations.

In the first analysis, 89YAR7a, a collophane sample collected from the top of Stratum S (or the bottom of T), gave uncorrected ages near 160 ka (Table 3.2). The plated U and Th discs were recounted in an attempt to improve the results, but this produced little change in the data (see 89YAR7b). By correcting the data for detrital ^{230}Th, assuming that the initial Th ratio in the detritus was $(^{232}\mathrm{Th}/^{230}\mathrm{Th})_0 = 1.25$, the ages changed little, still averaging 156^{+24}_{-20} ka. Regardless of the dating method used, single dates must always be viewed with skepticism. While the computed age for 89YAR7 may be valid, without confirming dates, it remains at best a preliminary estimate.

Calculating an isochron plot often proves a more reliable method to determine ages for samples with high detrital Th concentrations (Schwarcz and Latham 1989). This method assumes that all the samples dated derived from the same event, and hence, must be the same age, regardless of their detrital ^{230}Th contents. The strength of the isochron regression line demonstrates whether such an assumption is valid. With few points in the Yarımburgaz data set having reliable isotope ratios, however, the errors on such estimates become very large. Using all the data available produced regressions, with R^2 = 99.99% and 50% for the ^{234}U/^{238}U and ^{230}Th/^{234}U regressions respectively, that resulted in an age of 1.9 ± 0.8 ka. Using only the data from samples 89YAR37 and 90107 from Stratum X gave an age of 38^{+40}_{-30} ka (R^2 = 89%, 60%). Using 89YAR37 alone gave an age of approximately 18 ka. For 89YAR7 alone, the isochron gave approximately 38 ka. Given the low regression coefficients, none of these results can be considered reliable. The lack

Table 3.2. $^{230}Th/^{234}U$ dates for Yarımburgaz carbonate deposits

	Concentrations		Yield		Activity Ratios				$^{230}Th/^{234}U$ Age	
Sample	**U**	**Th**	**U**	**Th**					**Uncorrected**	**Corrected**[1]
(Stratum)	**(ppm)**	**(ppm)**	**(%)**	**(%)**	$\frac{^{234}U}{^{238}U}$	$\left(\frac{^{234}U}{^{238}U}\right)_0$	$\frac{^{230}Th}{^{234}U}$	$\frac{^{232}Th}{^{230}Th}$	**(ka)**	**(ka)**
89YAR4a	6.24	0.20	1.	< 1.	0.819	0.820	0.012	1.000	-	-
(Z)	1.04	0.47			0.033	0.028	0.034	0.784	-	-
89YAR4b	3.55	-	1.	0.	0.913	-	-	-	-	-
(Z)	0.01	-			0.093	-	-	-	-	-
89YAR9	5.08	0.01	2.	36.	0.807	0.806	0.021	0.019	-	-
(X top)	0.77	0.02			0.044	0.044	0.004	0.038	-	-
89YAR37a	0.93	0.04	< 1.	5.	1.174	1.207	0.437	0.026	-	-
(X top)	0.36	0.03			0.156	0.109	0.169	0.046	-	-
89YAR37b	0.76	0.06	< 1.	10.	1.100	1.154	0.772	0.032	-	-
(X top)	0.33	0.01			0.268	0.142	0.351	0.007	-	-
90107	0.16	0.03	< 1.	25.	2.474	2.727	0.446	0.056	-	-
(X)	0.06	0.01			0.839	1.158	0.179	0.015	-	-
89YAR7a	1.01	0.136	11.	16.	1.083	1.128	0.778	0.051	$157.8^{+22.1}_{-18.5}$	$152.4^{+22.7}_{-19.1}$
(S or ?T)	0.04	0.013			0.044	0.008	0.045	0.005		
89YAR7b	1.00	0.136	11.	16.	1.067	1.105	0.796	0.052	$167.3^{+25.1}_{-20.5}$	$160.5^{+26.0}_{-21.4}$
(S or ?T)	0.04	0.013			0.036	0.009	0.052	0.005		

[1] Corrected assuming $[^{230}Th/^{232}Th]_0 = 1.25$

of consistent isochron results suggests that some, or all, of the samples have been affected by secondary dissolution or reprecipitation, and that some reprecipitated calcite may well date to the Late Pleistocene or Holocene.

These two data sets do provide some limits for the travertine and collophane deposition within Yarımburgaz. The collophane deposition in Stratum S or T appears to date between 160 ± 25 ka and the present. The nature of the deposits, which occur as reaction rims on or draping over *éboulis* indicates that some or all collophane deposition postdates that of the overlying sediment. Until more carbonates or their associated sediment can be dated successfully, however, these dates and conclusions must remain preliminary.

ESR DATING

For standard ESR dating from Yarımburgaz, 17 cave bear molars from Strata Z through X, originally identified as *Ursus* cf. *deningeri*, were analyzed (Table 3.3). Owing to their small quantities of enamel, it was seldom possible to obtain more than one subsample from any tooth. Therefore, ESR isochron analyses (e.g., Blackwell and Schwarcz 1993a) could not be attempted. Two teeth, however, were dated by coupled ESR-^{230}Th/^{234}U to determine ages and to estimate suitable U uptake parameters to use to calculate the other ESR ages. Enamel in all the teeth showed some evidence for secondary vivianite replacement along cracks, but on the whole, the enamel was not very diagenetically altered.

Table 3.3. ESR enamel samples from Yarımburgaz

Sample Number Analysis	Catalogue	Field ID	Location Trench	Layer	Stratum	Depth (m)	Find Spot	Species	Tooth type
T57[1]	88109	V29	V	0	Z			*Ursus* cf. *deningeri*	right upper 3rd molar
T76	88111	T68	T	1	Y	7.20	b/4	*Ursus* cf. *deningeri*	left lower 3rd molar
T60	88115	S148	S	1	X	7.15	b/2	*Ursus* cf. *deningeri*	left lower 1st molar
QT49	90109	R51	R	1	X	7.78	e/2	*Ursus* cf. *deningeri*	left lower 3rd molar
T133	90110	R79	R	1	X	7.80	2	*Ursus* cf. *deningeri*	left upper 2nd molar
T134	90111	R99	R	1	X	7.80	2	*Ursus* cf. *deningeri*	left 3rd molar
QT50	90113	R169	R	1	X	7.80	c/2	*Ursus* cf. *deningeri*	left 3rd molar
QT51	90117	R624	R	1	X	7.82	b/2	*Ursus* cf. *deningeri*	lower 2nd molar
QT52	90119	R812	R	1	X	7.82	c/3	*Ursus* cf. *deningeri*	molar
QT53	90120	R913	R	1	X	7.90	c/1	*Ursus* cf. *deningeri*	left upper 2nd molar
T132	90103	P287	P	1	X	9.07[3]	d/1	*Ursus* cf. *deningeri*	left upper 2nd molar
QT48	90104	P345	P	1	X	9.33[3]	a/2	*Ursus* cf. *deningeri*	3rd molar
T58	88110	Y431	Y	1b	W	7.69	a/3	*Ursus* cf. *deningeri*	right lower 3rd molar
T131	90091	S104	S	4	W	7.78	b/3	*Ursus* cf. *deningeri*	left lower 2nd molar
T77	88112	S148	S	4	W	7.97	c/2	*Ursus* cf *deningeri*	left upper 3rd molar
QT47[1]	90092	S187	S	4	W	7.98	d/1	*Ursus* cf. *deningeri*	left lower 2nd molar
T59	88116	A71.172	A71[4]	4	W			*Ursus* cf. *deningeri*	lower 3rd molar

[1] Coupled ^{230}Th/^{234}U analyses performed on dentine.
[2] No find spot: collected from sieving screens.
[3] Corrected depth; original location incorrectly recorded.
[4] Equivalent to Trench V.

Theory

In well crystallized fossil tooth enamel, the intensity in the ESR signal at $g = 2.0018$ increases as a function of the total radiation absorbed by the tooth over time, although its intensity is zero in modern tooth enamel. Due to the signal's great thermal stability (Skinner *et al.* 2000), ESR can date samples as old as 4 Ma. Fossils are dated by measuring the equivalent accumulated radiation dose, $\mathcal{A}_\Sigma$, needed to produce the observed ESR signal and the total dose rate, $D_\Sigma(t)$, affecting the fossil during burial. The age is then calculated from:

$$\mathcal{A}_\Sigma = \mathcal{A}_{int} + \mathcal{A}_{ext} = \int_{t_0}^{t_1} D_\Sigma(t)\, dt = \int_{t_0}^{t_1} (D_{int}(t) + D_{ext}(t))\, dt$$

where $\mathcal{A}_S$ = the total accumulated dose in the sample,

$\mathcal{A}_{int}$ = the internally derived accumulated dose component,

$\mathcal{A}_{ext}$ = the externally derived accumulated dose component,

$D_S(t)$ = the total radiation environmental dose rate,

$D_{int}(t)$ = the sample's dose rate: from U, its daughters, and any other radioisotopes,

$D_{ext}(t)$ = the external dose rate: from sedimentary U, Th, and K, and the cosmic dose,

t_1 = the sample's age,

t_0 = today (Blackwell and Schwarcz 1993b).

$\mathcal{A}_S$ is determined by extrapolation from experimental irradiation using the additive dose method, while the dose rates are determined from analyzing the natural radioactivity in the tooth's immediate environment, $D_{ext}(t)$, and in its enamel and dentine, $D_{int}(t)$, assuming an U uptake model.

In calculating $D_{int}(t)$, the U uptake model is critical for samples that contain more than a few ppm of U. Where samples permit, coupled ESR-^{230}Th/^{234}U are now the norm for determining the uptake model. When the Yarımburgaz study began in 1989-1991, however, ^{230}Th/^{234}U dating had not been developed to the point that it would routinely give reliable ages for dentine or enamel. Consequently, the ESR dates were calculated by assuming three theoretical uptake models. Calculated by assuming that all the U now measured in the tooth was adsorbed very soon after its deposition, early uptake (EU) ages yield the minimum possible age given the $\mathcal{A}_S$. By assuming that U was continuously adsorbed at a constant rate since the tooth's deposition, linear uptake (LU) ages provide a median ESR age, whereas recent uptake (RU) ages, which assume that most U now in the tooth was acquired very recently, set the maximum ESR age limit. Which uptake model best predicts the true sample age remains somewhat site dependent (*cf.* McDermott *et al.* 1993; Grün and McDermott 1994), but the ESR dates were calculated by assuming all three theoretical uptake models.

The external dose rate, $D_{ext}(t)$, can strongly affect the calculated ESR ages. $D_{ext}(t)$ is calculated from the sedimentary concentrations from ^{40}K, the U and Th decay series isotopes, as determined by geochemical measurement or by direct measurement at the site using TL or γ ray dosimetry. To derive $D_{ext}(t)$ at Yarımburgaz, powdered sediment collected in conjunction with the sample was analyzed for its bulk geochemistry by NAA. The radioisotopic concentrations were used to mathematically calculate the current dose rates, $D_{ext,\beta}(t_0)$ and $D_{ext,\gamma}(t_0)$, whose calculation includes corrections for β and γ dose attenuation due to mineral density, water concentration, and backscattering (Brennan *et al.* 1997b). Because Yarımburgaz has thin and mineralogically inhomogeneous ("lumpy") layers, the inhomogeneity in the dose field requires volumetric analysis (Guibert *et al.* 1998). The contribution from each component is averaged depending on its volumetric abundance and radiation output, in order to accurately calculate the actual contribution to $D_{ext}(t)$ from each different components or layers within the β and γ "spheres of influence" (e.g., Appendix 3.1). Given the collophane deposits and problems with the ^{230}Th/^{234}U ages for the travertines, the $D_{ext}(t)$ calculations need to consider possible U uptake, U daughter isotope ingrowth, and potential radon loss within the sediment (Blackwell & Blickstein 2000). In caves such as Yarımburgaz, $D_{cos}(t)$, the cosmic dose rate, can be usually assumed to be 0 mGy/y, because the cave is roofed by > 10 m of limestone and the samples were collected from well inside the cave, away from the cave entrance.

In caves, karst processes and bioturbation can easily displace a tooth from its original depositional location to its final recovery location, which may be within a layer with different radioactivity from that of the original. Thus, its $D_{ext}(t)$ will differ. To ensure that reworked teeth do not occur, at least 2-3 independent ESR analyses for 4-5 teeth per layer are needed (Blackwell 1994). This study tested for reworked teeth by comparing dates for 5 teeth from Stratum W and 10 teeth from Stratum X, which yielded the majority of faunal remains and artefacts. From Strata Y and Z, which had comparatively less paleontological and archaeological material, only one tooth each was tested.

ESR Analytical Protocol

Before ESR analysis, each fossil was photographed, identified, and its diagenetic features fully described. For analysis, subsamples of enamel and attached dentine were carefully selected to minimize contamination by visibly altered tissue using a hand-held diamond-tipped Dremel drill. After the dentine was separated from the enamel, the outer 20 μm was removed from the enamel surfaces. Both enamel and dentine were powdered to 200-400 mesh. Neither the drilling nor the powdering affects the $g = 2.0018$ signal intensity, produces any spurious signals in that region, or alters the U concentrations in the tissues. For detailed sample preparation, signal processing, and age calculation, see Blackwell (1989) and Blackwell *et al.* (1992).

Individual enamel subsamples were divided into 10-15 identical aliquots for γ irradiation with a ^{60}Co γ radiation source to doses ranging from 0 to 128 krad. The samples were annealed for 3.0 days at 90°C to remove unstable signal interference (Skinner *et al.* 2000). Samples were analyzed on a Bruker ESR spectrometer, ED100, at room temperature using 20 dB microwave power at 9.76 GHz under 5 Gpp field modulation with a 0.25 G/s scan rate, and a 0.2 s time constant. Measurement gain was varied to maximize the signal intensity. ESR spectra were graphed as derivative peaks without deconvolution. Peak heights were measured from the graphs.

For neutron activation analysis (NAA), 50 sedimentary components (Appendix 3.1) and more than 50 tissue samples (Appendix 3.2) were analyzed. Approximately 1.0 g of each enamel and dentine subsample was analyzed using delayed neutron counting (DNC) for U. Samples were analyzed by instrumental NAA for U, Th, K (see Blackwell, 1989 for details). The enamel samples used here contained no insoluble residue minerals that could have contributed to the internal dose rate. Sediment samples weighing from 0.7-1.2 g were also analyzed for U, Th, and K. Each NAA data set was calibrated to a combination of two or more NIST standards, from among 1633A, 1633B, NIM-L, NIM-G, or S03.

Data Processing

For each subsample, $\mathcal{A}_{\Sigma}$ values were determined from the x-intercepts for curves plotting the ESR intensity (I) *versus* the experimentally added radiation dose, assuming an exponential fit to a saturating exponential curve weighted by $1/I^2$. $D_{ext}(t)$'s were calculated from the U, Th, and K concentrations in sediment, assuming 20% water content (Adamiec and Aitken 1998), while the internal dentine and enamel dose rates, $D_{int}(t)$'s, were calculated from U concentrations alone.

The ages, dose rates, and their respective errors were calculated using Data-HPS and Rosy v. 1.4 (Brennan *et al.* 1997b). The age calculations assume that the dose rates result from four sources:

attenuated penetration by β radiation into tooth enamel from the adjacent U-enriched dentine; α and β dose from U within the enamel; attenuated penetration by β and γ radiation into the enamel from U, Th, and K in the surrounding sediment; and, some absorption of the incident radiation by pore water in all phases. Since the radiation generated by the U within the tooth itself usually dominates the dose rate, the age calculation assumes a diffusional uptake history with no associated U or Rn loss, coupled with ingrowing U daughter isotopes. The a efficiency for tooth enamel was assumed to be 0.15 ± 0.02 (Grün 1989). Rn loss was assumed to be 0 vol% for the initial calculations, and its effect tested by recalculating the ages with varying Rn (see below). Because removing the outer 20-30 μm of the enamel during sample preparation should have eliminated any external α dose contributions, the calculations assumed that only internally generated α particles contributed to $D_{\text{int}}(t)$.

For ESR enamel ages, the most important error sources usually derive from the $\mathcal{A}_{\Sigma}$ and $D_{\text{ext}}(t)$ determinations, which can each contribute up to 5-10% uncertainty in the age (Blackwell, 1989). In interlaboratory calibrations, results from both McMaster University and Williams College fell in the centre of the distribution for the participating labs (Barabas *et al.*, 1993; Wieser *et al.*, 2005).

$D_{ext}(t)$ Determinations

Within Yarımburgaz, 50 sedimentary components were analyzed by NAA for U, Th, and K to assess the external dose rates, $D_{\text{ext}}(t)$. Significant variation exists within the U, Th, and K concentrations in different strata (see Appendix 3.1). Within a single stratum, considerable variation exists even within a single trench or between trenches. Also, many strata were noticeably inhomogeneous, containing large concentrations of bone or limestone *éboulis* detached from the roof. Some units contained some evidence for geochemical alteration, which may have leached U, Th, or K from some mineralogical components within given layers, redepositing these same elements in other, lower strata. Meanwhile, several layers contain secondary minerals, including U-rich collophane, possibly deposited long after the units' original deposition. These problems make it difficult to determine the time-integrated external dose rate, $\overline{D}_{\text{ext}}(t)$, experienced by any tooth. Lacking sufficiently large teeth for multiple subsampling, no isochrons could be derived to assess potential changes in $\overline{D}_{\text{ext}}(t)$ or U uptake rates (e.g., Blackwell and Schwarcz 1993a; Blackwell *et al.* 2002). Were the analyses being done today, advanced radiometric analyses of the sedimentary components might reveal $D_{\text{ext}}(t)$ variations not possible in 1991 (e.g., Guibert *et al.*, 2007).

Since some teeth were selected from the archaeological collections after the excavations had been completed, exact locations within the layers or trenches for a few teeth were difficult to assess. Hence, determining the best $\overline{D}_{\text{ext}}(t)$ to use in the calculations remains an educated guess. Even for those teeth collected specifically for ESR dating, the great variation in individual $D_{\text{ext}}(t)$ measurements within and across strata (see Appendix 3.1) complicates the $\overline{D}_{\text{ext}}(t)$ calculation. Given that some strata are very thin, the $\overline{D}_{\text{ext},\beta}(t)$ and $\overline{D}_{\text{ext},\gamma}(t)$ values were calculated by volumetrically integrating each layer's or sedimentary component's contribution to the sphere of influence for the radioactive particle in question (*e.g.*, Blackwell *et al.*, 2001). Given the heterogeneity in the layers, even teeth from the same layer may have experienced very different $D_{\text{ext}}(t)$'s (*cf.* Table 3.4).

Table 3.4. ESR dates for Yarımburgaz

ESR Sample	Stratum		Accumulated Dose, A_s (Gray)	External Dose Rates $D^{BG}_{ext,\beta}(t)$ (mGy/y)	$D^{BG}_{ext,\gamma}(t)^2$ (mGy/y)	Standard ESR Ages[1] EU (ka)	LU (ka)	RU (ka)	CU (ka)
T57e1	Z		276.0	0.163	0.387	199.8	289.7	430.3	270.4
		±	34.5	0.057	0.134	38.1	69.0	139.0	61.3
T57e2	Z		257.9	0.163	0.387	209.5	293.4	418.8	266.2
		±	56.2	0.057	0.134	56.5	90.8	158.7	78.6
T57e3	Z		266.3	0.163	0.387	191.9	278.6	420.2	256.7
		±	28.6	0.057	0.134	34.4	63.6	134.4	55.2
T57	Z		268.9	0.163	0.387	198.3	286.0	423.5	263.8
mean		±	20.5	0.057	0.134	23.3	41.6	82.6	36.4
T76e1[4]	Y		380.9	0.260	0.631	94.3	153.2	299.6	165.5
		±	50.9	0.106	0.264	16.3	31.7	96.4	35.4
T76e1[5]	Y		380.9	0.260	0.631	213.8	289.5	390.5	-
		±	50.9	0.106	0.264	53.2	90.4	157.1	-
T76e1[6]	Y		380.9	0.260	0.631	161.1	236.8	364.9	-
		±	50.9	0.106	0.264	33.9	63.7	128.2	-
T76e1[7]	Y		380.9	0.260	0.631	145.0	218.6	353.6	-
		±	50.9	0.106	0.264	29.0	55.6	130.3	-
T60e1	X		174.3	0.177	0.431	152.6	202.4	264.2	188.6
		±	15.8	0.075	0.201	39.8	67.7	113.0	59.1
QT49e1	X		236.5	0.191	0.533	200.0	251.5	307.0	218.1
		±	8.2	0.091	0.245	60.9	91.9	135.2	72.4
T133e1	X		291.0	0.178	0.737	169.2	224.9	292.9	225.4
		±	12.2	0.048	0.239	29.9	51.3	85.5	51.3
T134e1	X		219.3	0.171	0.498	121.6	179.2	276.6	181.0
		±	26.5	0.075	0.247	27.0	52.5	117.2	53.3
QT50e1	X		438.0	0.319	0.776	166.9	238.4	351.9	187.6
		±	21.0	0.103	0.239	23.4	45.8	97.6	29.3
QT50e2	X		392.8	0.319	0.776	135.0	198.2	298.7	157.2
		±	19.6	0.103	0.239	17.7	35.7	78.7	23.4
QT50	X		416.3	0.319	0.776	149.9	217.0	323.9	171.7
mean		±	14.4	0.103	0.239	14.5	28.7	62.1	18.5
QT53e1	X		377.9	0.308	0.788	161.7	223.2	310.7	187.3
		±	13.0	0.082	0.208	21.6	39.4	74.9	28.6
T132e1	X		274.3	0.217	0.619	249.2	287.1	319.0	288.4
		±	18.2	0.091	0.286	87.2	114.9	141.6	115.9
QT48e1+2	X		270.0	0.267	0.831	204.8	225.1	241.7	210.4
		±	8.7	0.082	0.257	53.5	64.2	73.8	56.5
Mean	X					160.9	220.7	299.2	190.9
		±				9.9	17.7	31.8	21.5
T58e1	W		331.0	0.295	0.637	91.9	145.7	265.4	123.0
		±	70.8	0.096	0.213	21.5	37.2	82.1	30.2
T58e2	W		328.6	0.295	0.637	94.1	148.3	269.7	125.7
		±	50.5	0.096	0.213	17.1	31.2	80.3	24.7

Table 3.4. ESR dates for Yarımburgaz (continued)

ESR Sample	Stratum		Accumulated Dose, $\mathcal{A}_S$ (Gray)	External Dose Rates $D^{BG}_{ext,\beta}(t)$ (mGy/y)	$D^{BG}_{ext,\gamma}(t)^2$ (mGy/y)	Standard ESR Ages[1] EU (ka)	LU (ka)	RU (ka)	CU (ka)
T58	W		329.4	0.295	0.637	93.3	147.3	267.6	124.7
mean		±	41.1	0.096	0.213	13.3	23.9	57.4	19.1
T58[8]	W		329.4	0.295	0.637	137.8	200.8	302.8	-
mean		±	41.1	0.096	0.213	25.5	46.0	94.2	-
T131e1	W		484.3	0.312	0.701	165.5	248.4	398.8	245.2
		±	41.5	0.087	0.213	20.6	41.5	101.1	40.2
T77e1	W		326.5	0.312	0.607	123.9	185.2	295.5	177.0
		±	25.8	0.087	0.227	18.5	36.8	87.2	33.6
QT47e1	W		291.5	0.251	0.585	161.5	223.2	313.4	187.0
		±	14.2	0.120	0.230	32.5	61.2	118.8	43.2
T59e1	W		427.5	0.226	0.582	158.8	245.4	430.0	210.3
		±	29.4	0.111	0.315	29.1	63.8	190.1	48.4
Mean	W					138.8	202.7	325.3	184.8
+ T58[9]		±				9.5	18.0	41.7	15.3
Mean	W					151.4	200.8	349.7	-
+ T58[8,9]		±				9.4	18.6	43.4	-
Mean	W					152.3	226.2	357.4	209.7
− T58[10]		±				11.9	23.9	57.1	20.7

1 Abbreviations:	EU	=	assuming early U uptake, p_{en} = -1.0, p_{den} = -1.0			
	LU	=	assuming linear (continuous) U uptake, p_{en} = 0.0, p_{den} = 0.0			
	RU	=	assuming recent U uptake, p_{en} = 10.0, p_{den} = 10.0			
	CU	=	assuming recent U uptake, p_{en} = 1.4, p_{den} = -0.8			
	Calculated using α efficiency factor, κ_α	=	0.15	±	0.02	
	initial U activity ratio, $(^{234}U/^{238}U)_0$	=	1.11	±	0.01	
	enamel density, ρ_{en}	=	2.95	±	0.02	g/cm^3
	dentine density, ρ_{den}	=	2.85	±	0.02	g/cm^3
	enamel water concentration, W_{en}	=	2.0	±	2.0	wt%
	dentine water concentration, W_{den}	=	5.0	±	2.0	wt%
	radon loss from the teeth, Rn_{mol}	=	0.	±	0.	vol%
	sedimentary water concentration, W_{sed}	=	20.0	±	5.0	wt%
2 Calculated using	sediment density, ρ_{sed}	=	2.66	±	0.02	g/cm^3
3 Calculated using	cosmic dose rate, $D_{cos}(t)$	=	0.0	±	0.0	μGy/y
4 Calculated using	enamel U concentration, U_{en}	=	10.29	±	0.02	ppm
5 Calculated using	enamel U concentration, U_{en}	=	1.00	±	0.02	ppm
6 Calculated using	enamel U concentration, U_{en}	=	3.00	±	0.02	ppm
7 Calculated using	enamel U concentration, U_{en}	=	4.00	±	0.02	ppm
8 Calculated using	enamel U concentration, U_{en}	=	3.00	±	0.02	ppm
	dentinal U concentration, U_{den}	=	25.00	±	0.02	ppm

9 Calculated including T58.

10 Calculated without T58.

The $\overline{D}_{ext,\beta}(t)$ and $\overline{D}_{ext,\gamma}(t)$ values (Table 3.4) were initially calculated by assuming 20 wt% sedimentary water concentrations. If the calculations doubled the water content to 40 wt%, this models the minimum possible $D_{ext}(t)$, and hence, the maximum possible age. Under a scenario with 40 wt% sedimentary water, $\overline{D}_{ext,\beta}(t)$ and $\overline{D}_{ext,\gamma}(t)$ would both be reduced by 27%, producing an overall increase in the calculated ages by 12% for the EU ages, 19% for LU ages, 31% for the

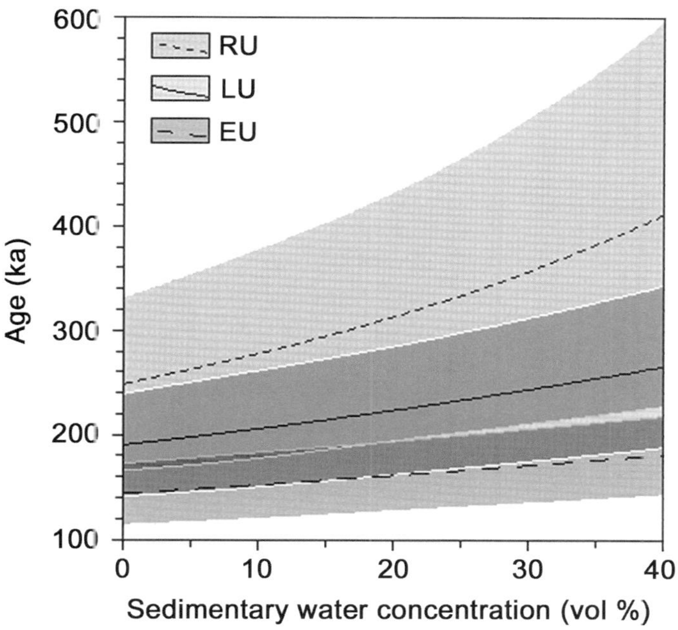

Figure 3.1
The variation in the calculated model ages with sedimentary water concentrations for QT47e1, Stratum W. Recalculating the ages for QT47e1 with different sedimentary water concentrations shows that as the water concentration increases from 20 wt% (as used in the calculation reported on Table 6), all the model ages increase. The shaded fields around each line show how the associated errors change with U concentration. At 40 wt% water, the calculated model EU ages would increase by 12%, the LU ages by 19%, and the RU ages by 31%. Assuming a value of $[U_{en}]$ = 4 ppm, gives an age of 219 ± 56 ka for T76en1. At 0 wt% water, the calculated model EU ages would decrease by 11%, the LU ages by 15%, and the RU ages by 21%. For most of the range of sedimentary water, any change in the age is not significant given the errors reported for the age.

RU ages (see Figure 3.1). Since 40 wt% water would exceed the maximum water concentration under total saturation for most sedimentary units near the teeth, the actual increase in the ages from even complete saturation would be less. On the hand, drying out the sediment totally (i.e., 0 wt%) increases $\bar{D}_{ext,\beta}$ (t) and $\bar{D}_{ext,\gamma}$ (t) by 29-31%, causing a decrease in the calculated ages by 11% for the EU ages, 15% for LU ages, 21% for the RU ages (see Figure 3.1).

When the cave bears or humans inhabited the cave, the sedimentary water saturation was not likely much higher than 5-10 wt%, because both species prefer living on dry substrates. The cave probably was not completely inundated at any time, except during MIS 5e, and for not more than a few thousand years. The collophane deposition, however, argues for wetter sedimentary conditions at least temporarily, necessary for bone dissolution. Collophane precipitation along with the secondary rim cements suggest rather drier sedimentary moisture conditions later on. Thus, the time-averaged sedimentary water concentrations were likely to be much closer to 10-20 wt% than 40 wt%. Although the error limits quoted on $D_{ext}(t)$ values (Appendix 3.1, Table 3.4) exclude the values possible under the fully saturated or the completely dry condition, the difference at all but the most extreme changes is insignificant given the associated errors. Using 10 wt% water, which is the most dramatic difference of any that could have likely occurred, would reduce the LU ages by 8%, which is well within the errors quoted for each model age calculation.

Looking at overall potential errors from any miscalculations in the actual $\bar{D}_{ext}$ (t) reveals that the analysis is more robust than might be expected given the previous discussion (Figure 3.2). Thanks to the relatively high dentinal U concentrations, the effect from inaccuracies in $\bar{D}_{ext}$ (t) are relatively minor. Considering the 2 σ error, any change in the calculated LU ages only becomes significantly different from those reported in Table 3.4, if the actual values for $\bar{D}_{ext}$ (t) fell below 450 μGy/y or exceeded 2200 μGy/y. Although surrounding the tooth with *éboulis* for 20 cm in all directions would reduce its $\bar{D}_{ext}$ (t) to ~ 450 μGy/y, no such extremely thick *éboulis* layers were found. To make $\bar{D}_{ext}$ (t) exceed 2200 μGy/y requires that the tooth be totally surrounded by collophane or bone in all directions for up to about 25 cm away from the tooth, which certainly would have been noted during the excavation. Thus, the errors reported for the ages include all uncertainty arising from any likely changes in $\bar{D}_{ext}$ (t).

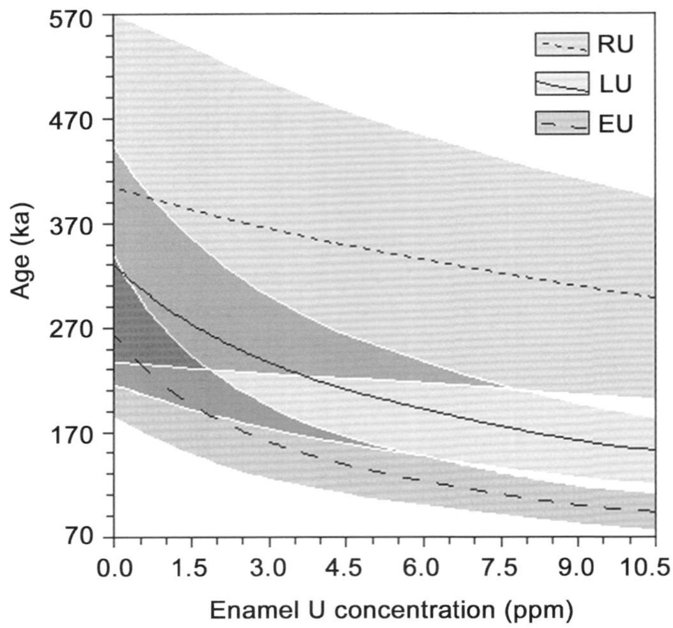

Figure 3.2.
The variation in the calculated model ages with U concentration for T76en1, Stratum Y.
Recalculating the ages for T76en1 with different U concentrations shows that as the U concentration decreases from the anomalously high value measured in T76en1, all the model ages increase. The shaded fields around each line show how the associated errors change with U concentration. Assuming a value of $[U_{en}]$ = 4 ppm, gives an age of 219 ± 56 ka for T76en1.

$D_{int}(t)$ Determinations

In *Ursus* (bear) teeth, like human teeth, the enamel forms a thin outer shell, but covers the dentine only on the crown. The crowns of many Yarımburgaz bear molars were considerably worn, thereby reducing the enamel available for ESR analysis. Since no cementum protects the enamel, it tends to dry out more easily, making enamel and dentine separation easier, but also making the enamel more prone to diagenesis and cracking than in bovid teeth. During preparation, the enamel tended to break into extremely small pieces. That, plus the enamel's thinness, made it difficult to clean the *Ursus* enamel properly for analysis. In T76en1, for example, the very high U concentration, $[U_{en}]$, and consequently, the high enamel dose rates, $D_{en}(t)$ (Table 3.7), might represent contamination within the enamel from adjacent sediment or dentine. If so, then the ages reported for T76en1 (Table 3.4) would be anomalously low.

This raises another concern. For several teeth reported here, ensuring that all the outer 20-30 μm doped by α particles had been removed from the enamel was impossible. Thus, some fragments in some samples may have not had the 20-30 μm shaved off either partially and some, possibly, not at all. ^{230}Th/^{234}U analyses have suggested that, in many fossil teeth, the U concentrations drop steeply from the outer uraniferous enamel surfaces (e.g., Eggins *et al.* 2003). This represents a diffusional gradient created as U slowly penetrates deeper into the enamel with time. Because the ESR dating procedure routinely removes that outer layer, but rarely analyzes its U concentration, samples from which the outer 20-30 μm of enamel have not been so completely removed, possibly such as T76e1, may show very high $[U_{en}]$. Indeed, T76el's high $[U_{en}]$ might reflect inclusion of the outer uraniferous zone, rather than dentine contamination. If so, the time-averaged dose contribution, $\overline{D}_{en}(t)$, arising from this thin uraniferous outer skin would be substantially less than that calculated by assuming that the total uraniferous zone had been present from the tooth's earliest history. Assuming that the U had been there from the initial burial would also result in ages that were anomalously low. Alternatively, T76el may not be abnormal at all, and its reported $[U_{en}]$ and $D_{en}(t)$ may be accurate. In which case, T76 would appear to be an extraneous tooth, possibly having been reworked from another younger layer, which has since been lost to erosion.

In several teeth, such as T58, T59, QT50, and QT52 (Appendix 3.2), the rather high dentinal U concentrations, $[U_{den}]$, suggest that considerable groundwater flow has affected this site to enable strong U migration, as does the presence of collophane in some layers. Consequently, for the $D_{ext}(t)$ and age calculations, we chose sedimentary water concentrations, W_{sed} = 20%, rather than the modern values. These high $[U_{den}]$ concentrations may also reflect minute amounts of secondarily deposited minerals, such as U-rich aragonite, calcite, or collophane in the dentinal porosity elements. Therefore, any simple uptake model may not accurately reflect the actual U uptake history, which might be better modeled using a stepped uptake model. Obviously, U could also have been leached from teeth during the events which led to the collophane deposition in the sediment. The very presence of collophane argues for some significant dissolution of hydroxyapatite or other phosphatic minerals, probably from overlying layers, although possibly from outside, upstream from the cave.

ESR Dates

Given the concerns outlined above, the ESR dates for the Yarımburgaz teeth do show much more consistency than might be expected, especially considering the marked variation in $\mathcal{A}_{\Sigma}$'s within Strata X and W (Table 3.4, Figure 3.3). Within single teeth, multiple analyses show good consistency.

From Stratum Z, only T57 was dated. All the model ESR ages for T57 (Table 3.4) appear older than those for teeth from the layers below, although its LU age, 286 ± 42 ka, still falls within 2 σ of the means for Strata X and W. If the tooth was found immediately adjacent to a large chunk of *éboulis*, its age would appear erroneously younger than it actually was (Brennan *et al.* 1997a), due to the lower $D_{ext}(t)$ emanating from the *éboulis*. Stratum Z did contain some large *éboulis* (Farrand and McMahon, this volume; Figure 2.5). With only one dated tooth from this layer,

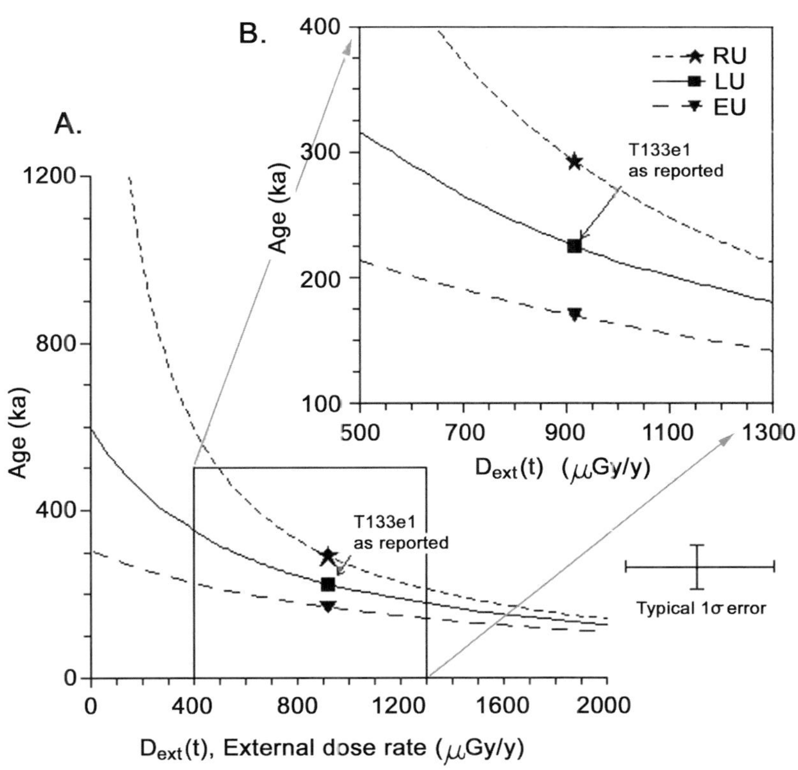

Figure 3.3.
The variation in the calculated model ages with external dose rate for T133e1, Stratum X. Recalculating the ages for T133e1 with different external dose rates, $D_{ext}(t)$, shows that $D_{ext}(t)$ increases, all the model ages decrease. Considering the 2 σ error, any change in the calculated LU age only become significantly different from those reported in Table 6, if the actual $D_{ext}(t) < 450$ μGy/y or $D_{ext}(t) > 2200$ μGy/y.

T57 cannot be proven unequivocally to have been *in situ* when collected. T57 could represent an extraneous tooth moved to its current location by bioturbation or left there during earlier excavations in Trench A71. Stratum Z certainly was disturbed by later occupants. Only dating more teeth from this layer would settle the question of T57's attribution to Stratum Z.

T76 was the only tooth dated from Stratum Y. The higher $[U_{en}]$ in T76 (Appendix 3.2) hints that it might represent an extraneous tooth. Nonetheless, its LU age, 153 ± 32 ka (Table 3.4), still falls within 2 standard deviations of the means for Strata X and W. While Stratum Y was not observed to have been disturbed near T76, minor disturbance, had it occurred, might not have been easily observed. If it is extraneous, T76 could be older or younger than reported here, but calculating its actual age would require some idea about the unit in which it was originally deposited (Blackwell 1994). Alternatively, if we were to assume that the currently measured $[U_{en}]$ for T76el is inaccurate, and recalculate the age for T76 using a value for $[U_{en}]$ more like those seen in the other teeth, say $[U_{en}]$ = 2-4 ppm, the calculated ages for T76 then agree better with those for the other units (Figure 3.1; Table 3.4). The post-depositional collophane and weathered travertine in several units, including at the top of Stratum X, argue for significant U migration, and may suggest recent uptake of some U into T76. If we can believe that T76 is *in situ* rather than extraneous, and that the ESR ages for the teeth are reasonably accurate, then the similarity in the ages for T76 using $[U_{en}]$ = 4 ppm to the ages for the teeth in Stratum X and W, might suggest that for much of T76's history, $[U_{en}]$ was approximately 4 ppm, but that $[U_{en}]$ increased to its present high concentration within the last few millennia. The $^{230}Th/^{234}U$ isochron ages for the collophane (see above) support this supposition. If for most of its history, T76's $[U_{en}]$ was really close to 4 ppm, T76's age would be very close to 215-220 ka (LU).

From Stratum X, seven teeth were dated. Although T134 appears somewhat young, while QT49 appears somewhat older than the rest, both these had large uncertainties due to small sample sizes. Considering all the samples, their ages averaged 162 ± 10 ka, assuming EU, 221 ± 18 ka, assuming LU, or 299 ± 32 ka, assuming RU.

From Stratum W, five teeth were dated. Although the ages for T58 do not differ by more than 2 σ from those for the other teeth within Stratum W (Table 3.4), it does appear young compared to the others. Like T57, this age difference could arise from adjacent *éboulis* or bioturbation. Including T58 in the group, the teeth average between 133 ± 7 ka, assuming EU, or 197 ± 12 ka, assuming LU, and 389 ± 31 ka, assuming RU. Examining T58's dentine, however, reveals that $[U_{den}]$ = 74-78 ppm (Appendix 3.2), almost double that in any other tooth from Stratum W. This hints that T58 is an extraneous tooth reworked from another layer. If we consider T58 to have been reworked, and hence, an outlier to the group, the other four teeth average 152 ± 12 ka, assuming EU, 226 ± 24 ka, assuming LU, or 356 ± 57 ka, assuming RU. Alternatively, T58, the only tooth analyzed from Trench Y, may have experienced late U uptake that substantially boosted its $[U_{den}]$ in recent millennia. Under that scenario, T58's age would be closer to 200 ka (LU), and the average ages for Stratum W closer to 226 ka (LU).

COUPLED ESR-^{230}TH/^{234}U DATING

With the technology currently available (2009), ESR dates are normally combined with $^{230}Th/^{234}U$ analyses for both the enamel and dentine within the tooth to calculate a coupled ESR-$^{230}Th/^{234}U$ age (hereafter, "coupled" age). In 1989, however, when the Yarımburgaz dating began, such analyses were neither technologically, nor financially, feasible. In 2000, we attempted several

^{230}Th/^{234}U with the Yarımburgaz samples that had been archived following their NAA irradiations, to see if coupled ages were possible. Of several samples attempted, only two dentines, T57 and QT47, had enough powdered sample with a sufficiently high U concentration to enable calculated ^{230}Th/^{234}U ages (see Tables 3.1, 3.2). No enamels were directly datable.

Theory

Both ^{230}Th/^{234}U and ESR dating use the U concentration in the teeth to determine an age. Standard ^{230}Th/^{234}U dating uses the U and Th isotopic concentrations to calculate a radiometric age using the decay rates for ^{230}Th, ^{234}U, and ^{238}U, but the standard method assumes closed system behavior, particularly that no U or Th have been added to or removed from the system except through radioactive decay. ESR, however, assumes that the U and its daughters within the teeth generate the radiation that produces the ESR signals detectable by ESR spectrometry. Standard ESR dating does not necessarily assume closed system behavior, since it uses several U uptake models to calculate the ages. Only the EU model assumes a closed system.

Coupled dating assumes that one unique age exists and can be calculated with the data sets from the enamel ^{230}Th/^{234}U, dentine ^{230}Th/^{234}U, and ESR data. The ^{230}Th/^{234}U data are used to calculate U uptake rate parameters for the enamel, p_{en}, and dentine, p_{den}, that enable the ESR ages to agree with the enamel and dentine ^{230}Th/^{234}U ages. These p values are used to recalculate a new ESR age and to calculate open-system ^{230}Th/^{234}U ages for the enamel and dentine. These three new ages are then used to recalculate two more p values, and the process is repeated iteratively until two p values are determined that generate a unique date that satisfies the ESR and both ^{230}Th/^{234}U raw data sets. Without ^{230}Th/^{234}U analyses for the enamel, one must assume a value for p_{en} in order to calculate the coupled age.

Coupled ESR-^{230}Th/^{234}U Age Calculations

For both the Yarımburgaz teeth, the coupled age calculations initially assumed that $p_{en} = p_{den}$ (Table 3.5). Then, the calculations were repeated with other values to assess how the coupled ages changed with the value assumed for p_{en}.

For T57, the dentine data showed that $p_{den} = -0.80 \pm 0.08$, but the coupled age calculation would only give a finite age, if $p_{en} \geq 1.39 \pm 0.34$ (Table 3.5). This indicates that the U uptake into the dentine occurred primarily early in the tooth's history, namely that the dentine's actual U uptake fell between the values for the EU ($p = -1.0$) and LU ($p = 0.0$) models. In the enamel, however, some U uptake occurred later in the tooth's history, with the actual U uptake falling between the values for LU ($p = 0.0$) and the typical RU ($p = 10.0$) model. In both laboratory simulations and fossils, U uptake rates into dentine produce concentrations that exceed those in enamel by roughly an order of magnitude (Grün and Taylor 1996; Blackwell *et al.* 2002). Moreover, the magnitude of the differences between the U concentrations in dentine and enamel bear no obvious relationship to the fossil age. Hence, the differences in p values in T57 do not appear aberrant. Therefore, assuming that no U loss from T57 has occurred, its minimum age is 211 ± 22 ka ($p_{en} = 1.39$). As p_{en} increases, T57's calculated age becomes older.

For QT47, the dentine data indicated that $p_{den} = -0.93 \pm 0.06$, suggesting that the U uptake into the dentine occurred very early in the tooth's history, and is best modelled with an EU model (Table 3.5). Using $p_{en} = -0.93 \pm 0.06$, the coupled calculation gave an age of 154^{+13}_{-12} ka, an age which appears rather young compared to that for T57, and an age that agrees well with that for

Table 3.5. Coupled ESR-^{230}Th/^{234}U dates for Yarımburgaz teeth

Sample (Stratum)		Concentrations U (ppm)	Th (ppm)	Activity Ratios $\frac{^{234}U}{^{238}U}$	$\left(\frac{^{234}U}{^{238}U}\right)_0$	$\frac{^{230}Th}{^{234}U}$	$\frac{^{232}Th}{^{230}Th}$	^{230}Th/^{234}U Age Uncorrected (ka)	Coupled ESR-^{230}Th/^{234}U Age[1,2] p_{den}	if $p_{en} = p_{den}$ (ka)	if $p_{en} > p_{den}$ (ka)	p_{en}
T57d1		36.69	0.634	1.074	1.116	0.777	0.000	$159.1^{+0.5}_{-0.5}$	-0.80	-	211^{+22}_{-22}	1.39
(Z)	±			0.001	0.001	0.001	0.025		0.08			0.34
QT47d5		20.75	0.586	1.121	1.179	0.733	0.000	$138.8^{+1.1}_{-1.1}$	-0.93	154^{+13}_{-12}	178^{+23}_{-22}	1.40
(W)	±			0.003	0.003	0.003	0.048		0.06			0.30

[1] Abbreviations: p_{en} = U uptake parameter in the enamel
p_{den} = U uptake parameter in the dentine

[2] Calculated using other assumptions for standard ESR as listed in Appendix 3.2.

the secondary collophane deposition in Stratum S (Table 3.1). This hints that most of the U in the enamel was acquired some time after its deposition. As p_{en} increases, the coupled age for QT47 becomes older. While no *a priori* reason exists to assume that p_{en} for QT47 should equal that for T57, having one tooth with p_{en} = 1.39 makes it more likely that others could have similar values, since all experienced reasonably similar environmental conditions. Assuming p_{en} = 1.4 ± 0.3, the age for QT47 increases to 178^{+23}_{-22} ka. This age, however, still does not agree well with the stratigraphy. If T57 has not been reworked from an older layer, then QT47's age should ≥ 211 ka. To produce such an age, given the constraint of the dentine data, requires that p_{en} ~ 10 (RU).

Given the lack of actual ^{230}Th/^{234}U dates for the enamel, however, these ages are at best minimum age estimates. While it is unlikely that the teeth could be much younger than the dates here, they could be considerably older, because the value for p_{en} remains an estimate. Therefore, the coupled dating suggest that all the layers at Yarımburgaz date to ≥ 154^{+13}_{-12} ka, and may well be ≥ 211 ± 22 ka.

Using the *p* values from the coupled analyses to recalculate the standard ESR ages generates standard ESR ages calculated with a coupled U uptake (CU) model (Table 3.4). This gives an average CU age for Stratum X at 191 ± 22 ka, while for Stratum W, the standard CU ages average 185 ± 15 ka, including T58 in the group, and 210 ± 21 ka, without T58. These CU ages do not different significantly from the LU model ages.

CONCLUSIONS

All three dating methods used at Yarımburgaz suggest that the deposits had formed by, and more likely well before 150-160 ka (Table 3.6). Among the different methods, however, the standard LU model ages appear to offer the best internal stratigraphic consistency, and do not differ significantly from the CU model ages. While ^{230}Th/^{234}U dates on stalagmitic deposits were not conclusive, they do not disagree with these ESR ages. Moreover, the presence of the secondary collophane, for which the mean dates average 156 $^{+}$25/-20 ka, argues that the deposits predate that time, and that secondary U was likely added to the teeth at that time, necessitating a CU or

LU model uptake scenario. From the standard LU ESR ages, primary deposition of Strata W to Z likely were deposited between 226 ± 24 ka and 211 ± 22 ka. This age correlates most closely with the cold interval MIS 7d (Figure 3.4; *cf.* Martinson *et al.* 1987; Lisiecki and Raymo, 2005), which some estimates would have been as cold as the start of MIS 2. The associated errors do not rule out the possibility that these beds were deposited during MIS 6e, or MIS 8a. Nonetheless, these ages suggest relatively rapid deposition, averaging about 6.7 cm/ka for these strata. An RU model assumption would produce more gradual deposition rates, while an EU model produces extremely high sedimentation rates.

Among the Yarımburgaz fauna, the cave bears provide the strongest biochronological markers. Since remains that fit the morphological ranges for both *Ursus deningeri* and *U. spelaeus* occur (Tsoukala, this volume), the Yarımburgaz bears, and hence, the deposits which contain them, should correlate with the period in which *U. spelaeus* replaced *U. deningeri*. *U. deningeri* made its last appearance in late MIS 8 at 250 ka (Séronie-Vivien and Tillier 2002), but transitional forms intermediate between the two apparently persisted until ~ 150 ka in mid OIS 6d (Bordes 1984). In Abri Suard, La Chaise-de-Vouthon, Charente, *U. spelaeus* bones (Bordes 1984) pre-dated 112 ± 12 ka (Blackwell *et al.* 1983). At Pradayrol, Lot, France, *U. deningeri* teeth dated to 330 ± 5 ka (Blackwell *et al.* 2007). At Abri Vaufrey, Dordogne, France, for example, the transition occurred from slightly before 250 ± 70 ka to somewhat after 170 ± 10 ka (Blackwell and Schwarcz 1989). At Pech de la Azé, Dordogne, bones attributed to *U. deningeri* (Bordes 1972) and the transitional forms (Bordes 1984) dated between 240 ± 30 and 123 ± 15 ka (Schwarcz and Blackwell 1983), while at Montgaudier, Dordogne, *U. deningeri* bones (Bordes 1984) pre-date 296 ± 54/-37 ka (Blackwell 1980). At Kent's Cavern, England. *U. deningeri* dated at > 340 ka (Proctor *et al.* 2005. At Treugol'naya in the Russian Caucasus, *U. deningeri* and *Vulpes vulpes* dated to 366 ± 12 to 375 ± 9 ka (Blackwell *et al.*, 2005). More generally, Middle Pleistocene

Table 3.6. Dates for stratigraphic events at Yarımburgaz

Stratum Deposition Type	Method (Assumptions)	Age (ka)
Z (primary)	coupled ESR-^{230}Th/^{234}U[2] (p_{en} = 1.4 ± 0.3; T57 not reworked)	211 ± 22
Y (primary)	standard ESR[2] (LU; $[U]_0$ = 4 ppm)	219 ± 56
X (primary)	standard ESR[2] (LU)	221 ± 18
W (primary)	standard ESR[2] (LU)	226 ± 24
S collophane (secondary)	 detritally corrected ^{230}Th/^{234}U ($[^{230}Th/^{232}Th]_0$ = 1.25)	156^{+24}_{-20}

[1] Abbreviations: p_{en} = U uptake parameter in the enamel
p_{den} = U uptake parameter in the dentine
LU = assuming linear (continuous) U uptake, p_{en} = 0.0, p_{den} = 0.0

[2] Calculated using other assumptions for standard ESR as listed is Appendix 3.2.

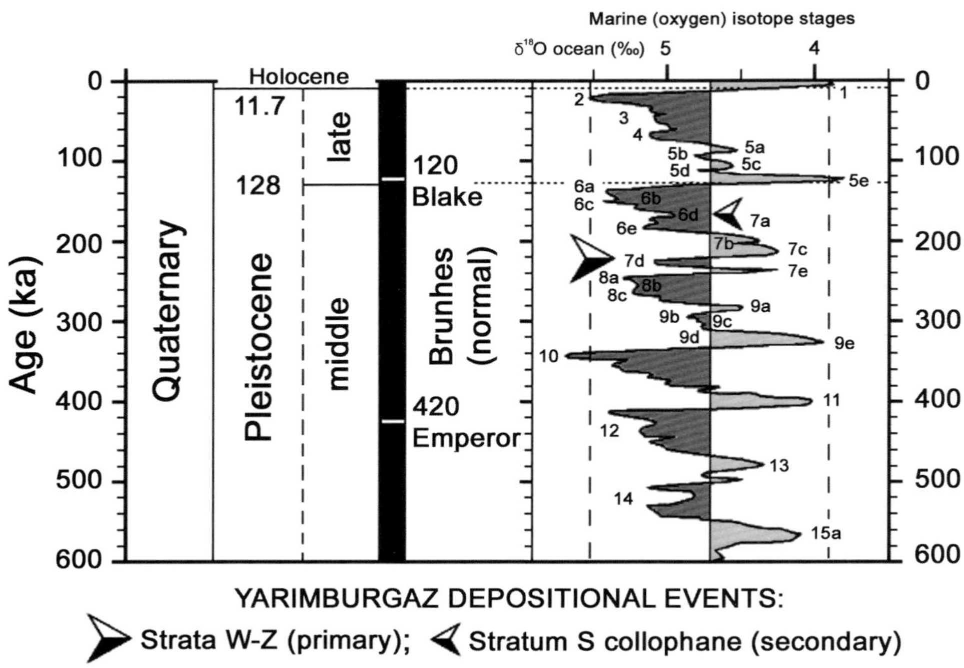

Figure 3.4. The timing for the depositional events in the Lower Chamber of Yarımburgaz cave. The dates for the primary deposition of Strata W through Z, between 226 ± 24 ka and 211 ± 22 ka, correlates best with Marine Isotope Stage (MIS) 7d. The secondary deposition of the collophane found in Stratum S, dated at 156^{+24}_{-20} ka, correlate with MIS 6d. Other secondary deposition of travertine and collophane probably occurred more recently.

faunas including *U. deningeri, U. arctos, Dicerorhinus kirchbergensis, Megaceros giganteus,* and *Equus mosbachensis,* range from 200 to 350 ka. Examples include > 320 ± 50 ka at Belle Roche, Belgium (Gascoyne and Schwarcz 1985), ~ 300 ± 50 ka in Pontnewydd, Wales (Schwarcz 1984), 190 to 220 ± 20 ka for the Layer I-J, Bau de l'Aubesier, France (Blackwell *et al.*, 2001), and > 150 to 240 ka in the Lower Travertine, Ehringsdorf, Germany (Blackwell and Schwarcz 1986; Grün and Stringer 1991).

In sum, the time range for the *U. deningeri-U. spelaeus* transition at other absolutely dated European sites agrees well with our determinations here. As noted, however, choosing the uptake model is critical here. The EU ages, ranging from 130-160 ± 10-20 ka for Strata X and W, seem young when compared with other faunal estimates and the occurrence of the marine deposits in the Upper Chamber correlated with MIS 5e. Alternatively, the RU ages ranging from 270-390 ± 40-60 ka, MIS 8-11, agree better with some estimates for the antiquity of *U. deningeri,* but not the appearance of *U. spelaeus* or transitional forms. While it seems unlikely that all, or even the greater part of, the U in these teeth, was absorbed only recently, the good agreement between the age for T76 when corrected for recent U uptake and the teeth below it suggests that its actual enamel uptake rate may lie somewhere between the LU and RU, like for T57's enamel. The sedimentary water concentrations for the teeth, however, appear to have been fairly high at selected times, such as 150-160 ka and at 120 ka when the beach was deposited, which should have provided the ideal conditions for transferring U from the surrounding sediment into the teeth. Those same wet conditions would also have favored dissolution of bone, other phosphates, and carbonates, including any primary travertine, and possibly generating secondary carbonate or phosphocarbonate precipitation as collophane, calcite, or aragonite.

Wet conditions also promote progressive exponentially increasing degradation of the organic fraction in the teeth, which would act as a locus for U uptake. The combined $^{230}Th/^{234}U$ and ESR data argue for a CU or LU-like model.

In conclusion, all the $^{230}Th/^{234}U$ and ESR data for Yarımburgaz suggest that the fossil-bearing Strata W to Z were deposited between 226 ± 24 ka and 211 ± 22 ka during MIS 7d (Figure 3.4), and suffered secondary collophane deposition at 156^{+24}_{-20} ka during MIS 6d, and possibly at other times thereafter.

ACKNOWLEDGEMENTS

US National Science Foundation (ILI 9151111, SBR 9709912, SBR 9896289) grants to BABB and ARS, City University of New York Research Foundation support to BABB, NSF (SBR 8801699) and NSERC grants to HPS, and NSF grant BNS 8801699 to F.C. Howell, provided funding for the ESR analyses. HPS and the archaeological crew at Yarımburgaz collected the dating and sediment samples. K. Runnels and Ch. Falguères assisted with the computational analyses. K. Goodger performed the NAA analyses. ^{60}Co irradiations and NAA analyses were performed at the McMaster Nuclear Reactor facilitated by M. Butler and A. Pidruczny. A. Wieser helped to calibrate the γ source. McMaster University, Purdue University, the University of Windsor, Queens College of City University of New York, and Williams College provided logistical support.

REFERENCES CITED

Adamiec, G., Aitken, M.J., 1998. Dose-rate conversion factors: update. *Ancient TL* 16, 37-50.

Barabas, M., Walther, R., Wieser, A., Radtke, U., Grün, R., 1993. Second interlaboratory-comparison project on ESR dating. *Applied Radiation & Isotopes* 44, 119-129.

Blackwell, B.A.B., 1989. *Laboratory Procedures for ESR Dating of Tooth Enamel.* McMaster University Department of Geology Technical Memo 89.2, 234 pp.

Blackwell, B.A.B., 1994. Problems associated with reworked teeth in electron spin resonance dating. *Quaternary Geochronology (Quaternary Science Reviews)* 13, 651-660.

Blackwell, B.A.B., 2001. Electron spin resonance (ESR) dating in lacustrine environments, in: Last, W.M., Smol, J.P. (Eds.), *Tracking Environmental Change Using Lake Sediments: Basin Analysis, Coring, and Chronological Techniques*, vol. 1. Kleuwer, Dordecht, pp. 283-368.

Blackwell, B.A.B., 2006. Electron spin resonance (ESR) dating in karst environments. *Acta Carsologica* 35, 10-48.

Blackwell, B.A.B., Blickstein, J.I., 2000. Considering sedimentary U uptake in external dose rates determinations for ESR and luminescent dating. *Quaternary International* 68-71, 329-343.

Blackwell, B.A.B., Schwarcz, H.P., 1986. U-series analyses of the Lower Travertine at Ehringsdorf, DDR. *Quaternary Research* 25: 215-222.

Blackwell, B.A.B., Schwarcz, H.P., 1989. Datation des spéleothems de la grotte Vaufrey par la famille de l'uranium, in: Rigaud, J.-Ph. (Ed.), *La Grotte XV (Grotte Vaufrey) á Cénac-et-St-Julien (Dordogne): Paléoenvironnements, Chronologie, et Activités humaines.* Mémoires de la Société Préhistorique Française 19. CNRS, Paris, pp. 365-380.

Blackwell, B.A.B., Schwarcz, H.P., 1993a. ESR isochron dating for teeth: A brief demonstration in solving the external dose calculation problem. *Applied Radiation & Isotopes* 44, 243-252.

Blackwell, B.A.B., Schwarcz, H.P., 1993b. Archaeochronology and scale, in: Stein, J.K., Linse, A. (Eds.), *The Effects of Scale on Archaeological and Geoscientific Perspectives.* Geological Society of America Special Paper 283, 39-58.

Blackwell, B.A.B., Schwarcz, H.P., 1995. The uranium series disequilibrium dating methods, in: Rutter, N.W., Catto, N.R. (Eds.), *Dating Methods for Quaternary Deposits.* Geological Association of Canada, St. John's, Geotext 2, 167-208.

Blackwell, B.A.B., Schwarcz, H.P., Debénath, A., 1983. Absolute dating of hominids and Paleolithic artefacts from the cave of La Chaise-de-Vouthon (Charente), France. *Journal of Archeological Science* 10, 493-513.

Blackwell, B.A.B., Porat, N., Schwarcz, H.P., Debénath, A., 1992. ESR dating of tooth enamel: Comparison with $^{230}Th/^{234}U$ speleothem dates at La Chaise-de-Vouthon (Charente), France. *Quaternary Science Reviews* 11, 231-244.

Blackwell, B.A.B., Skinner, A.R., Blickstein, J.I., Lebel, S., Leung, H.Y., 2001. ESR isochron analyses at Bau de l'Aubesier, Provence: Clues to U uptake in fossil teeth. *Geoarchaeology* 16, 719-761.

Blackwell, B.A.B., Skinner, A.R., Brassard, P., Blickstein, J.I., 2002. U Uptake in tooth enamel: Lessons from isochron analyses and laboratory simulation experiments, in: Whitehead, N.E., Ikeya, M. (Eds.), *Proceedings of the International Symposium on New Prospects in ESR Dosimetry and Dating,* Society of ESR Applied Metrology, Osaka. Advances in ESR Applications 18, 97-118.

Blackwell, B.A.B., Liang, S.S., Golovanova, L.V. Doronichev, V.B., Skinner, A.R., Blickstein, J.I.B., 2005. ESR at Treugol'naya Cave, northern Caucasus Mt., Russia: Dating Russia's oldest archaeological site and paleoclimatic change in Oxygen Isotope Stage 11. *Applied Radiation and Isotopes* 62, 237-245.

Blackwell, B.A.B., Long, R.A., Séronie-Vivien, M.R., Montoya, A.C., Skinner, A.R., Blickstein, J.I., Tillier, A.M., 2007. ESR dating at Pradayrol, Lot, France: Constraining the age for the Middle Pleistocene bone-bearing layer, its Middle Paleolithic artefacts, and hominid incisor. *Geological Society of America Annual Meeting Program with Abstracts* 39, Abstract 205-2.

Bordes, F., 1972. *A Tale of Two Caves.* Harper Row, New York, 202 pp.

Bordes, F., 1984. *Léçons sur le Paléolithique, I: Notations de Géologie Quaternaire.* Éditions du Centre National de la Recherche Scientifique, Paris.

Brennan, B.J., Schwarcz, H.P., Rink, W.J., 1997a. Simulation of the γ radiation field in lumpy environments. *Radiation Measures* 27, 299-305.

Brennan, B.J., Rink, W.J., McGuirl, E.L., Schwarcz, H.P., 1997b. β doses in tooth enamel by "one-group" theory and the Rosy ESR dating software. *Radiation Measures* 27, 307-314.

Eggins, S., Grün, R., Pike, A.W.G., Shelley, M., Taylor, L., 2003. ^{238}U, ^{232}Th profiling, and U-series isotope analysis of fossil teeth by laser-ablation ICPMS. *Quaternary Science Reviews* 22, 1373-1382.

Farrand, W.R., McMahon, J.P., 1997. History of the sedimentary infilling of Yarımburgaz Cave, Turkey. *Geoarchaeology* 12, 537-565.

Gascoyne, M., Schwarcz, H.P., 1985. Uranium-series dates for the Lower Paleolithic site of Belle Roche, Belgium. *Current Anthropology* 26, 641-642.

Gascoyne, M., Schwarcz, H.P., Ford, D.C., 1978. *Uranium series dating and stable isotope studies of speleothem: Part 1, Theory and techniques.* British Cave Research Association, Transactions 5, 91-111.

Grün, R., 1989. Electron spin resonance (ESR) dating. *Quaternary International* 1, 65-109.

Grün, R., McDermott, F., 1994. Open system modelling for U-series and ESR dating of teeth. *Quaternary Science Reviews* 13, 121-125.

Grün, R., Stringer, C.B., 1991. Electron spin resonance dating and the evolution of modern humans. *Archaeometry* 33, 153-199.

Grün, R., Taylor, L., 1996. Uranium and thorium in the constituents of fossil teeth. *Ancient TL* 14, 21-25.

Guibert, P., Bechtel, G., Schvoerer, M., Müller, P., Balescu, S., 1998. A new method for γ dose rate estimation of heterogeneous media in TL dating. *Radiation Measurements* 29, 663-671.

Guibert, P., Lehaye, C., Veillevigne, E., 2007. Datation par thermoluminesence de Niveau B et I: Rapport de recherche octobre 2006, in: Sandgathe, D., Turq, A., Dibble, H., McPherron, S., Goldberg, P. (Eds.), Roc de Marsal (Campagne de Brugue, Dordogne): Rapport de fouille programmée triannuelle, 2005-2007. Report to French Ministry of Culture, University of Pennsylvania, Philadelphia, pp. 241-266.

Ivanovich, M., Harmon, R.S. (Eds.), 1982. *Uranium Series Disequilibrium: Applications to Problems in Earth, Marine, and Environmental Sciences.* Clarendon, Oxford.

Lisiecki, L.E., Raymo, M.E., 2005. A Pliocene-Pleistocene stack of 57 globally distributed benthic $\delta^{18}O$ records. *Paleoceanography* 20, 1003-1071.

Martinson, D.G., Pisias, N.G., Hays, J.D., Imbrie, J., Moore, T.C. Jr., Shackleton, N.J., 1987. Age dating and the orbital theory of the ice ages: Development of a high-resolution 0 to 300,000-year chronostratigraphy. *Quaternary Research* 27, 1-29.

McDermott, F., Grün, R., Stringer, C.B., Hawkesworth, C.J., 1993. Mass-spectrometric U-series dates for Israeli Neanderthal/early modern hominid sites. *Nature* 363, 252-255.

Özdoğan, M., Koyunlu, A., 1986. Yarımburgaz Mağarası. 1986 Yılı Çalışmalarının ilk sonuçları ve bazı gözlemler. *Arkeoloji ve Sanat* 32/33, 4-17.

Proctor, C.J., Berridge, P.J., Bishop, M.J., Richards, D.A., Smart, P.L., 2005. Age of Middle Pleistocene fauna and Lower Palaeolithic industries from Kent's Cavern, Devon. *Quaternary Science Reviews* 24, 1243-1252.

Rae, A.M., Ivanovich, M., 1986. Successful application of uranium series dating of fossil bone. *Applied Geochemistry* 1, 419-426.

Schwarcz, H.P., 1984. Uranium-series dating and stable isotope analyses of calcite deposits, in: Green, H.S., (Ed.), *Pontnewydd Cave: A Lower Palaeolithic Hominid Site in Wales.* National Museum of Wales, Cardiff, pp. 88-97.

Schwarcz, H.P., Blackwell, B.A., 1983. $^{230}Th/^{234}U$ age of a Mousterian site in France. *Nature* 301, 235-236.

Schwarcz, H.P., Blackwell, B.A., 1992. Archaeological applications, in: Ivanovich, M., Harmon, R.S. (Eds.), *Uranium-series Disequilibrium: Applications to Earth, Marine, and Environmental Sciences.* Clarendon, Oxford, pp. 513-552.

Schwarcz, H.P., Latham, A.G., 1989. Dirty calcites 1: Uranium series dating of contaminated calcites using leachates alone. *Isotope Geoscience* 80, 35-43.

Séronie-Vivien, M.R., Tillier, A.M., 2002. La dent humaine de Pradayrol (Caniac-du-Causse, Lot) dans son contexte stratigraphic et paléontologique. *Paléo* 14, 257-264.

Skinner, A.R., Blackwell, B.A.B., Chasteen, D.E., Shao, J.M., Min, S.S., 2000 Improvements in dating tooth enamel by ESR. *Applied Radiation & Isotopes* 52, 1337-1344.

Wieser, A., Debuyst, R., Fattibene, P., Meghzifene, A., Onori, S., Brik, A., Scherbina, O., Bayankin, S.N., Ivanov, D., Pivovarov, S., Trompier, F., Bugay, A., Ivannikov, A., Stepanenko, V., Skvortsov, V., Ciesielski, B., Schultka, K., Hoshi, H., Imata, H., Miyazama, C., Tielewuhan, E., Toyoda, S., Chumak, V., Sholom, S., Romanyukha, A., Romanyukha, L., Schauer, D., Thomas, J.A., 2005. The 3rd international intercomparison on EPR tooth dosimetry: Part 2, Final analysis. *Applied Radiation & Isotopes* 62, 163-171.

CHAPTER 4

ECOLOGICAL SIGNIFICANCE OF THE SMALL MAMMAL FAUNA FROM YARIMBURGAZ CAVE (TURKISH THRACE)

WIGHART V. KOENIGSWALD – CHRISTA LINDENAU – WINFRIED SANTEL

INTRODUCTION

Bone assemblages found in cave sediments may have very different taphonomic backgrounds. Bones of larger mammals found in archaeological layers are often regarded as being mainly the result of human hunting activities. Often, however, it is not easy to distinguish bones resulting from human activities from those brought in by carnivores or originating from carcasses of animals that died within the caves. For the most part the bones of small mammals found in caves accumulated independently from human activities. Most of these remains originated from at least two very different sources. One source consists of animals that lived and died in the cave. The second includes animals that were brought into the cave by birds of prey and other carnivores. In either case, these small animals represent a sampling of the fauna of the surrounding area, and as a result they are important indicators of ecological conditions. Small mammals have the additional advantage that they tend to occur in larger numbers than macrofauna. Thus, provided that the documentation is good enough, small animal remains are an important component to interpretations of the environment and of the chrono-stratigraphic position of the various layers.

Small mammal remains found during the excavation of the Lower Chamber of Yarımburgaz Cave were unevenly distributed among different stratigraphic units. For the most part there are too few bones from specific layers in the various trenches allow for independent interpretations. Although the level numbers were assigned independently for each trench, the stratigraphic sequences could be correlated with each other, except for square U (see Farrand and MacMahon, this volume). Consequently, the faunal elements from various layers and trenches have been grouped according to the three major sedimentary cycles found in the cave. The correlations of the various units and the sedimentary cycles are given in Table 4.1.

Square U (not layer U) (see Figure 1.6) represents a unique case. It produced a very rich microfaunal assemblage: almost 70% of the bats and the majority of the other small mammals (78%) were concentrated within this square. Unfortunately the deposits in this square are very shallow and the stratigraphy could not be tied in with that of other squares. Thus, the microfauna from square U cannot be securely attributed to any single stratigraphic cycle.

The small mammal fauna from the Yarımburgaz Cave was studied in several stages. A preliminary study was carried out by v. Koenigswald (Koenigswald 1989). Later, Santel conducted a more detailed study of the insectivores and rodents as a diploma thesis at the University of Bonn (Santel 1994); results were partially published (Santel and Koenigswald 1998). Lindenau analyzed the bat remains for another diploma thesis at the University of Bonn (Lindenau 2000); a condensed version of the thesis was also published (Lindenau 2005). Although these results

Table 4.1. Stratigraphic correlation of sedimentary cycles and square containing faunal remains

Sedimentary Cycle	Layer	Square
III	Z	
	Y	T-1
	X	T-2, S-1, S-2, S-3, P-1, R-1, and Y-1a
	W	T-3, T-4, S-4, Y-1b
II	V	Y-1c, d
	U	
	T	
I	S	Y-5
	R	
uncorrelated	-	U

are accessible in the German journal *Eiszeitalter und Gegenwart* we repeat major parts of these studies here, in order to make all relevant data on the small mammal fauna available in this volume.[1]

Studies of bats and of other small mammals are presented separately below. This approach is justified because of the different taphonomic background and a slightly different aims of the interpretation. The bats were using the cave as a shelter while most of the other small mammals were bought in by birds of prey.

THE BAT FAUNA FROM YARIMBURGAZ

The Yarımburgaz caves provided a rich Middle Pleistocene chiropteran fauna. The bat fauna from the site is especially significant in light of the very limited number of fossil localities providing Pleistocene bats in the Mediterranean. So far fossil bats are known mainly from some Mediterranean islands, but their diversity is very limited and the faunas contain mostly endemic species (Storch 1974).

In contrast to the rodents, which were most likely accumulated in owl pellets, the bats probably used Yarımburgaz as a roost where they were protected from bad weather and predators. Bats use caves and other shelters in several different ways. The most significant uses are as roosts, which may be used daily, as transient roosts, mating roosts, maternity roosts and as winter roosts (Gebhard 1997; Schober and Grimmberger 1998). Generally, the first three kinds of roots, also called summer roosts, can occur in tree hollows or caves, whereas hibernation normally occurs only in caves. Some of the chiropterans certainly used the cave seasonally, in summer, whereas others came there for hibernation. For climatic reconstruction it would be very useful to separate species occurring only during summer times from those that were present in the area all year long. Because Yarımburgaz Cave probably provided a shelter during both summer and winter such discrimination is impossible using the fossil record alone.

1 We thank the owner and publisher of the journal "Eiszeitalter und Gegenwart" for the permission to proceed in this way.

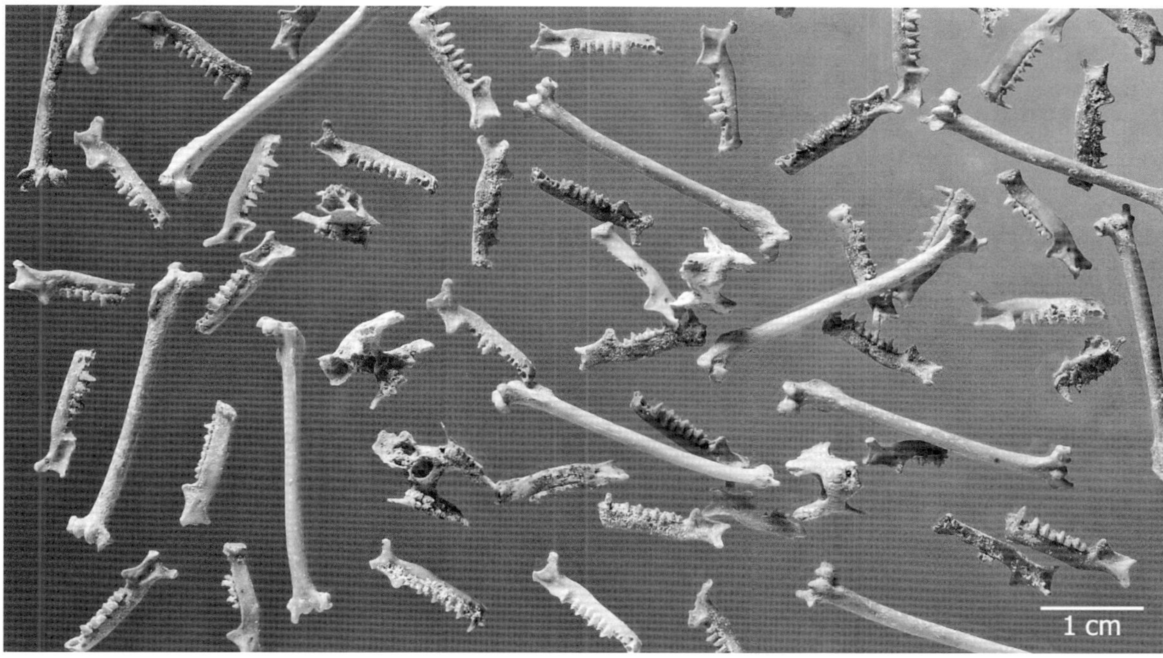

Figure 4.1. Skull fragments, mandibles and humeri from the middle Pleistocene bat *Rhinolophus mehlei.* The sample is typical for the bone preservation in Yarımburgaz Cave. (Photo Georg Oleschinski, Bonn).

The chiropteran taxa

From the rich fossil sample, 229 fragmentary rostra and maxillae, 1562 mandibles, and 717 humeri of chiropterans were taxonomically identifiable. The typical preservation of the material is documented in Figure 4.1. Several bones are covered with a light sedimentary incrustation, which could be partially removed with an insect needle for better identification. In addition to the identifiable material, many remains are damaged in such a way that they cannot be identified to a taxonomic level beyond Chiroptera. For identification of skull elements the following literature was used: Schober and Grimmberger (1998), Görner and Hackethal (1988), and Kowalski and Ruprecht (1981). A special key for jaws can be found in Ruprecht (1987). From among the postcranial material only the humeri were analysed. This element - especially the distal epiphysis - shows important differences in size and morphological characteristics between the various taxa (Felten et al. 1973). Most other elements of bat skeletons are very fragile and do not generally preserve well.

THE DIVERSITY OF THE BAT FAUNA

The various species occurring in the Yarımburgaz fauna are described in Lindenau (2005). All of the bats from Yarımburgaz Cave are Microchiroptera. Recent microchiropterans are distributed worldwide, with 17 families, 150 genera, and 808 species. They are absent only from polar regions and on some isolated oceanic islands (Nowak 1999). Most are insectivorous, but some tropical species are specialists, feeding on fruits, leaves, nectar, pollen, fishes, other small vertebrates, or blood (Koopman 1994; Nowak 1999). The microchiropteran bats are represented by two infraorders, both of which are present in the Yarımburgaz fauna. The Yinochiroptera (premaxillaries almost never fused with maxillary) are represented by the family Rhinolophidae, including four species of one genus. The Yangochiroptera (premaxillaries always fused with maxillaries in adults) are represented by three genera and ten to twelve species of the Vespertilionidae (Figure 4.4). Table 4.2 describes the systematic array and Table 4.3 shows the stratigraphic distribution of the species represented in the middle Pleistocene sediments of Yarımburgaz Caves.

Table 4.2. Chiropteran Taxonomy

Yangochiroptera KOOPMAN 1984
Rhinolophidae GRAY 1825
Rhinolophus blasii PETERS 1866
Rhinolophus euriale BLASIUS 1853
Rhinolophus ferrumequinum (SCHREBER 1774)
Rhinolophus mehelyi MATSCHIE 1901
Yinochiroptera KOOPMAN 1984
Vespertilionidae GRAY 1821
Vespertilioninae GRAY 1821
Myotini TATE 1942
Myotis bechsteini (KUHL 1817)
Myotis blythii (THOMES 1857)
Myotis capaccinii (BONAPARTE 1837)
Myotis daubentoni (KUHL 1817)
M. emarginatus (GEOFFROY 1806)
M. mystacinus (KUHL 1817)
M. brandti (EVERSMANN 1845)
Myotis myotis (BORKHAUSEN 1797)
Myotis nattereri (KUHL 1817)
Plecotini Gray 1866
Plecotus auritus (LINNAEUS 1758)
Plecotus austriacus (FISCHER 1829)
Miniopterinae DOBSON 1857
Miniopterus schreibersi (KUHL 1817)

Determination of species was based on measurements of teeth and humeri as well as on morphological characteristics. The skulls and mandibles of two species, *Myotis myotis* (greater mouse-eared bat) and *Myotis blythii* (smaller mouse-eared bat), are nearly the same size. Nevertheless the taxa can be distinguished by comparing the lower tooth rows to extant material (4.2). The p4 from *M. myotis* is clearly relatively much larger than the p4 of *M. bythini*. Not all bones or teeth can be differentiated by measurements. Because many bats are very similar in size, it is necessary to search for other characteristics for some taxonomic distinctions. For example, the mandibles of *Plecotus auritus* and *Plecotus austriacus* differ in the morphology of the processus angularis. The dorsal aspect of the processus angularis in *Plecotus auritus* is smooth, whereas in *Plecotus austriacus* it bears a small dorsal projection (Figure 4.3 C,D).

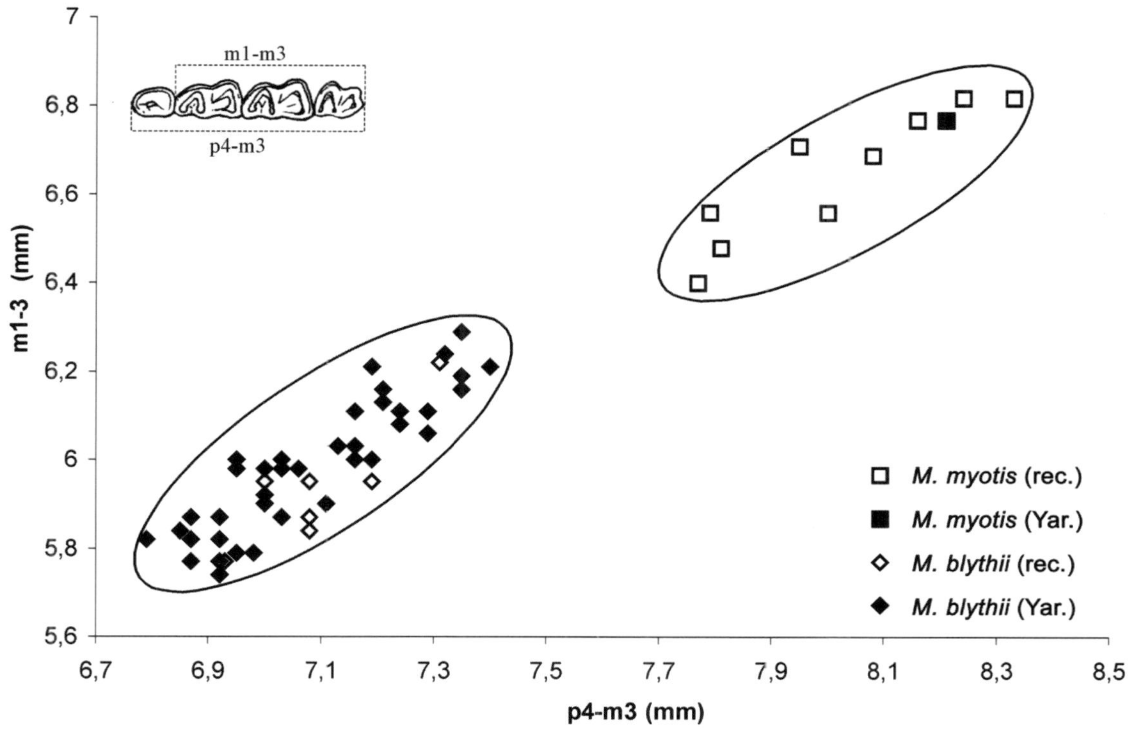

Figure 4.2. Separation of the fossils of *Myotis myotis* and *Myotis bl ythi* based on the proportions of the lower dentition in comparison with extant material. (Drawings CL).

Table 4.3. Stratigraphic distribution of the Chiroptera within the sedimentary cycles I, II, III, and the unstratified unit U of the Yarımburgaz Cave. The numbers refer to the number of mandibles or humeri, respectively.

		Rhinolophus ferrumequinum	*Rhinolophus euryale*	*Rhinolophus blasii*	*Rhinolophus mehelyi*	*Myotis myotis*	*Myotis blythii*	*Myotis bechsteini*	*Myotis natteri*	*Myotis emarginatus*	*Myotis mystacinus-brandti-daubentoni group*	*Myotis capaccinii*	*Plecotus auritus*	*Plecotus austriacus*	*Miniopterus schrebersi*
Cycle III	Z	1					5								6
	Y						7								
	X				6		27	2			1				7
	W				1		25								7
Cycle II	V														
	U				1										
	T														
Cycle I	S	4	4	1	214		102	33				1			16
	R														
Square U		14	1		567	1	260	94	2	5	4	11	1	4	54
Total		19	5	1	789	1	430	129	2	5	5	12	1	4	90

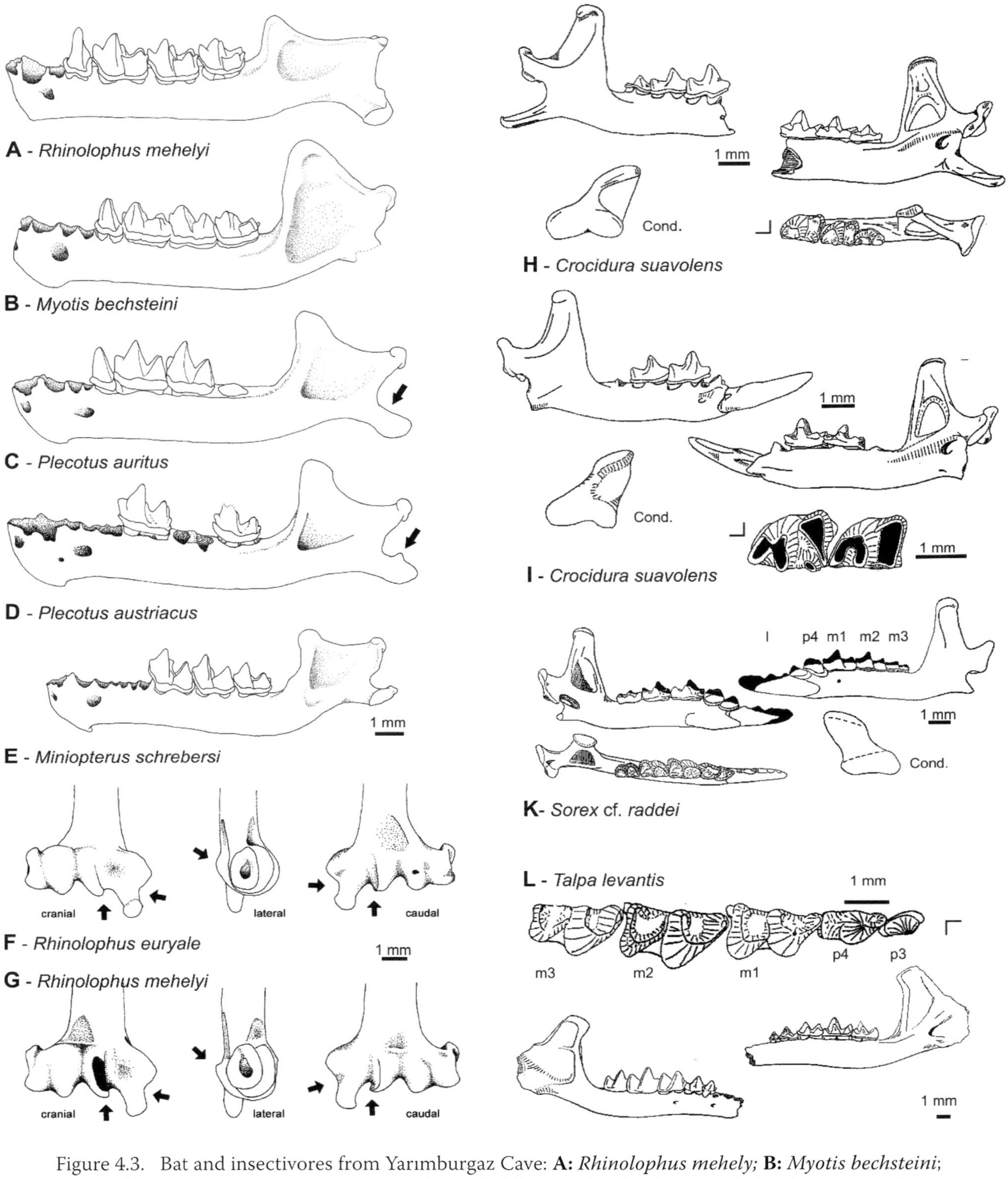

Figure 4.3. Bat and insectivores from Yarımburgaz Cave: **A:** *Rhinolophus mehely;* **B:** *Myotis bechsteini;* **C:** *Plecotus auritus;* **D:** *Plecotus austricaus;* **E:** *Miniopteris schreibersi;* **F-G**: distal humerus of *Rhinolophus euryale* and *Rhinolophus mahely* showing the features for identification; **H:** *Crocidura suavolens*, right mandible with m1-m3, buccal, lingual, and occlusal aspect and condylus (YSM 67); **I:** *Crocidura suavolens*, right mandible with i, m1-m2, buccal, lingual, and occlusal aspect and condylus (YSM 68); **K:** *Sorex* cf. *raddei*, left mandible with i-m3, buccal, lingual and occlusal aspect and condylus (YSM 64); **L:** *Talpa levantis*, right mandible with p3-m3, buccal, lingual and occlusal aspect (YSM 63). (Drawings CL and WS).

Ecological implications of the bat faunas from the three sedimentary cycles

Three sedimentary cycles were distinguished by Howell and Arsebük (1989), Farrand (1989, 1990) and Farrand and McMahon (1997) in the Lower Chamber of Yarımburgaz. Cycle I is equivalent to sedimentary sequence 1 (strata R and S) as described by Farrand and McMahon in this volume; cycle II is equivalent to sedimentary sequence 3 (strata T, U and V); cycle II is equivalent to sedimentary sequence 3 (strata W, X and Y). Layer Z (recent and disturbed infilling) is not considered part of this sequence. Only cycles I and III produced rich bat faunas. Additionally, many remains come from square U, but sediments in this unit cannot be clearly linked to any one of the cycles.

Cycle II produced only a single mandible of *Rhinolophus mehely*, which is clearly insufficient for any ecological reconstruction. The frequencies of the taxa are based on the numbers of mandibles identified. For *Rhinolophus euryale* the humeri were counted because there were no mandibles (Table 4.3). *Myotis mystacinus, Myotis brandti, and Myotis daubentoni* were combined as a single taxon due to the fact that there is much overlap in the characteristics of these species.

The bat fauna of cycle I, represented by 375 mandibles, contains eight species:

Rhinolophus blasii, Rhinolophus euryale, Rhinolophus ferrumequinum, Rhinolophus mehelyi, Myotis bechsteini, Myotis blythii, Myotis capaccinii, and *Miniopterus schreibersi.* The frequency distribution (Table 4.3 and Figure 4.4) is dominated by *Rhinolophus mehelyi* (57%) and *Myotis blythii* (27%), followed by *Myotis bechsteini* (9%). The abundance of *Rhinolophus mehelyi* is ecologically significant because today this species occurs only in warm Mediterranean climates (Lanza and Finotello 1985; Mitchell-Jones et al. 1999). Its abundance indicates that the climate during cycle I was not colder than today, but instead was typically Mediterranean. The coexistence of the second species *Miniopterus schreibersi,* another typical thermophilic taxon, supports this inference. Its recent distribution includes the Mediterranean as well as parts of south Asia, Australia, and Africa (Rabeder 1972; Lanza and Finotello 1985). Most of the species in cycle I, except for *Myotis bechsteini,* prefer open habitats. The tree-dweller, *Myotis bechsteini,* probably used Yarımburgaz Cave for hibernation. Based on the bat species an open landscape with bushes and some trees can be postulated for most of the time represented by cycle I.

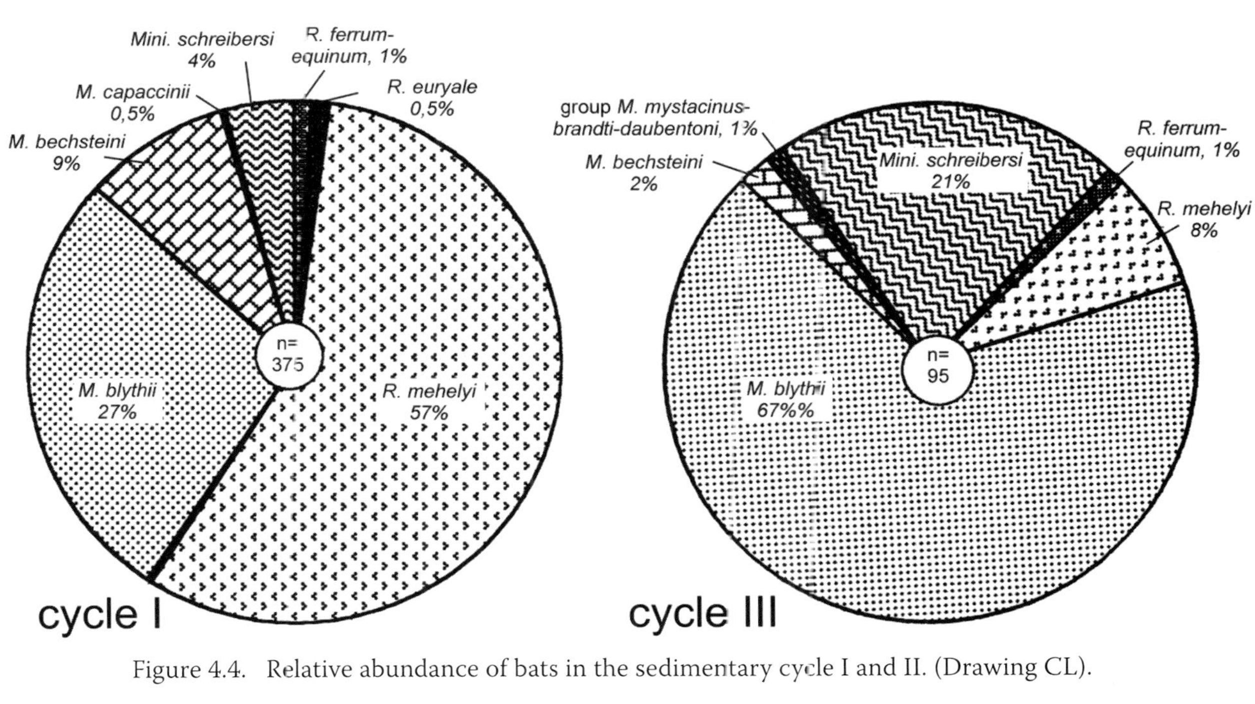

Figure 4.4. Relative abundance of bats in the sedimentary cycle I and II. (Drawing CL).

The cycle III produced 95 mandibles belonging to six taxa: *Rhinolophus ferrumequinum, Rhinolophus mehelyi, Myotis bechsteini, Myotis blythii, Myotis mystacinus-brandti-daubentoni* group, *Miniopterus schreibers.* The fauna is strongly dominated by *Myotis blythii* (67%). *Miniopterus schreibersi* (21%) and *Rhinolophus mehelyi* (8%) are the next most abundant taxa, but are distinctly less common (Table 4.3 and Figure 4.4). Since all of the documented taxa are present in the extant chiropteran fauna of Turkish Thrace no difference from the present-day climate is indicated by the bat fauna of cycle III.

The unstratified square U yielded remains of *Rhinolophus euryale, Rhinolophus ferrumequinum, Rhinolophus mehelyi, Myotis bechsteini, Myotis blythii, Myotis capaccinii, Myotis myotis, Myotis mystacinus-brandti-daubentoni* group, *Myotis nattereri, Plecotus auritus, Plecotis austriacus,* and *Miniopterus schreibersi* (Table 4.3). In this assemblage the most frequent species is *Rhinolophus mehelyi* which represents 52% of the total sample. The high percentage of *Rhinolophus meheyli* might indicate a warm climate, similar to cycle I, but the unit U presumably represents a long time period which could not be subdivided during excavation. Unfortunately, this very rich layer without a clear stratigraphic position cannot be used for the reconstruction of the climatic history of Yarımburgaz Cave.

Summary of the bat fauna

The assemblage of chiropteran remains from Yarımburgaz Cave is the most diverse bat fauna of the Middle Pleistocene in the Eastern Mediterranean. Although the antiquity of the deposits is unquestioned, the taxonomic composition of the assemblage does not differ significantly from the extant chiropteran fauna of the Mediterranean. Obviously a fossil bat fauna in which *Rhinolophus meheli* is dominant can only indicate relative mild climatic conditions. The fossil record of the bats does not indicate any mayor ecological difference from the present day, a strong contrast with the rodent fauna, a part of which indicates much colder and drier climatic conditions. One possible explanation for the seeming lack of concordance is that bats did not populate the area of Turkish Thrace during the colder periods at all, so that most fossils represent the much longer warm periods. The seasonal migration common for various bats may also have an effect, in that bats may have occupied the area only during summer so that the colder seasons are not represented in the fauna.

SORICIDS, TALPIDS, RODENTS AND LAGOMORPHS

A diverse fauna of soricids, talpids, rodents, and lagomorphs was identified from the various trenches in Yarımburgaz Cave (Table 4.4). Unfortunately, almost 80% of the identified material comes from the unstratified square U, and only about 20% can be attributed to the sedimentary cycles I and III. No identifiable small mammals are available from cycle II. Table 4.4 presents the systematic array of the observed taxa and Table 4.5 shows their stratigraphic occurrence. In contrast to the chiropteran fauna, several taxa present in the cave sediments of the Yarımburgaz Cave do not occur in the contemporary fauna of the area. Typical specimens representing the various groups are illustrated in Figures 4.3 and 4.5-4.7).

Table 4.4. Systematic of insectivores, rodents and lagomorphs found in the Yarımburgaz Cave

LIPOTYPHLA	
Soricidae	
	Crocidura suaveolens (Pallas 1811) ve *C. leucodon* (Hermann 1780)
	Sorex cf. raddei Satunin 1895 – **not present in extant fauna**
Talpidae	
	Talpa levantis Tomas 1906
RODENTIA	
Sciuridae	
	Spermophilus cf. *citellus* (Linnaeus 1766)
Gliridae	
Myomimus roachi (Bate 1937)	
Muridae	
Murinae	
	Apodemus (Sylvaemus) sylvaticus (Linnaeus 1758)
	Apodemus sp.
	Apodemus mystacinus (Danford and Alston 1877) – **not present in extant fauna**
	Rattus rattus (Linnaeus 1758)
Cricetinae	
	Cricetus cricetus (Linnaeus 1758) – **not present in extant fauna**
	Mesocricetus cf. *newtoni* (Nehring 1898) – **not present in extant fauna**
	Cricetulus migratorius (Perres 1773)
Arvicolinae	
	Arvicola terrestris (Linnaeus 1758)
	Microtus arvalis (Pallas l788)
	Microtus guentheri Danford and Alston 1880
	Microtus (Terricola) sp.
	Clethrionomys glareolus (Schreber 1780) **not present in extant fauna**
	Lagurus transiens/lagurus: – **not present in extant fauna**
Spalacinae	
	Nannospalax cf. *leucodon* Nordmann 1840
Dipodidae	
	Sicista subtilis (Pallas 1773) **not present in extant fauna**
LAOMORPHA	
	Ocbotona pusilla (Pallas 1769) – **not present in extant fauna**
Lepus sp. – Cycle I: S; Cycle III: U-1.	
	cf. *Oryctolagus cuniculus* (Linnaeus 1758) – **not present in extant fauna**

Paleoecological evaluation of the small mammal fauna in Cycle I

All small mammals from cycle I come from layer S. The fauna consists of *Apodemus mystacinus* (Figure 4.5 H), *Apodemus* sp., *Cricetus cricetus, Cricetulus migratorius, Microtus (Terricola)* sp., and *Nannospalax* cf . *leucodon*. Most species of this assemblage occur as part of the contemporary fauna of the general region. Currently, *Apodemus mystacinus* occurs only on the eastern side of the Bosporus and on the Balkan Peninsula. *Cticetus cricetus* is the only species indicative a somewhat more continental and drier environment for layer S in Yarımburgaz Cave, since its nearest present day occurrence is in the Russian plains north of the 44°. A more detailed stratigraphic assignment can not be provided for this faunal assemblage. Its composition indicates a middle or upper Pleistocene age due to the presence of *Cricetus cricetus* and *Cricetulus migratorius*.

Table 4.5. Stratigraphic distribution of insectivore, rodents and lagomorphs in the Yarımburgaz Cave

		Crocidura suavoles vel leucodon	*Sorex cf. raddei*	*Talpa levantis*	*Spermophilus cf. citellus*	*Myomimos roachi*	*Apodemus sylvaticus*	*Apodemus mystacinus*	*Apodemus sp.*	*Rattus rattus*	*Cricetur cricetus*	*Mesocroiicetus cf. newtoni*	*Cricetulus migratorius*	*Arvicola cantinaus/terrestris*	*Microtus arvalis*	*Microtus guenther*	*Microtus (Terricola) sp.*	*Clethrionomys glareolus*	*Lagurus transiens*	*Nannospalax cf. leucodon*	*Sicista subtilis*	*Ochotona pisillus*	*Lepus sp.*	*Oryctolagus cf. cuniculus*
Cycle III	Z																							
	Y																							
	X				2		6				2	4	1	2	1	1	2	1		15				
	W				4					3	3	4	4	1	5	1	2					5		1
Cycle II	V																							
	U																							
	T																							
Cycle I	S							2	1		3		9				5			13				
	R																							
	?	2	4		4		1					5		4	5		2			3	3			
Square U		6		3		10	114					22	206		15	11	52	14	10	20		4	1	
Total		8	4	3	10	10	123	2	1	3	8	34	221	7	26	13	63	15	10	51	3	9	1	1

Paleoecological evaluation of the fauna in Cycle III

The stratigraphic younger cycle III yielded the major part of the faunal remains. Most of the archeological material belongs to this sedimentary cycle as well (Howell and Arsebük 1989). Most of the small mammal remains come from strata W and X. The following species are present: *Spermophilus* cf. *citellus* (Figure 4.5 C-G), *Apodemus (Syvaemus) sylvaticus* (Figure 4.5 I-K), *Rattus rattus* (Figure 4.5 L), *Cricetus cricetus* (Figure 4.5 M-N), *Mesocrocetus* cf. *newtoni* (Figure 4.5 O), *Cricetulus migratorius* (Figure 4.5 P), *Arvicola cantianus/terrestris* (Figure 4.6 A), *Clethrionomys glareolus* (Figure 4.6 D-E), *Microtus (Terricola)* sp. (Figure 4.6 F), *Microtus arvalis* (Figure 4.6 G), *Microtus guentheri* (Figure 4.6 H), *Nannospalax* cf. *leucodon* (Figure 4.7 A-C), *Sicista subtilis* (Figure 4.6 D-E), *Ochotona pusilla* (Figure 4.6 F-H), and cf. *Oryctolagus cuniculus*. The species *Microtus arvalis* may represent the extant *Microtus rossiameridionalis*. But the latter taxon is based on karyological characters, which cannot be examined in the fossil material. Since we do not want to make inferences about the antiquity of the karyotype modification we used the traditional term *Microtus arvalis* based on morphological characters.

The fauna of cycle III contains species characteristic of a range of different environments such as steppe, dry and open country, and open landscapes with bushes and some trees (Niethammer and Krapp 1978, 1982, 1990). This is not necessarily a non-analog fauna, since similar patterns of coexistence can be observed in the extant fauna of Turkish Thrace. However, the percentage of steppic elements is distinctly higher in cycle III of Yarımburgaz Cave, indicating a landscape more open and dryer than today. In particular the greater steppic influence is indicated by taxa that no longer occur in the area, including *Cricetus, Mesocricetus, Cricetulus, Lagurus, Sicista subtilis, and Ochotona pusilla.* Most of these species today are found in the Ukraine and the Russian plains north of Turkish Thrace. Although the steppic species occur even in semi-des-serts, other species such as *Sorex, Apodemus sylvaticus* and *Clethrionomys glareolus* signal the presence of at least restricted areas with denser vegetation. In contrast to the many immigrants from the north, *Rattus rattus* definitely immigrated from the south, most probably from the

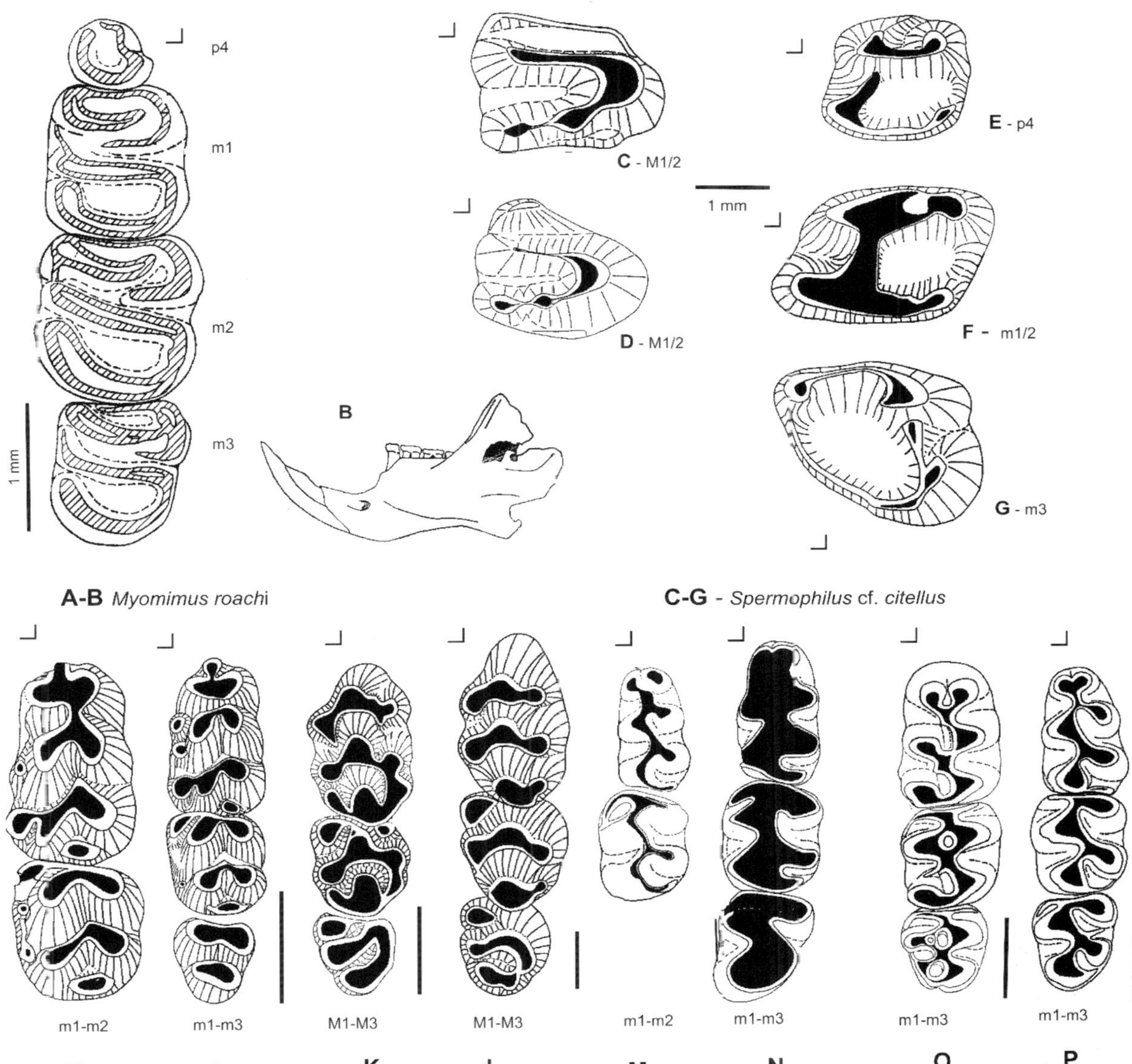

Figure 4.5. Fossil rodents from the Yarımburgaz Cave. **A-B**: *Myomimus roachi* occlusal view of the lower dentition (p4-m3) and left mandible (YSM 7); **C-G:** *Spermophilus* cf. *citellus*, isolated teeth; **C:** right p4 D:right m1/2; E left m3(invers); **F** upper M1/2; G: left M3 (invers); **H:** *Apodemus mystacinus*; left lower m1 and m2 (YSM 8). **I:** *Apodemus sylvaticus*, left lower m1-m2 (YSM 9) and left m3 (YSM 10); **K:** *Apodemus* sp. right upper toothrow (M1-M3) (YSM 12); **L:** *Rattus rattus* , right upper toothrow (M1-M3) (YSM 11); **M:** *Cricetus cricetus,* right m1 (YSM 15) and m2 (YSM 16) (both teeth are only slightly worn); **N:** *Cricetus cricetus,* right m1-m3 (YSM 14) (heavily worn teeth); **O:** *Mesocricetus* cf. *newtoni; N:* right m1-m3 (YSM 18); **P:** *Cricetulus migratorius,* left m1-m3 (YSM21). (Drawings WS).

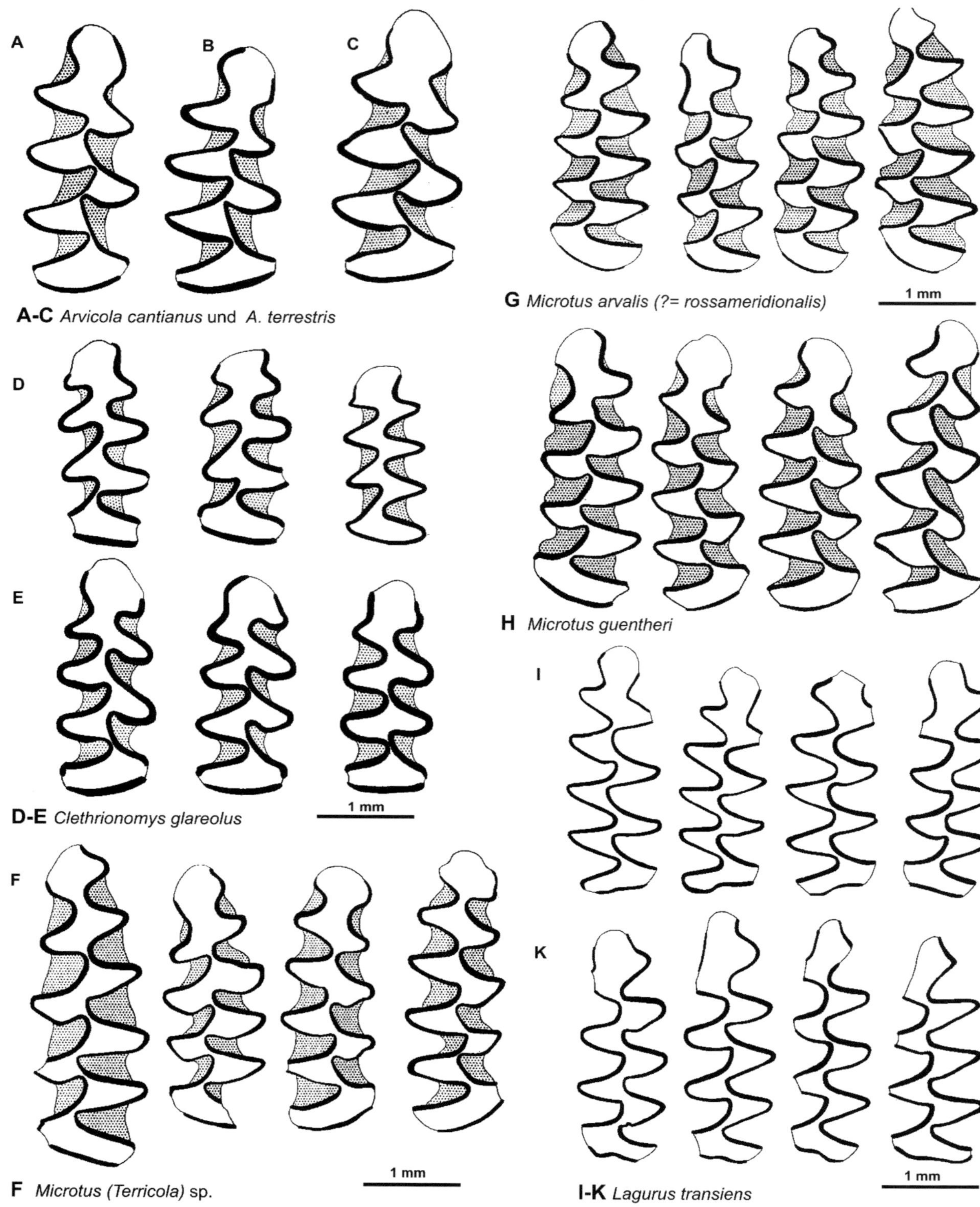

Figure 4.6. Fossil voles from Yarımburgaz Cave. The occlusal surfaces of lower m1 is figured to show the variability. **A:** *Arvicola cantianus/terrestris*, Yarımburgaz (note the similar thickness of the enamel band on leading and trailing edges) (YSM 22); **B:** *Arvicola cantianus*, Middle Pleistocene Petersbuch, Germany (for comparison); **C:** *Arvicola terrestris persicus*, recent, Turkey (Zoological Museum Alexander Koenig, Bonn 68.283) (for comparison); **D-E:** *Clethrionomys glareolus* (YSM 38 - 43); **F:** *Microtus (Terricola)* sp. (YSM 31-33); **G:** *Micotus arvalis* (?=*rossiaemeridionalis*) (YSM 23-26); **H:** *Microtus guentheri*, YSM 27-30); **I-K:** *Lagurus transiens*, variation of the m1 occlusal surface (YSM 44-51). (Drawings WS).

Levant where the earliest *Rattus* is found at Oumm Quatafa (Tchernov 1994). Because of the wide range of *Rattus*, however, this genus is not a good ecological indicator. In Central Europe, especially *Lagurus* and *Ochotona* are regarded as indicators for cool and dry climatic conditions (Heinrich 1983; Sutcliffe and Kowalski 1976). Among the soricids, *Crocidura leucodon* and/or *Crocidura suavolens* are more sensitive to cold climates and are absent during glacial periods in Central Europe.

Taxa of the uncorrelated Square U

The taxa found in square U generally correspond those of Cycle III, but includes several additional species as well: *Crocidura suaveolens* vel *leucodon* (Figure 4.4 H-I), *Sorex* cf. *raddei* (Figure 4.4 J), *Talpa levantis* (Figure 4.4 K), *Myomimus roachi* (Figure 4.5 A-B), *Lagurus transiens/lagurus* (Figure 4.6 I-J). The presence of these taxa, and of *Myomimus roachi*, a typical species of the Mediterranean area, provides the lower limit for temperatures for the fauna of square U. The southern shift of the steppic area to the Bosporus region as documented by the fauna of Square U was probably caused by a general decrease in temperatures and a reduction of rainfall. A cold phase with such significant ecological shifts most probably reflects a period with distinctly lower temperatures and less precipitation. As discussed above, however, the chiropteran fauna contains typical Mediterranean species.

TAXA OF SPECIFIC CHRONO-STRATIGRAPHIC SIGNIFICANCE

In addition to the ecologically informative species, the small mammal assemblages of cycle III and the unstratified square U contain some species often regarded as significant for the purposes stratigraphic correlation. These species are *Arvicola terrestris, Lagurus transiens/lagurus, Ochotona pusilla*, and *Rattus rattus*.

Arvicola cantianus/terrestris (Figure 4.6 A-C)

The genus *Arvicola* is of great significance for the biostratigraphy of the middle Pleistocene in central Europe (Koenigswald 1973; Heinrich 1982, 1987; Koenigswald and Kolfschoten 1996; Koenigswald and Heinrich 1999). Changes in the relative thickness of the enamel band between the leading and trailing edges of the dentin triangles can be used as time-stratigraphic indicators. The less derived form, characterized by a relatively thick trailing edge, characterizes the middle Pleistocene *Arvicola cantianus*, while in the upper Pleistocene *Arvicola terrestris* shows a significant reduction of the thickness of trailing edge. The transition occurs about the last interglacial period. In the Mediterranean, however, especially in Spain and Turkey, the underived condition with a thick trailing edge is still present in the extant fauna (Roettger 1987). The Spanish form is identified as *Arvicola sapidus* (Miller 1908), the Turkish one as *Arvicola terrestris persicus* (De Filippi 1865). *Arvicola* from Yarımburgaz Cave, represented by 7 molars, shows somewhat thickened trailing edges like *A. cantianus*, but in the Mediterranean this character obviously cannot be used in age assignments. Neither can the absence of the primitive *Mimomys*-fold in two lower M1 (length 3.84 and 3.77 mm) be used as a stratigraphic indicator, since this trait is rare in *A. cantianus*. In Turkey it is difficult to distinguish between *A. cantianus* and *A. terrestris*. As Kolfschoten (1990) accepts both taxa as subspecies of *A. terrestris*, the Yarımburgaz *Arvicola* is not identified to a finer level.

Lagurus transiens/lagurus (Figure 4.6 I-K)

Molars of *Lagurus*, the gray lemming, are characterized by the lack of cementum and the typical dentine tract in the anterior lobe. Two species, the Middle Pleistocene *L. transiens* and the extant *L. lagurus*, have to be considered in the identification of the fossil material of Yarımburgaz, which consists of eight specimens of the lower M1. The length of the M1 (2.35-2.58 mm with a mean of 2.45 mm) is intermediate between the two species if measurements from Central Europe are used (Chaline 1972; Heinrich 1990; Jánossy 1962). *Lagurus lagurus* from Chios described by Storch (1975) is slightly larger than the Yarımburgaz specimens. The morphology of the anterior lobe can be studied in seven specimens of the M1. Two of them fall within the range of variability of *Lagurus lagurus*, but four others show the pattern typical of *Lagurus transiens*, which is characterized by missing or fairly small re-entrant folds of the cap of the anterior lobe. If the material, which came exclusively from square U, is regarded as representative, the Yarımburgaz population most likely was transitional between the two species.

Rattus rattus (Figure 4.5 L)

A left upper maxilla with all molars in place morphologically matches *Rattus rattus* so well that identification is certain. It differs from *Rattus norvegicus* in several morphological characters of the M1. Although the measurements of the molars fall into the range of extant *Rattus rattus* from western Anatolia (Felten et al. 1971), contamination by recent material is not probable, since the maxilla is well fossilized like all other bones from layer W. *Rattus* is rare in the Mediterranean Middle Pleistocene and missing in Central Europe. *Rattus haasi* was described from Oumm Qatafa in Israel (Tchernov 1968). However, as it is represented by lower molars only it is difficult to compare with our material. Storch (1975) attributes upper and lower molars from Chios to *Rattus rattus*. These teeth are somewhat larger than the Yarımburgaz material but are still in the range of *Rattus rattus* in Anatolia.

Ochotona pusilla (Figure 4.7 F-H)

This small species is the most common lagomorph in the Yarımburgaz fauna. It is represented by several isolated molars, one P3, and seven mandibular fragments. The species *O. pusilla* can be identified by the morphology of the lower P3, especially by the deep lingual reentrant fold separating an enlarged posterior part from the medial part and by the rhomboid shape of the anterior part (Figure 4.1c). This shape is known in *O. pusilla* from the Middle Pleistocene from Central Europe, from Stránská Skála (Tobien 1972) and Hunas (Heller et al. 1983). The length of the alveolar tooth row in the lower jaw is very similar in Middle and Upper Pleistocene material attributed to *O. pusilla*. *Ochotona* is known in the Mediterranean Middle Pleistocene in Central Anatolia (Montuire et al. 1994) and from Oumm Qatafa in Israel (Tchernov 1994) but is missing in Chios (Storch 1975). *Ochotona pusilla* which occurs today in the Russian plains did not reach the Eastern Mediterranean during the cold periods of the Late Pleistocene.

The stratigraphic position of the fauna of Yarımburgaz Cave is marked best by the transition of *Lagurus transiens* to *Lagurus lagurus* in square U (Koenigswald and Tesakov 1997). In the Ukraine and in the plains of southern Russia, the transition between *L. transiens* and *L. lagurus* occurs in the Gungki-Complex, containing the Oka-Glacial and the Likhvin interglacial period of the Middle Pleistocene (Markova 1990, 1992). The lower Middle Pleistocene, starting at the Brunhes-Matuyama boundary, can be excluded because at this time *Mimomys savini* was still present. *Arvicola* occurs only in the latest interglacial periods of the Cromerian complex. The

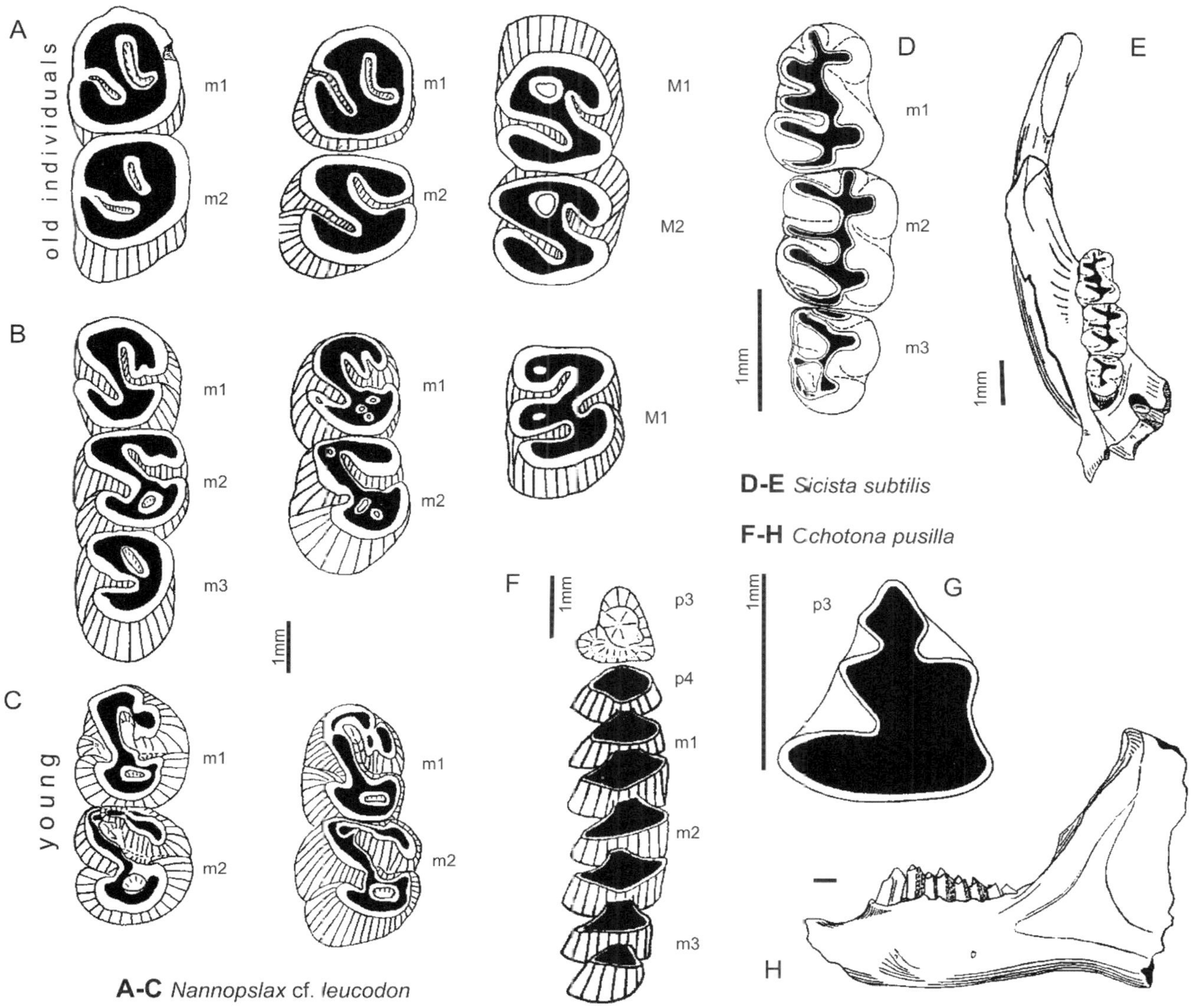

Figure 4.7. Exotic rodents and lagomorphs in the fossil fauna of Yarımburgaz Cave. **A-C:** *Nannospalax* cf *leucodon,* molars in various stages of wear, **A:** young, **B**: medial, **C:** old individuals; **D-E**: *Sicista subtilis,* left mandible with m1-m2 and characteristic foramen besides m2/3; E: lower m1-m3; F-I: *Ochotona pusilla*, **F-H:** left mandible and tooth row (YSM 62); **H:** enlarged right p4. (Drawings WS).

upper part of the Middle Pleistocene in Russia is characterized by the presence of the derived *Lagurus lagurus*. The Yarımburgaz fauna appears to document the transition from *Lagurus transiens* to *L. lagurus*. This allows us to conclude that the fauna in Yarımburgaz Cave dates to a cold period in the middle part of the Middle Pleistocene.

THE CHRONO-STRATIGRAPHIC POSITION OF YARIMBURGAZ CAVE RELATIVE TO SOME MIDDLE PLEISTOCENE LOCALITIES IN THE REGION AND ITS BIOGEOGRAPHIC SIGNIFICANCE

The fissure filling of Emirkaya-2 in Central Anatolia was published by Montuire et al. (1994). The fauna seems definitely older than that of Yarımburgaz Cave, cycle III, if the molar of *Mimomys savini* belongs to this assemblage. Among the many species this fauna shares with the Yarımburgaz cycle III fauna, *Ochotona* and three cricetine species are of significance. As an additional element of the steppe environment, *Allactaga*, occurs in the Anatolian fauna. Therefore the ecological interpretation of Emirkaya-2 as representing an open and humid environment near to a forest must be questioned. Furthermore, the assignment of this fauna to the Holsteinian, corresponding to the Elsterian-Saalian interglacial period, is not reliable because for the time being no vertebrate fauna in Central Europe can be attributed with certainty to the Holsteinian. Even the Holsteinian of northern Germany can not be correlated with sufficient confidence to the glacial sequence of the Alpine region (Koenigswald 1992, 2006; Koenigswald and Heinrich 1999).

Storch (1975) described a fauna from the fissure filling Latomi-1 on the Aegean Island of Chios, which was part of the mainland during the Middle Pleistocene. It is assumed to be not older than the middle part of the Middle Pleistocene. Storch (1975) attributes the *Arvicola* material from Chios to *A. praeceptor* which is regarded as a younger synonym of *Arvicola cantianus* (Koenigswald 1973). In the thickness of the enamel band, the trailing edges surpass the leading edges, as is typical for *A. cantianus* in Central Europe. As discussed above, this character is of little chronological significance in the Mediterranean. Therefore we do not consider the identification of *Arvicola cantianus* at Chios a reliable indication that this fauna is older than the cycle III fauna from Yarımburgaz Cave. *Lagurus* offers a better correlation. *Lagurus lagurus* that Storch (1975) identified in the fauna of Chios is more derived than the transitional form found at Yarımburgaz Cave. Therefore we argue that the Chios fauna is somewhat younger than the fauna of cycle III and square U in Yarımburgaz Cave.

Tchernov (1968, 1994) described a fauna from Oumm Qatafa in the Judean desert near Bethlehem in Israel. So far the chrono-stratigraphic position for Oumm Qatafa is not clear. It is difficult to compare this fauna with Yarımburgaz Cave since it is situated almost 10° further south. The Israeli fauna contains several African elements. Nevertheless, it shares two important taxa with the fauna of Yarımburgaz Cave. These are *Ochotona*, a northern immigrant, and *Rattus*, an immigrant from the southeast. On the other hand, Oumm Qatafa contains more primitive cricetines than occur in the Yarımburgaz Cave assemblages. The species at Yarımburgaz Cave occur in Israel much later, at Hayonim Cave for example (Tchernov 1994). This may indicate that the fauna of Oumm Qatafa is older than the cycle III and square U from Yarımburgaz Cave, even if such a correlation is tentative.

BIOGEOGRAPHICAL IMPLICATIONS

Even if Yarımburgaz Cave occupies a geographically marginal position in Europe, the occurrence of several taxa predates the estimated immigration time into this continent, as reported by Hosey (1982). While *Mimomys roachi, Apodemus mystacinus, Microtus guentheri*, and *Mesocricetus* were assumed to have crossed the Bosporus at the lowest sea level stand during the last glaciation, the finds from Yarımburgaz prove that these species were present west of the Bosporus before this time. The occurrence of *Talpa levantis* in the Middle Pleistocene of Yarımburgaz Cave may support Vohralik's (1991) postulation that *T. levantis* originated in Asia Minor and penetrated the Balkan region. The very small number of Middle Pleistocene localities makes it impossible to securely identify the routes of immigration or their exact timing, however. Because the Pleistocene encompasses several climatic oscillations and multiple sea level fluctuations, several waves of immigration must be postulated. Climatic oscillations may even have caused partial extinctions between phases of immigration. It is almost impossible to reconstruct the faunal movement without a large set of densely-spaced and well-dated localities. The small mammal fauna from Yarımburgaz Cave is biogeographically important since it is almost the only fauna in the region from this time period. Together with other localities to be discovered in the future, it should significantly contribute to improved biogeographical reconstructions for the Middle Pleistocene.

CONCLUSIONS

The small mammal fauna of the Yarımburgaz Cave is highly diverse. It consists of bats that lived within the caves and other small mammals which most probably were brought in by predatory birds. The distribution of the small mammalian fauna is very uneven in the three sedimentary cycles. Cycle II is almost void of fossils. Most of the material comes from square U which can not be correlated any one of the sedimentary cycles. Overall, the fauna contains steppic elements as well as those typical of the Mediterranean. Cycle I and III differ somewhat, but not significantly enough to interpret details of the climatic conditions when they were deposited. Among the rodents and lagomorphs are several species that no longer occur in Turkish Trace. Their current ranges are more to the northeast, indicating distinctly more steppic conditions in the area around the cave than can be found there at present. The gray lemming (*Lagurus transiens*) is also of stratigraphic significance, as it clearly indicates a Middle Pleistocene age. All bats identified from the Yarımburgaz paleontologial assemblages occur today in Turkish Trace. Some taxa are typically Mediterranean (e.g., *Rhinolophus mehelyi*). Others have ranges that reach much further south. Thus the bat fauna does indicate significant climatic differences from the present. This apparent discordance between bats and other small mammals may be explained by seasonal migrations of the bats, bringing thermophilic species into the area during the summer, or to an absence of bats from Turkish Thrace during the colder glacial periods.

REFERENCES CITED

Chaline, J., 1972. Les rongeurs du Pléistocéne moyen et supérieur de France (Systematique-Biostratigraphie-Paléoclimatologie). *Cahier de Paléontologie*, CNRS, Paris.

Farrand, W.R., 1989. Geological observations at Yarımburgaz Cave, in: Howell, F.C., Arsebük, G. (Eds.), Yarımburgaz Cave, report on investigations and current status of researches in the cave of Yarımburgaz (Marmara,Turkey). 1989 Field season. Committee on Research and Exploration, National Geographic Society, Washington D.C., pp. 15-20.

Farrand, W.R., 1990. Geochronology of Yarımburgaz Cave, Turkey, in: Bermudes de Castro, J.M., Arsuaga, J., Carbonell, E. (Eds.), *Human Evolution in Europe and the Atapuerca evidence.* Junta de Castilla y Lèon, Burgos, pp.19-36.

Farrand, W.R., McMahon, J.P., 1997. History of the sedimentary infilling of Yarımburgaz Cave, Turkey. *Geoarchaeology* 12, 537-565.

Felten, H., Helfricht, A., Storch, G., 1973. Die Bestimmung der europäischen Fledermäuse nach der distalen Epiphyse des Humerus. *Senckenbergiana Biologica* 54, 291-297.

Felten, H., Spitzenberger, F., Storch, G., 1971. Zur Kleinsäugerfauna des Bey-Gebirges, SW-Anatolien. *Natur und Museum* 101, 21-25.

Gebhard J., 1997. *Fledermäuse.* Birkhäuser, Basel.

Görner, M., Hackethal, H., 1988. Chiroptera., in: Görner, M., Hackethal, H. (Eds), *Säugetiere Europas.* Enke, Stuttgart, pp. 72-129.

Heinrich, W.D., 1982. Zur Evolution und Biostratigraphie von Arvicola (Mammalia: Rodentia) im Pleistozän Europas. *Zeitschrift für Geologisceh Wissenschaften* 10, 683-735.

Heinrich, W.D., 1983. Skelettreste von *Lepus europaeus* PALLAS, 1778 (Lagomorpha, Mammalia) aus dem fossilen Tierbautensystem von Pisede bei Malchin. Wiss. Z. Humboldt-Universität Berlin, *Math.-Nat. R.* 32, 713- 117.

Heinrich, W.D., 1987. Neue Ergebnisse zur Evolution und Biostratigraphie von Arvicola (Mammalia: Rodentia) im Quartär Europas. *Zeitschrift Geologischer Wissenschaften* 15, 389-406.

Heinrich, W.D., 1990. Nachweis von *Lagurus lagurus* (Pallas, 1773) für das Pleistozän von Neumark-Nord, Kreis Merseburg. *Veröffentlichungen des Landesmuseums für Vorgeschichte Halle* 43, 167-175.

Heller, F., Brunnacker, K., Freund, G., Groiss, J.T., Jánossy, D., Schoch, W., Schweingruber, F. Stadie, C., Streit, R., 1983. Die Höhlenruine Hunas bei Hartmannshof (Landkreis Nürnberger Land). Eine paläontologische und urgeschichtliche Fundstelle aus dem Spät-Riß. *Quartär-Bibliothek* 4. Röhrscheid, Bonn.

Hosey, G.R., 1982. The Bosporus land-bridge and mammal distributions in Asia Minor and the Balkans. *Säugetierkundliche Mitteilungen* 30, 53-62.

Howell, F.C., Arsebük, G. (Eds.), 1988. Yarımburgaz Cave 1988: Report on investigations in the cave of Yarımburgaz (Marmara, Turkey) 1988 Field season. Committee on Research and Exploration, National Geographic Society, Washington D.C.

Howell, C., Arsebük, G., (Eds.) 1989. Yarımburgaz Cave, Report on investigations and current status of researches in the cave of Yarımburgaz (Marmara, Turkey) 1989 Field season. Committee on Research and Exploration, National Geographic Society, Washington D.C.

Jánossy, D., 1962. Vorläufige Mitteilung über die mittelpleistozäne Vertebratenfauna der Tarkö-Felsnische (NO-Ungarn, Bükk-Gebirge). *Ann. Hist.-Natur. Mus. Nat. Hungarici* 54, 155-176.

Koenigswald, W.v., 1973. Veränderungen in der Kleinsäugerfauna von Mitteleuropa zwischen Cromer und Eem (Pleistozän). *Eiszeitalter und Gegenwart* 23/24, 159-167.

Koenigswald, W.v., 1989. Preliminary remarks on the small mammals from Yarımburgaz Cave (Marmara, Turkey). Appendix 1, in: Howell, C., Arsebük, G. (Eds.), Yarımburgaz Cave (1988) - Report on investigations in the cave of Yarımburgaz (Marmara,Turkey) 1988 Field season. Committee on Research and Exploration, National Geographic Society, Washington D.C.

Koenigswald, W.v., 1992. Various aspects of migrations in terrestrial mammals in relation to Pleistocene faunas of Central Europe, in: Koenigswald, W.v., Werdelin L. (Eds.), *Mammalian Migration and Dispersal Events in the European Quaternary.* Courier Forschungs-Institut Senckenberg 153, pp. 39-47.

Koenigswald, W.v., 2006. Climatic changes, faunal diversity and the environment of the Neanderthals in Central and Western Europe during the middle and upper Pleistocene. *Terra Nostra* 2006/2, 35-40.

Koenigswald, W.v., Heinrich, W.D., 1999. Mittelpleistozäne Säugetierfaunen aus Mitteleuropa - der Versuch einer biostratigraphischen Zuordnung. *Kaupia* 9, 53-112.

Koenigswald, W.v., Kolfschoten, T.v., 1996. The *Mimomys-Arvicola* boundary and the enamel thickness quotient (SDQ) of *Arvicola* as stratigraphic markers in the Middle Pleistocene, in: Turner, C. (Ed.), *The Early Middle Pleistocene in Europe*. Balkema, Rotterdam, pp. 211-226.

Koenigswald, W.v., Tesakov, A., 1997. The evolution of the schmelzmuster in Lagurini (Arvicolinae, RODENTIA). *Palaeontographica* B 245, 45-61.

Kolfschoten, T.v., 1990. The evolution of the mammal fauna in the Netherlands and the middle Rhine area (Western Germany) during the late middle Pleistocene. *Mededelingen Rijks Geoloische Dienst* 43/3, 1-69.

Koopman, K.F., 1994. *Chiroptera: Systematics.* Walter de Gruyter, Berlin.

Kowalski, K., Ruprecht, A.L., 1981. Order: Bats – Chiroptera, in: Pucek, Z. (Ed.), *Keys to Vertebrates of Poland, Mammals.* Polish Scientific Publishers, Warschau, pp. 101-154.

Lanza, B., Finotello, P.L., 1985. Biogeografia dei Chirotteri italiani. *Bolletino Museo Regionale Sciencia Naturae Torino* 3, 389-420.

Lindenau, C., 2000. Die mittelpleistozänen Chiroptera (Mammalia) aus der Yarımburgaz -Höhle, westlich von Istambul, Türkisch Thrazien.Unpublished diploma thesis, Bonn University.

Lindenau, C., 2005. Middle Pleistocene bats (Mammalia:Chiroptera) from the Yarımburgaz Cave in Turkish Thrace (Turkey). *Eiszeitalter und Gegenwart* 55, 122-148.

Markova, A.K., 1990. Pleistocene microtheriofauna of the European part of the USSR. *International Symposium On the Evolution, Physiology and Biostratigraphy of the Arvicolids (Mammalia, Rodentia)*, Prague, pp. 313-338.

Markova, A.K., 1992. Influence of palaeoclimatic changes in the Middle and Late Pleistocene on the composition of small mammal faunas: data from Eastern Europe. *Courier Forschunsinstitut Senckenberg* 153, 93-100.

Mitchell-Jones, A.D., Amori, G., Bogdanowicz, W., Krystufek, B., Reijnders, P.J.H., Spitzenberger, F., Stubbe, M., Thissen, J.B.N., Vohralik, V., Zima, J., 1999. *The Atlas of European Mammals.* Academic Press, London.

Montuire, S., Sen, S., Michaux, J., 1994. The Middle Pleistocene mammalian fauna from Emirkaya-2, Central Anatolia (Turkey); systematics and paleoenvironment. *Neues Jahrbuch für Geologie und Paläontologie, Abhandlungen* 193, 107-144.

Niethammer, J., Krapp, F. (Eds.), 1978. *Handbuch der Säugetiere Europas.* Band 1: Rodentia I (Sciuridae, Castoridae, Gliridae, Muridae). Akademische Verlagsgesellschaft, Wiesbaden.

Niethammer, J., Krapp, F. (Eds.), 1982. *Handbuch der Säugetiere Europas.* Band 2/I: Rodentia II (Cricetidae, Arvicolidae, Zapodidae, Spalacidae, Hystricidae, Capromyidae). Akademische Verlagsgesellschaft, Wiesbaden.

Niethammer, J., Krapp, F., (Eds.), 1990. *Handbuch der Säugetiere Europas.* Band 3/I: Insectivora - Primates. AULA-Verlag, Wiesbaden.

Nowak, R.M., 1999. *Walker's Mammals of the World.* Johns Hopkins University Press, Baltimore.

Rabeder, G., 1972. Die Insectivoren und Chiropteren (Mammalia) aus dem Altpleistozän von Hundsheim (Niederösterreich). *Annales Naturhistorisches Museum* Wien 76, 375-474.

Roettger, U., 1987. Schmelzbandbreiten an Molaren von Schermäusen (*Arvicola* Lacepede, 1799). *Bonner zoologische Beiträge* 38, 95-105.

Ruprecht, A.L., 1987. A key for mandible identification of Polish bats (Polish). *Przegl. Zool.* 31, 89-105.

Santel, W.T., 1994. Die mittelpleistozänen Insectivora, Chiroptera, Rodentia und Lagomorpha (Mammalia) aus der Yarımburgaz Höhle, westlich von Istanbul (Türkisch Thrazien). Unpublished diploma thesis Bonn University.

Santel, W.T., Koenigswald, W.v., 1998. Preliminary report on the middle Pleistocene small mammal fauna from Yarımburgaz Cave in Turkish Thrace. *Eiszeitalter und Gegenwart* 48, 162-169.

Schober, W., Grimmberger, E., 1998. *Die Fledermäuse Europas*. Kosmos Verlag, Stuttgart.

Storch, G., 1974. Quartäre Fledermaus-Faunen von der Insel Malta. *Senckenbergiana Lethaea* 55 (1/5), 407-434.

Storch, G., 1975. Eine mittelpleistozäne Nager-Fauna von der Insel Chios, Ägäis. *Senckenbergiana Biologica* 56 (4/6), 165-189.

Sutcliffe, A.J., Kowalski, K., 1976. Pleistocene rodents of the British Isles. *Bulletin of the British Museum of Natural History* (Geology) 27 (2), 31-147.

Tchernov, E., 1968. *Succession of rodent faunas during the Upper Pleistocene of Israel.* (Morphologie, Taxonomie und Systematik der Nagetierfauna in Israel während des jüngeren Pleistozäns.) Mammalia Depicta. Parey, Hamburg.

Tchernov, E., 1994. New comments on the biostratigraphy of the Middle and Upper Pleistocene of the Southern Levant, in: Bar-Yosef, O., Kra, R.S. (Eds), *Late Quaternary Chronology and Paleoclimates of the Eastern Mediterranean. Radiocarbon (Special Issue).* Dept. of Geosciences, University of Arizona, Tucson, pp. 333-350.

Tobien, H., 1972. *Citellus* (Rodentia) und Lagomorpha aus den älteren Aufsammlungen an der pleistozänen Lokalität Stránská Skala bei Brno (CSSR). *Studia Musei Moraviae Anthropos* 20 (12), 137-147.

Vohralik, V., 1991. A record of the mole *Talpa levantis* (Mammalia: Insectivora) in Bulgaria and the distribution of the species in the Balkans. *Acta Univ. Carolinae Biol.* 35, 119-127.

CHAPTER 5
THE YARIMBURGAZ LITHIC ASSEMBLAGE

STEVEN L. KUHN

INTRODUCTION

This section describes the technological, morphological and typological characteristics of the lithic assemblages from the Lower Chamber of Yarımburgaz Cave.

All available artifacts from the 1986 through 1990 excavations in the "Lower Chamber" (LC) of Yarımburgaz Cave are included in this analysis. Small samples of artifacts housed at the University of Istanbul and the Istanbul Archaeological Museum that could not be taken off display are excluded from the sample, as are artifacts from earlier excavations in the upper chamber (UC). Although there is indisputable evidence for a Paleolithic occupation in the UC, the assemblages appear to have been extensively mixed as a result of the activities of later inhabitants of the cave.

The entire collection from the lower cave is treated as a single unit in the following discussions, except for some analyses in the concluding section. The great majority of lithic artifacts are from layers V, W and X (see Farrand and McMahon, this volume). A very small number of specimens are from layer Z at the top of the sequence, and from layers T and U, lower in the stratigraphic column. As Table 5.1 demonstrates, there is a high degree of similarity among the assemblages derived from the three main artifact-bearing stratigraphic units. The most pronounced contrast is a difference of about 11% in the frequency of formal cores between layer V and layer X: otherwise, the assemblages from the various layers are remarkably homogeneous. If the assemblages defined by the layers actually represent distinct occupations, they are essentially indistinguishable from the standpoint of the stone tools. A more likely explanation is that the sample as a whole derives from a series of occupational episodes occurring within a relatively

Table 5.1. Percentages of major artifact categories among stratigraphic units.

		LAYER	
	V	W	X
Artifact classes			
flake tools	33.60%	31.00%	38.60%
flakes and debris	53.70%	55.7	47.6
cores & core tools	12.70%	13.3	13.8
Cores			
tested pebbles	20.00%	22.50%	27.30%
formal cores	53.3	45.7	42.4
choppers	26.7	32.6	30.3
Raw Materials			
flint	67.50%	66.00%	62.40%
quartz	15.2	18.9	20.4
quartzite	17.3	15	17.2

narrow interval of geologic time. Some of the artifact concentrations observed and mapped during the excavations may represent such episodes. Artifacts and bones were subsequently dispersed within a larger portion of the stratigraphic column as a result of digging by denning bears and other processes of disturbance. As a result, the assemblages have been homogenized to some degree.

RAW MATERIALS

There are sharp differences in the ways various raw materials were exploited and transformed into tools. Each major material type was exploited using a different modal method of flake production or shaping. The majority of the Yarımburgaz lithics are made of one of three raw materials: in order of abundance these are flint, quartz, and quartzite. Small quantities of jasper, silicified wood, and unidentified metamorphic rocks are also present in the assemblage. These rarer rock types, grouped under the heading of "other" raw materials in all tabulations, are too scarce to be treated as separate sub-samples. The major rock types are described below.

The Yarımburgaz flint appears to be of high quality. Its grain is somewhat variable, however, and many specimens are fossiliferous. It is possible that more than one type of flint is represented within the collections. However, more than 95% of the flint artifacts are heavily patinated to a uniform whitish-gray color, and a significant number of these are almost completely desilicified. This chemical alteration makes it very difficult to establish the original appearance of most specimens. The few unpatinated pieces, which might even be intrusive from later (Middle, Upper, and post-Paleolithic) occupations, are opaque and medium brown in color.

The quartz used by Yarımburgaz toolmakers is a semi-translucent, milky white crystalline variety. It has a strong tendency to fracture along crystal boundaries, producing flat surfaces and rhomboidal cross-sections in flakes. Despite its problematic fracture properties, the quartz used at Yarımburgaz is suitable for producing robust and quite regular working edges. The most common variety of quartzite appears to be relatively coarse grained, but it is well indurated and is quite homogeneous in texture. It is clear from the artifacts present in the collections that large, comparatively thin flakes could be struck from quartzite cobbles without hinging or transverse fracture.

Little is known about the natural distribution of raw materials in the vicinity of Yarımburgaz Cave. Archaeological specimens that retain some portion of the outer surface exhibit cortex typical of heavily rolled stream cobbles. None of the flint artifacts preserves unaltered nodular or tabular cortex. At least some, and quite possibly most, of the raw materials, may have been found quite close to the site in the past. Flint and quartz pebbles do occur naturally in the fluvial deposits underlying the archaeological strata within the cave (see Farrand and McMahon, this volume), suggesting that there is or was a source of pebbles nearby.

Although there is little reliable information on the origins of the raw materials utilized at Yarımburgaz, it does seem that some types of stone were available in larger "packages" than others. Unbroken quartzite cores and core tools average 9.8 cm in length, with a maximum of 17.1 cm. In contrast, flint cores average 6.3 cm (maximum = 16.2 cm), and quartz cores 5.8 cm in length (maximum = 11.3 cm). Even considering the less extensive exploitation of quartzite cores and core tools (see below), it is very likely that unmodified pebbles of quartzite were originally somewhat larger than pebbles of the other two frequently-used materials.

The stone artifacts from Yarımburgaz show few traces of post-depositional alteration aside from the desilicification of many of the flint artifacts. There is little evidence for movement by water. Even chemically-altered pieces show sharp, fresh-looking edges. Macroscopic evidence of burning is so rare as to be essentially absent from the assemblage.

ASSEMBLAGE COMPOSITION AND ARTIFACT FORMS

Table 5.2 shows the basic composition of the Yarımburgaz lithic assemblage. The dominant shaped elements are retouched flake tools: these outnumber core tools by approximately 9 to 1. One noteworthy feature of the assemblage is the high frequency of intentionally modified or utilized pieces. Modified flake tools far outnumber unbroken, unretouched flakes, and modified or utilized artifacts of all sorts comprise almost one-third of the total assemblage. Even though sediments were sieved through fine-mesh screens, quantities of small flaking debris are relatively limited. This may indicate that much primary stone flaking was done outside of the cave itself. Alternativly, the paucity of small debris could be a byproduct of the particular methods of flake production employed by Middle Pleistocene tool makers (see below). Table 5.2 also provides a first indication of the differential exploitation of raw materials at Yarımburgaz. Flint and quartz were apparently preferred for the manufacture of small (flake) tools, as modified and/or utilized flakes far outnumber core tools in these raw materials. In contrast, core tools (choppers and chopping tools) were most often manufactured from quartzite.

It is difficult to draw clear typological distinctions among the flake tools from the LC of Yarımburgaz Cave. The majority of modified tools have irregular or denticulate edges (Figures 5.1, 5.2). Artifacts with regular, continuously-retouched edges, which would be classified as that possesscrapers (e.g., Figures 5.3, 5.4) (Table 5.3) are less abundant. Several specimens possessing converging, denticulate margins resemble “Tayac points” (e.g., Fig. 5.1, n. 8; Fig. 5.2, n. 2, 3). The bec/perçoir group (Fig. 5.2, n. 4, 7, 8) consists of artifacts with pointed working edges formed either by the intersection of two retouched margins, or by the intersection of a retouched margin with a fracture or steep natural flake margin. These are considerably less abundant than denticulates and scrapers. The edges of many of the thinner flakes possess small, irregular, flake scars—often on both the ventral and dorsal faces—which could represent either use-wear or post-depositional damage. These are not included in the tool counts presented in Tables 5.2 and 5.3. Steep, alternating marginal “retouch” (e.g., Figure 5.5, n. 8), generally thought to be the result of post-depositional factors, is comparatively rare in the collections from the LC at Yarımburgaz.

Table 5.2. Basic assemblage composition.

artifact category	flint	quartz	quartzite	other	total
cores	60	41	35	3	139
core tools	10	5	48	1	64
flake tools	398	94	40	6	538
whole flakes	147	22	32	12	213
broken flakes	111	16	20	6	153
chips, chunks & debris	373	119	62	14	568
Total	1099	297	237	41	1674

Table 5.3. Artifact type frequencies.

artifact category	flint	quartz	quartzite	other	sum	%
sidescraper	83	30	7	1	121	22.60%
denticulate	162	17	11	4	194	36.20%
notch	38	6	6	0	50	9.30%
bec/perçoir	26	15	1	0	42	7.80%
burin	2	1	0	0	3	0.60%
composite	25	5	2	0	32	6.00%
rabot	2	1	4	0	7	1.30%
partially retouched	60	19	9	1	89	16.60%
Total	398	94	40	6	538	

The majority (61.3%) of the retouched flake tools from Yarımburgaz Cave are modified using simple, scalar retouch. Stepped and undercut (14.9%), abrupt (11.6%), and Clactonian retouch (8.9%) are the next most abundant forms of modification: the Clactonian technique was most often used to make notched pieces. Slightly fewer than 5% of the Yarımburgaz flake tools exhibit a form of edge modification resembling Quina retouch (Fig. 5.3, n. 7). This resemblance may be accidental, the result of multiple episodes of resharpening, rather than representing the systematic form of retouch (as well as flake production) found in the Quina Mousterian (e.g., Bourguignon 1996). Retouch is predominantly confined to the dorsal faces of tool blanks, but some specimens, especially notches and denticulates, exhibit ventral retouch (e.g., Fig. 5.1, n. 5, 12; Fig. 5.2, n. 6). The ventral surfaces of some artifacts also exhibit scattered, flat, highly invasive flake removals reminiscent of the Kombewa technique (e.g., Fig. 5.3, n. 7; Fig. 5.5, n. 1, 10). Such ventral modification could represent intentional thinning of tool blanks, but as the morphology of these tools is generally inconsistent with hafting, such intentional thinning is unlikely. Instead, as in increasingly recognized in both the Lower and Middle Paleolithic (e.g., Ashton 2007; Dibble and McPherron 2006; Hovers 2007), the ventral removals may reflect the use of large flakes or retouched tools as cores for the production of small flakes.

Consistent with the high frequency of modified artifacts, many flake tools have been heavily utilized and extensively modified and/or resharpened (e.g., Fig. 5.2, n. 2; Fig. 5.3, n. 1, 2, 4, 9, 11; Fig. 5.4, n. 1, 2, 9, 10). Retouched edges tend to be quite steep: the mean angle of retouch is approximately 70 degrees for all artifact forms and raw material types.

Even the very simple typological distinctions used above seem somewhat arbitrary when applied to the Yarımburgaz artifacts. The differences between "scraper" and "denticulate" edges are not nearly so clear cut as in some later Middle or Upper Paleolithic assemblages. Instead, there is a continuous range of variation in edge profile from deeply notched or serrated to regular and evenly retouched. Many of the so-called "scrapers" possess rather irregular edges (Fig. 5.4). Denticulates, notches and becs/perçoirs also seem to grade into one another. A tool with a single retouched concavity may be classed as a notch if the modified area is located near the center of a flake margin, but may resemble a bec/perçoir if the notch is adjacent to a fracture or steep natural edge (e.g., Fig. 5.1, n. 3; Fig. 5. 2, n. 1). Similarly an artifact with two or more adjacent notches might be called a notch, a denticulate, or a bec/perçoir, depending on the breadth and depth of the concavities (cf. Holdoway et al. 1996).

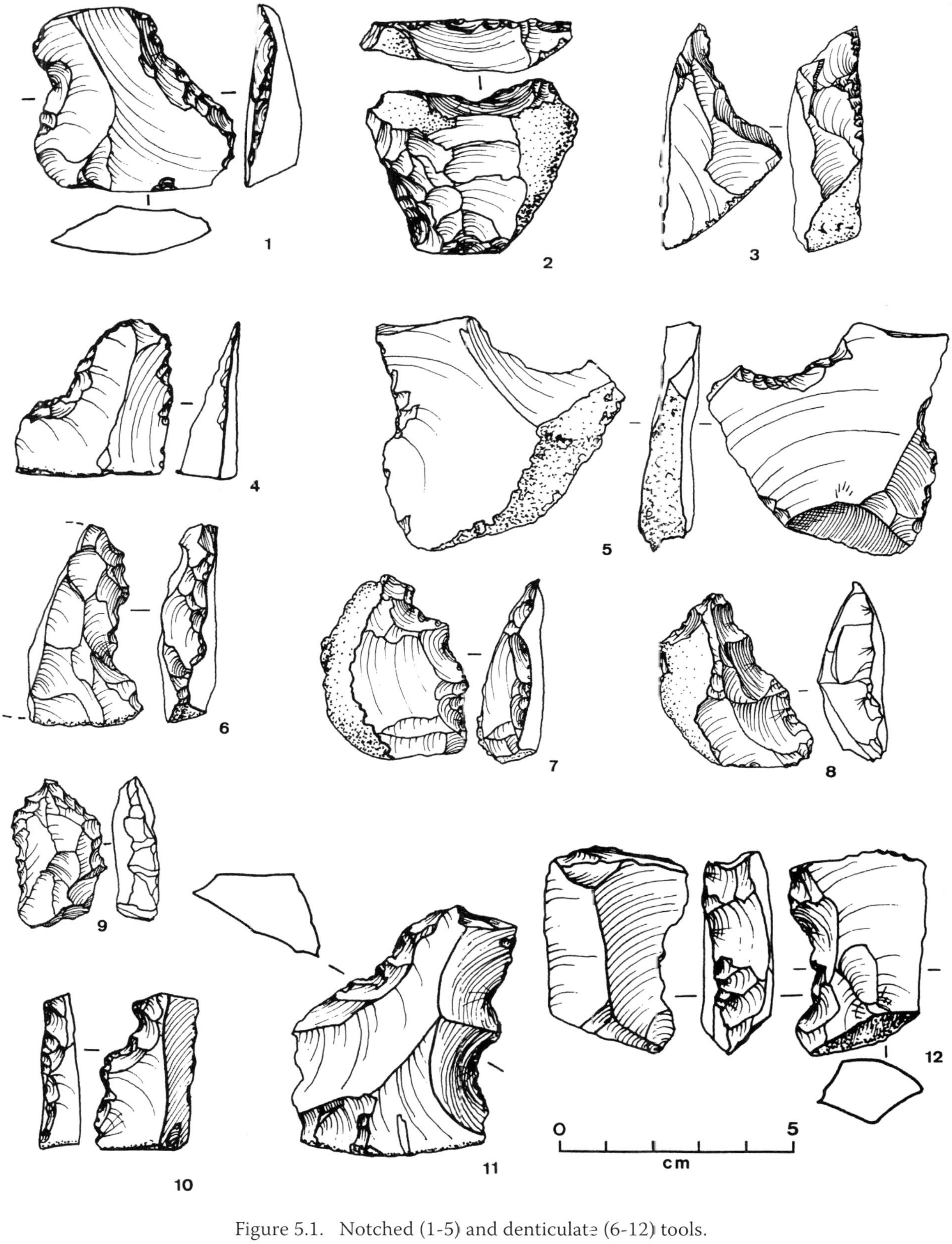

Figure 5.1. Notched (1-5) and denticulate (6-12) tools.

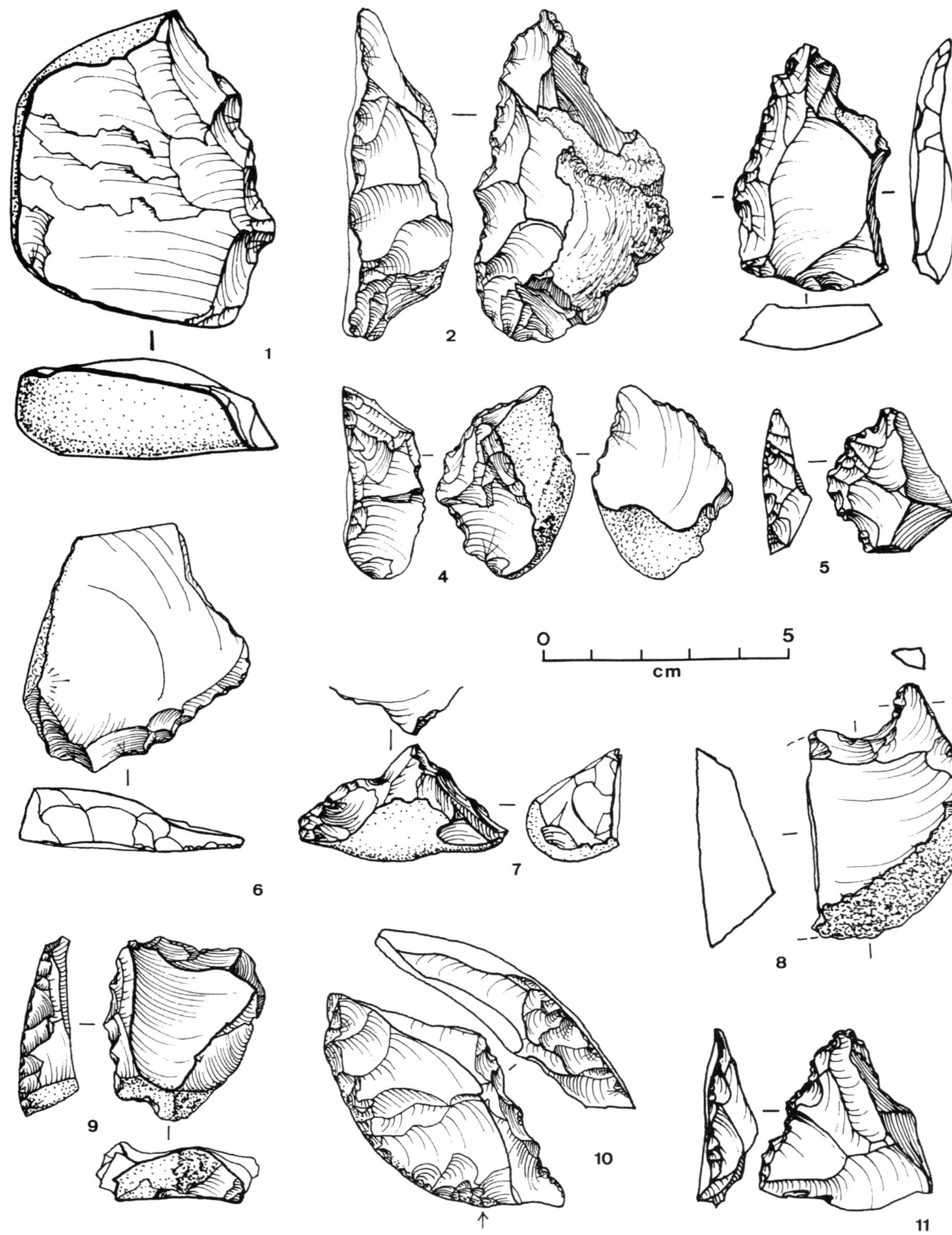

Figure 5.2. Denticulate (1, 5, 6, 9-11) and pointed (2-4, 7, 8) tools.

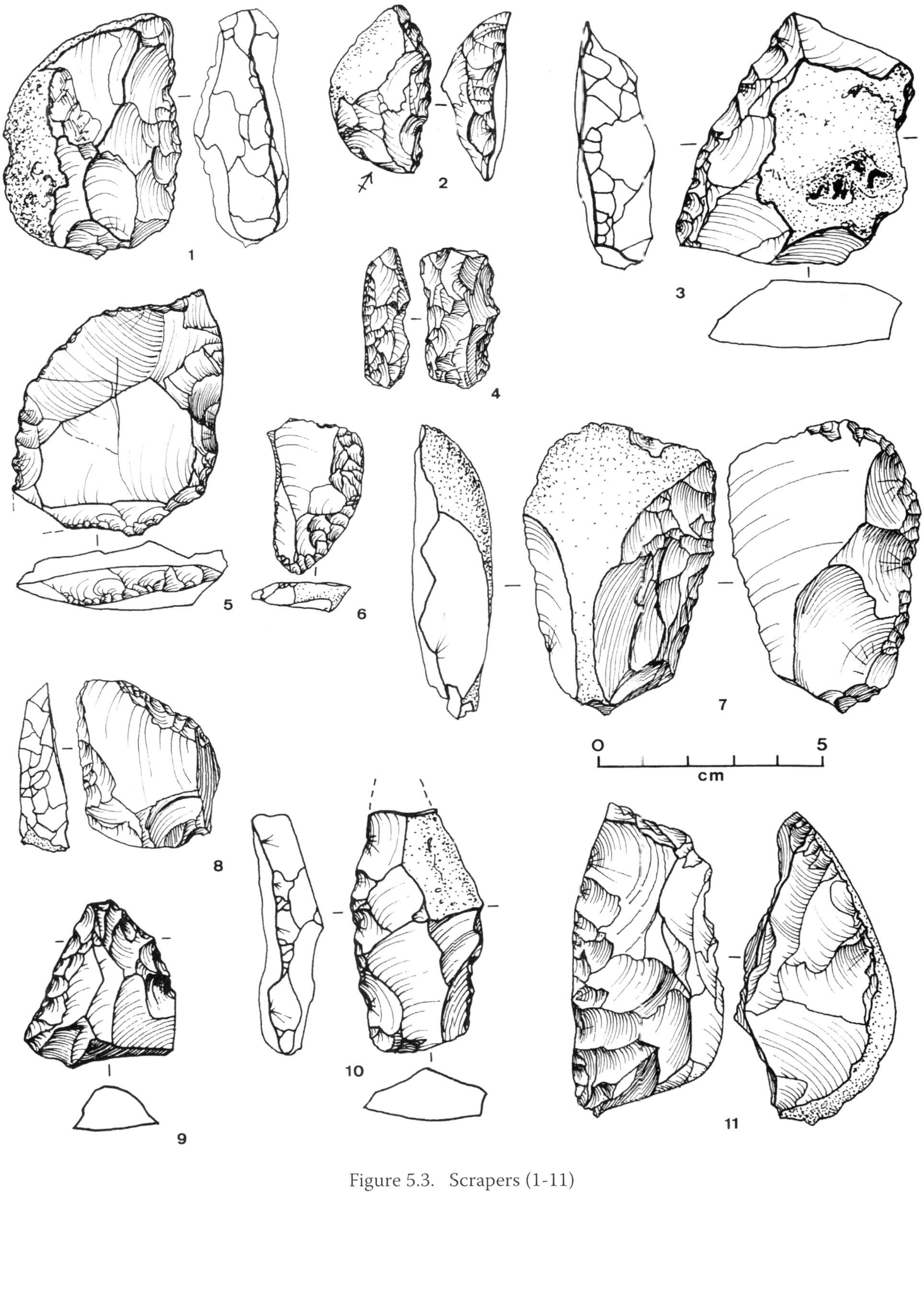

Figure 5.3. Scrapers (1-11)

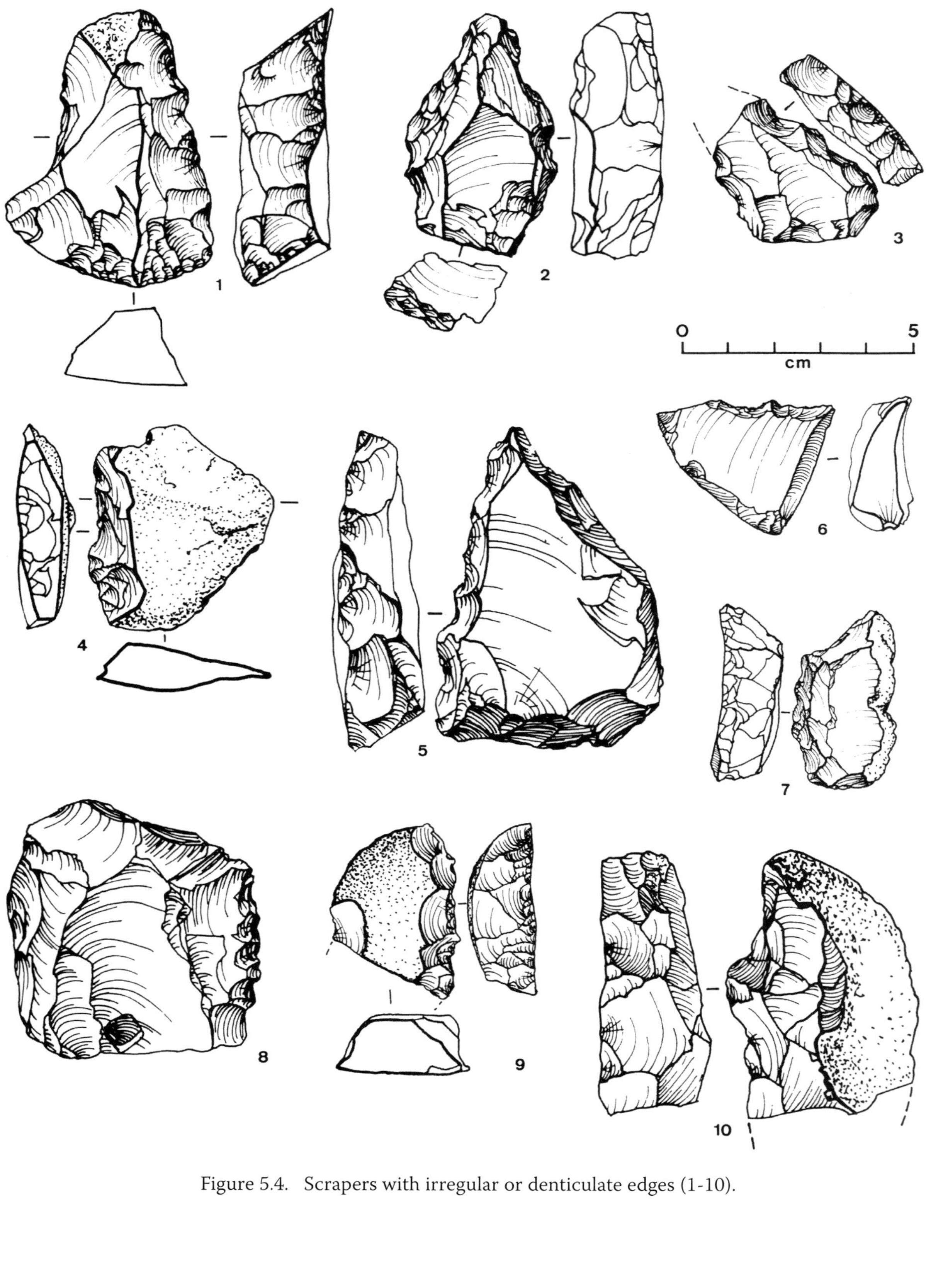

Figure 5.4. Scrapers with irregular or denticulate edges (1-10).

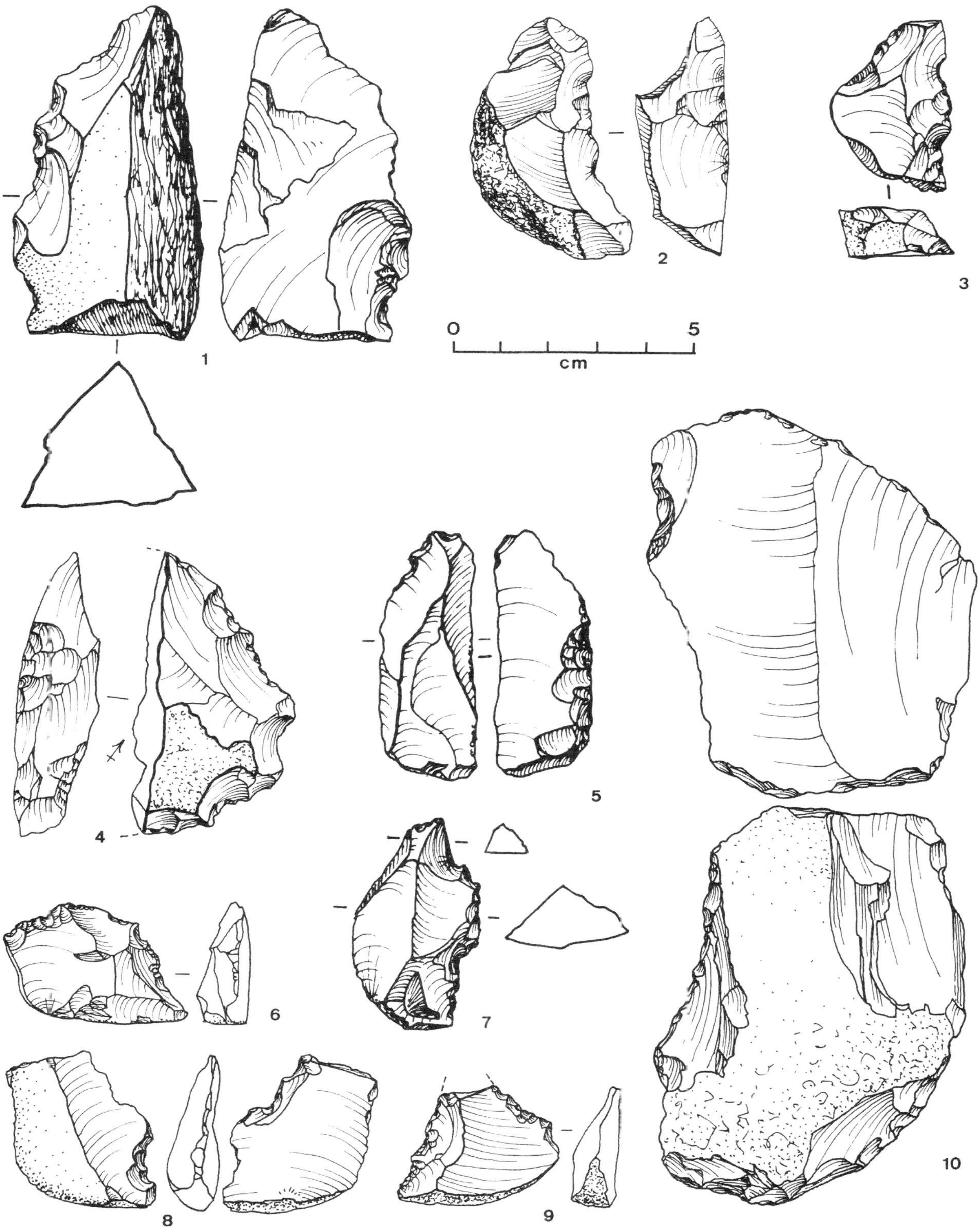

Figure 5.5. Denticulates (1-4, 6, 7, 9), partially-retouched pieces (5, 10) and pieces with alternating retouch (8).

In several cases, retouched flake tools appear to have been split by a blow directed either at a cortical platform or perpendicular to the retouched edge (Fig. 5.1, n. 3; Fig. 5.4, n. 6). Clearly-marked bulbs of percussion on the fracture surfaces demonstrate that breaks were caused by percussion, not by torsion or bending. The intersection of fractures with retouch scars further shows that the pieces were split after edges were modified. Because it normally results in a sharp, robust point formed by the intersection of the fracture with a retouched margin, this splitting of flake tools could represent a technique for the production of edges. Alternatively, the tools could have been split accidentally as a result of being struck with a percussor of some sort during use or while being modified.

Flakes and tool blanks in the Yarımburgaz assemblage tend to be thick and blocky, with relatively obtuse natural edges. The majority preserve at least some dorsal cortex. End-struck flakes—pieces longer than they are wide—outnumber side-struck pieces by almost four to one but the degree of elongation is very restricted. The length/width ratios of both flakes and tool blanks average around 1.2/1, and only four specimens can be characterized as technical blades (length equal to or greater than 2X width). Flakes and flake tools average between four and five cm in length. The largest plurality of flakes and tool blanks possess either cortical or plain butts (Table 5.4). Faceted platforms are extremely rare, and most of the dihedral platforms actually have one cortical facet. Importantly, neither Levallois flakes (typical or atypical) nor typical biface thinning flakes (*éclat de taille de biface*) are present in the Yarımburgaz assemblage. The forms of flakes and tool blanks of different raw materials do differ somewhat, particularly with respect to platforms, dorsal scar patterns, and the distribution of cortex. These differences are discussed in greater detail below.

Core tools comprise only 10.6% of the modified tools at Yarımburgaz. Of the 64 core tools, 61 are choppers and chopping tools (Figures 5.6, 5.7). Two of the remaining specimens were classified as "pseudo-bifaces". Both of these are ovoid pointed artifacts with irregular, flat, invasive retouch along both lateral margins: one is made on a flake, the other on a flat cobble (Fig. 5.8, n. 1). In reality, the "pseudo-bifaces" resemble true bifacial handaxes only superficially, as both are minimally modified, with no evidence for systematic bifacial shaping or thinning. One very large, round cortical flake of quartzite has bidirectional retouch or heavy use-damage on one edge (Fig. 5.8, n. 2). This artifact could be classified as an atypical cleaver flake, as a chopping tool on a flake, or simply as a very large utilized flake.

Table 5.5 shows a morphological breakdown of choppers and chopping tools. Unidirectionally flaked choppers are significantly more common than chopping tools with bidirectional flaking. Choppers also tend to have more regular "working edges" and appear better made than chopping

Table 5.4. Platform types, retouched tools and unretouched flakes.

platform type	flint	quartz	quartzite	other	sum	%
no platform	139	56	19	6	220	***
cortical	284	20	55	6	365	53.90%
plain	167	41	13	9	230	34.00%
dihedral	28	4	2	0	34	5.00%
facetted	10	0	1	0	11	1.60%
collapsed/crushed	24	10	2	1	37	5.50%

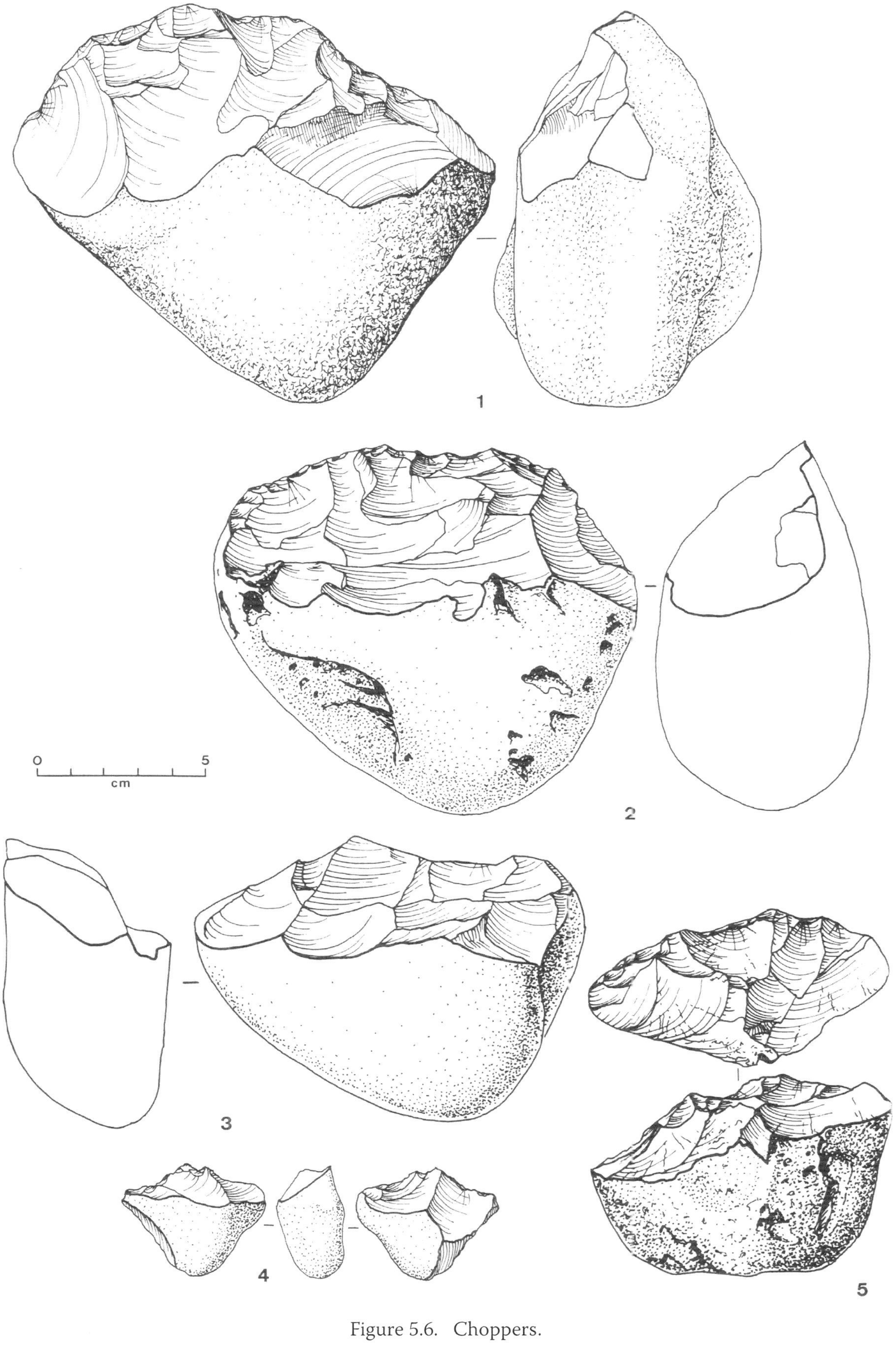

Figure 5.6. Choppers.

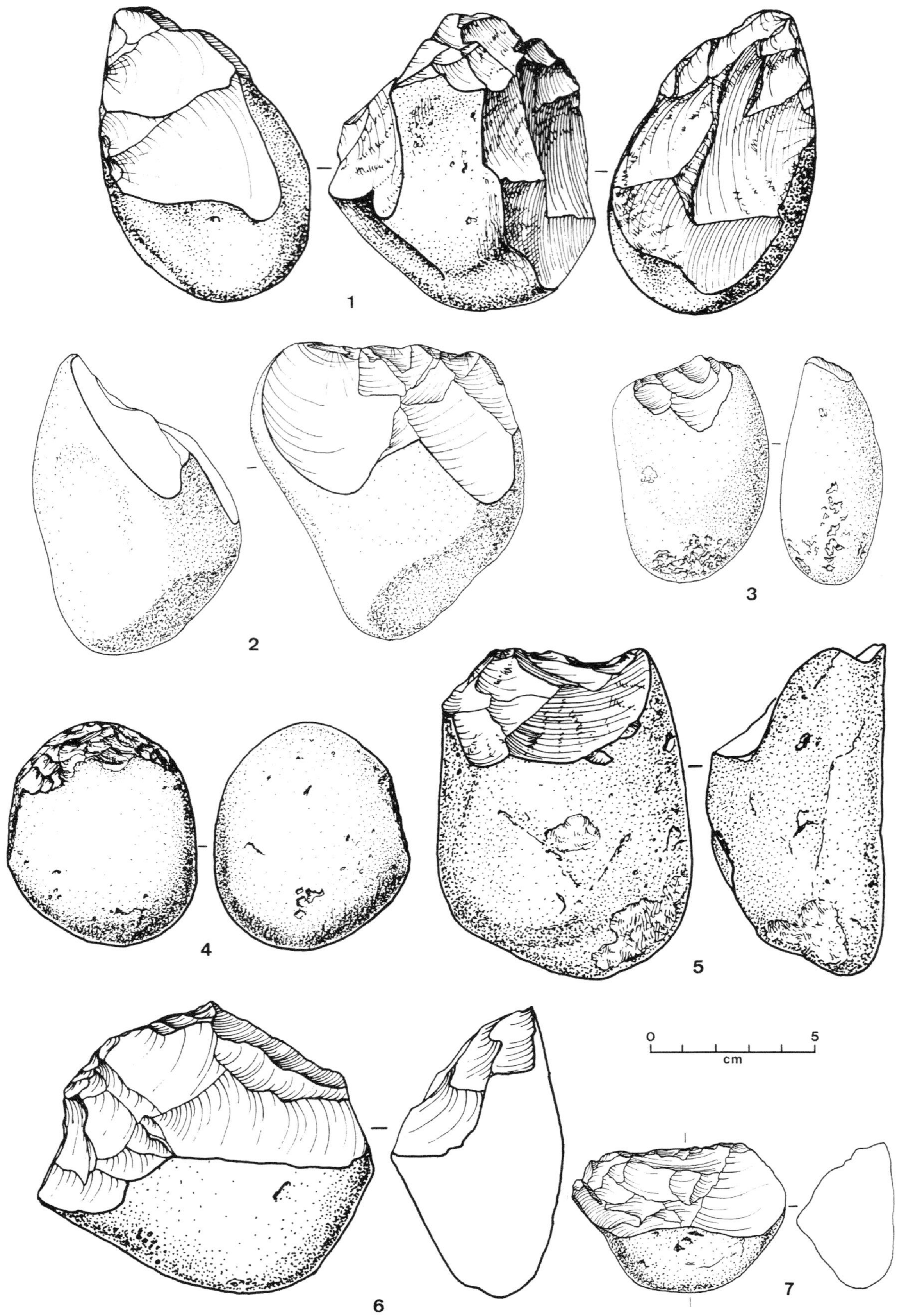

Figure 5.7. Choppers (1, 2, 5-7) and hammerstones (3, 4).

tools, which often exhibit irregular, sinuous margins. Aside from the preponderance of unidirectional modification, there is little morphological standardization among the core tools. Side choppers—retouched along the long edge of a cobble—are slightly, but not significantly, more abundant than end choppers. A few end-modified core tools are chisel-like in appearance, while some side choppers have very broad edges. The modal length for choppers and chopping tools is between 8.0 and 10.0 cm, but they vary considerably is size (Figure 5.9). The smallest, a flint chopping tool (Fig. 5.6, n. 4), is only 3.3 cm long, while the largest chopper (of quartzite) is 17.1 cm in maximum dimension. Edge angles of core tools display a somewhat bimodal distribution, with one peak at 70-75°, and a second at 80-85° (Figure 5.10).

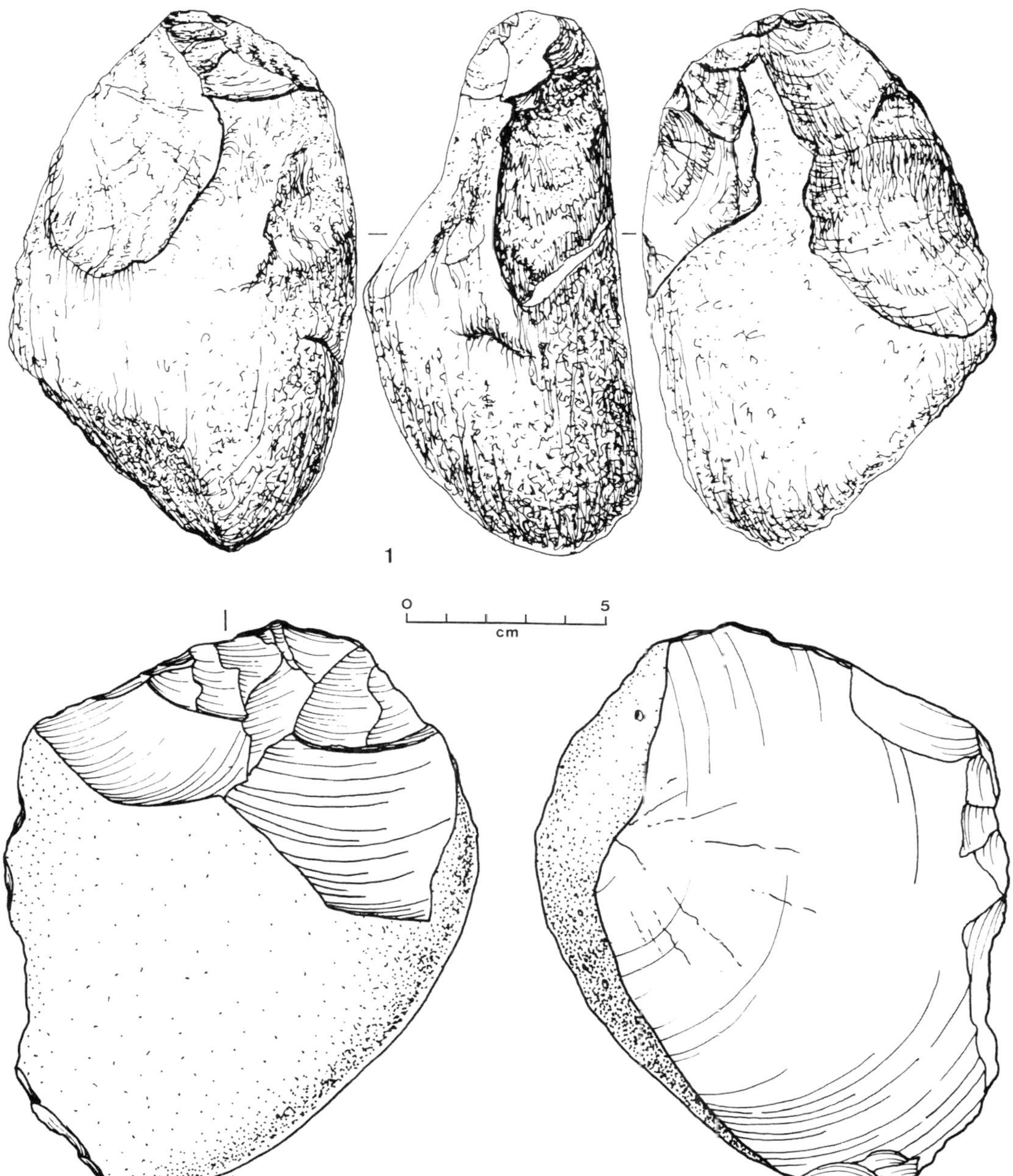

Figure 5.8. "Pseudo-biface" (1) and flake cleaver or utilized flake (2).

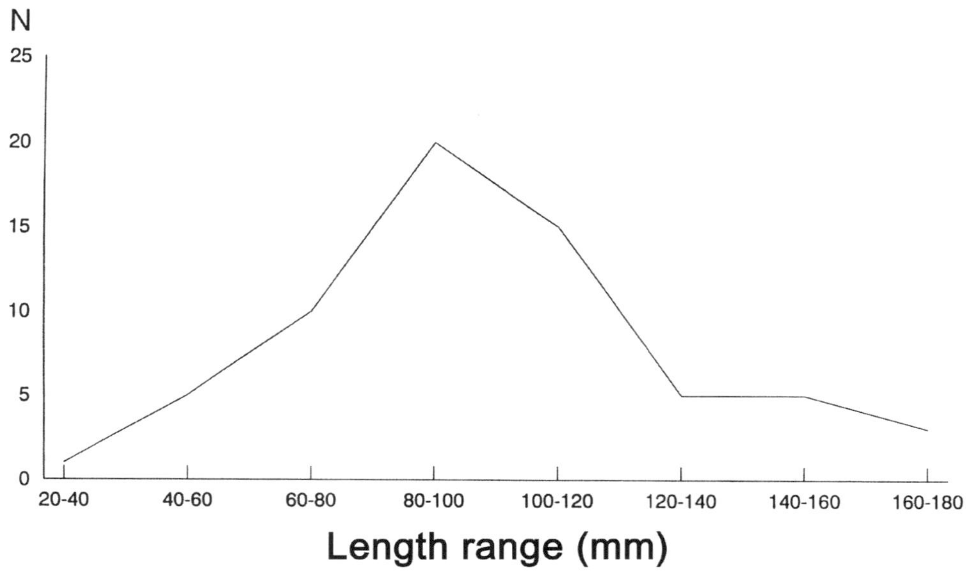

Figure 5.9. Maximum sizes of core tools (mm).

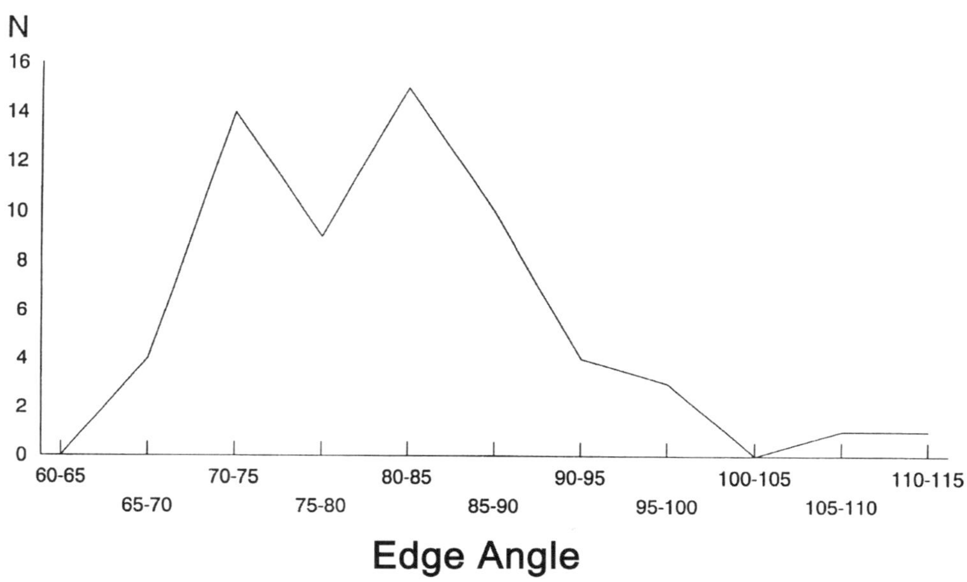

Figure 5.10. Edge angles of choppers.

The Yarımburgaz assemblage contains a wide range of core forms (Table 5.6, Figures 5.11-5.14). The most abundant single class is that of "tested pebbles", here defined as natural cobbles with three or fewer non-systematic removals. The most abundant variety of formal core consists of centripetally-worked or discoid specimens (Figures 5.11-5.13). Most of the centripetally exploited cores from Yarımburgaz have relatively flat faces of detachment, and the majority have been worked on one face only (e.g., Fig. 5.11, n. 1, 2, 5; Fig. 5.12, n. 2). A few are quite thick, with steep, conical faces of detachment, resembling the classic "pyramidal" discoid cores (Fig. 5.13, n. 1, 2). Striking platforms may be either plain or cortical. Consistent with the scarcity of faceted platforms on flakes and tool blanks, few of the Yarımburgaz cores preserves any traces of platform shaping, and there are no true Levallois cores. One specimen (Fig. 5.12, n.1) does appear to

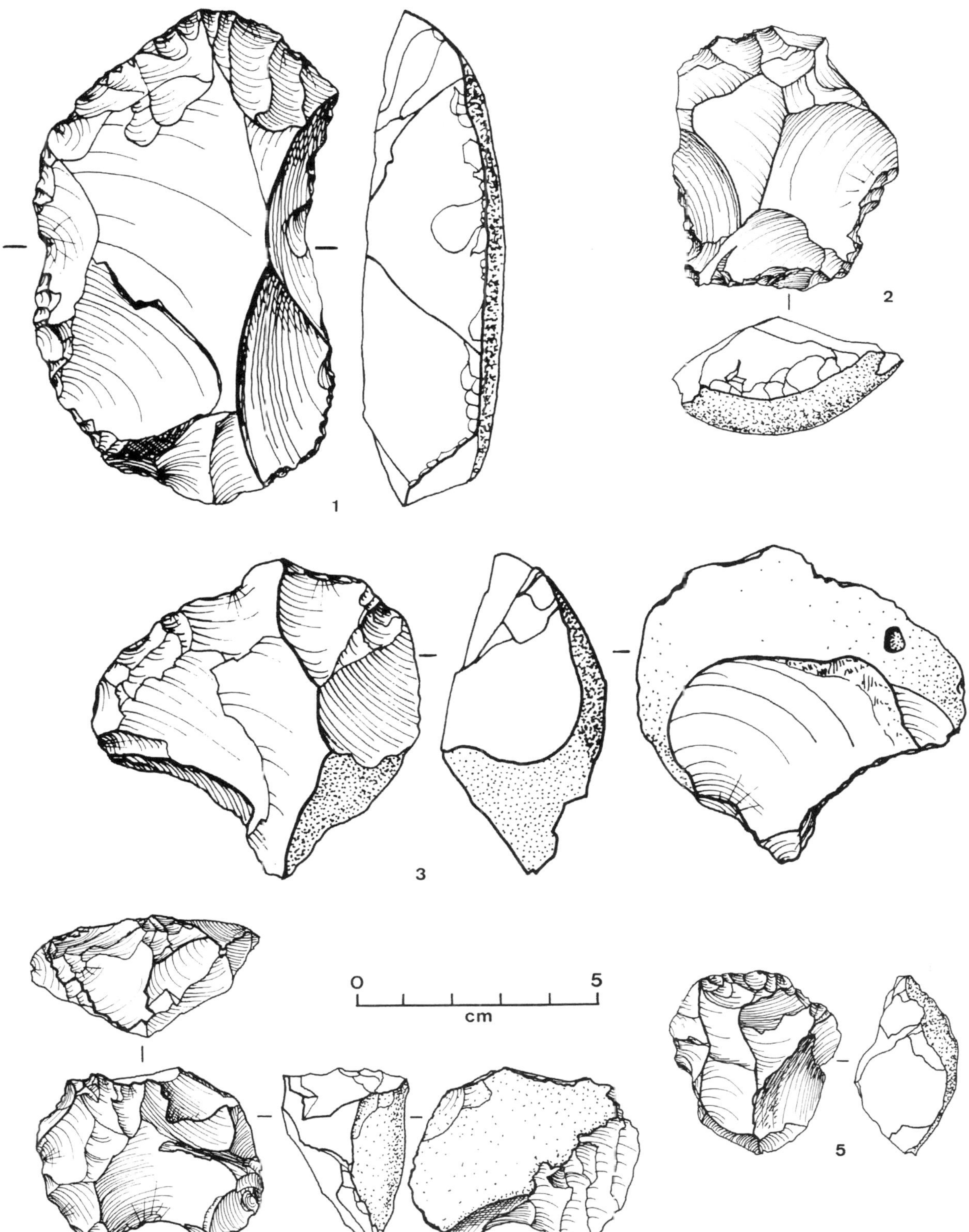

Figure 5.11. Unifacial discoid or centripetal cores.

Table 5.5. Forms of core tools. (Three specimens were too fragmentary to categorize)

form	side	end	side+end
chopper	27	22	1
chopping tool	5	3	0
Total	32	25	1

Table 5.6. Core forms.

core form	flint	quartz	quartzite	other	sum	%
tested (1-3 scars)	4	7	32	3	46	33.10%
centripetal	37	3	3	0	43	30.90%
1-2 platforms	5	7	0	0	12	8.60%
polyhedral	6	5	0	0	11	7.90%
bipolar	0	8	0	1	9	6.50%
amorphous	7	11	0	0	18	12.90%
Total	59	41	35	4	139	

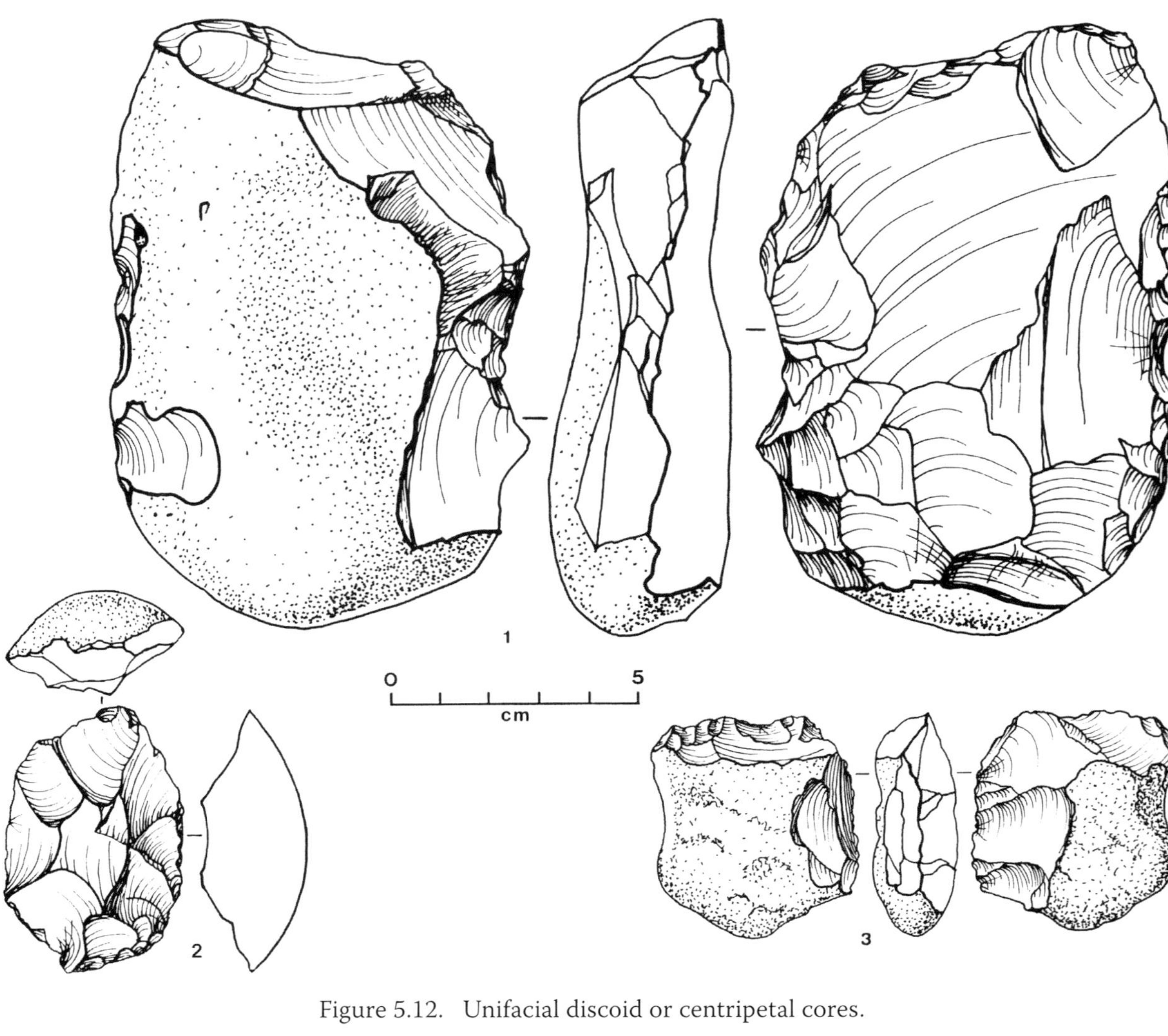

Figure 5.12. Unifacial discoid or centripetal cores.

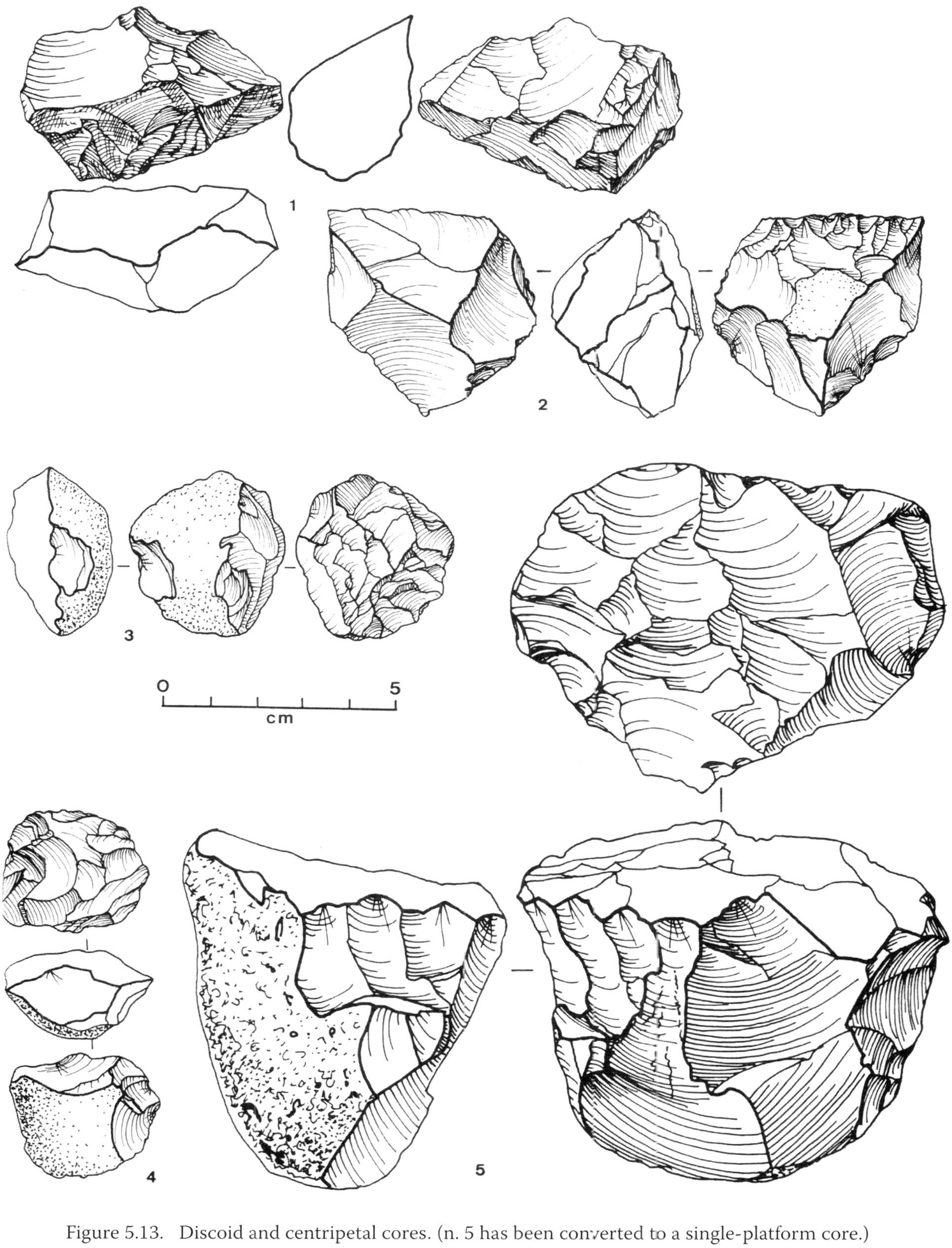

Figure 5.13. Discoid and centripetal cores. (n. 5 has been converted to a single-platform core.)

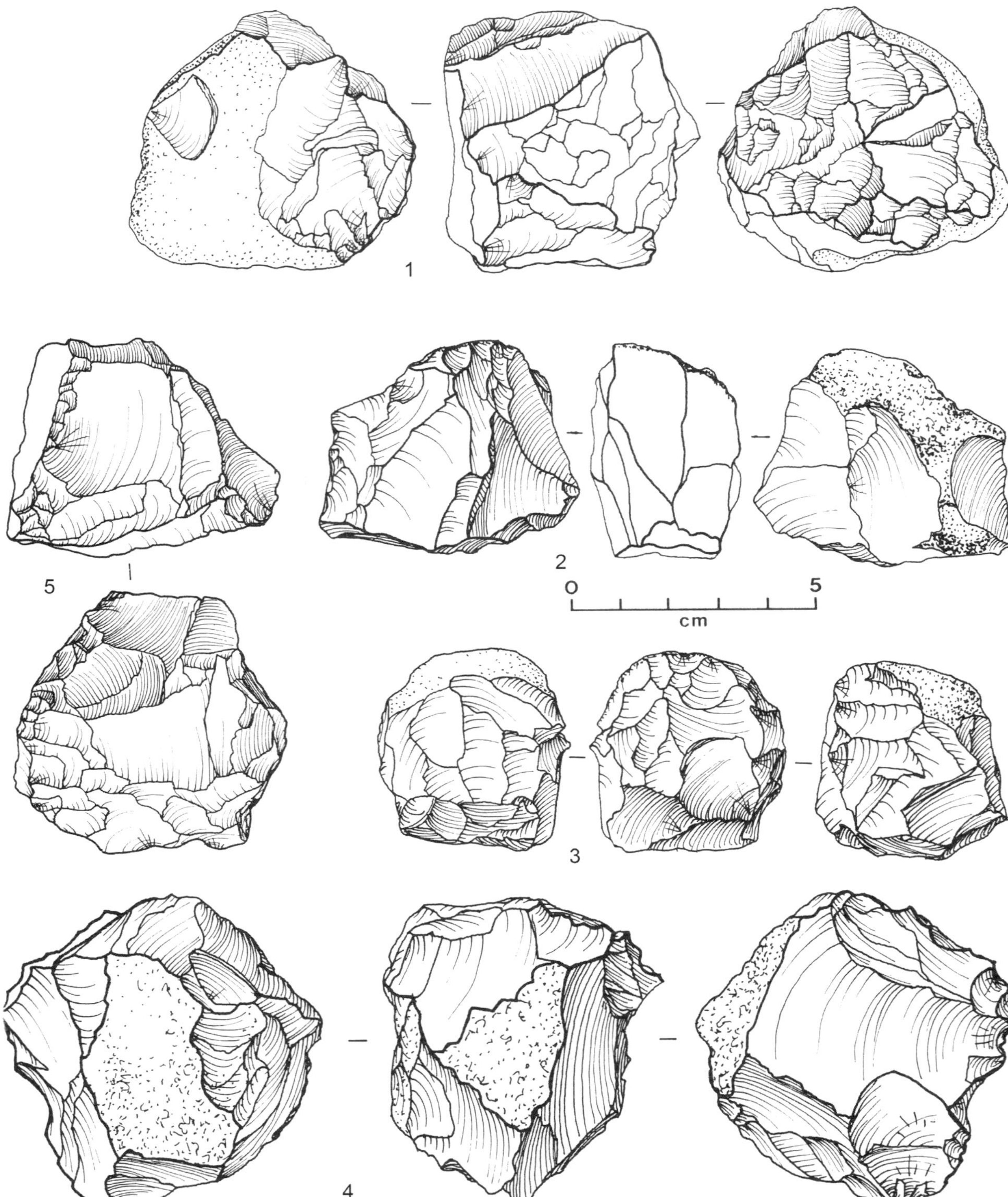

Figure 5.14. Globular and polyhedral cores.

show evidence for differential exploitation of two flat faces, with one face used to remove large, flat flakes and the other used primarily to establish striking platforms: however, this core does not show any of the other defining characteristics of Levallois, such as platform preparation or "predetermining" flakes on the removal surface.

Less common core forms include globular or polyhedral forms (Figure 5.14) with multiple striking platforms oriented at a variety of angles (e.g., Figure 5.14, n. 1-5). Depending on the shape of the original pebble or nodule and sequences of platform use, centripetal or discoidal exploitation can lead to the production of polyhedral core forms (Terradas 2003: 23-25). Informal flake/ blade cores with one or two platforms, bipolar (splintered) cores, and amorphous pieces are also present in small numbers. There are no spheroids in the collections from Yarımburgaz. The collection does contain 11 "hammerstones", unflaked pebbles with localized percussion damage (Figure 5.7, n. 3, 4).

DIFFERENTIAL EXPLOITATION OF RAW MATERIALS

In discussing raw material economy, flake production technology, and *chaînes opératoires* at Yarımburgaz Cave it is necessary to treat the three principal raw materials—flint, quartz, and quartzite—separately. Although there is some overlap, prehistoric toolmakers exercised distinct ranges of technological options in working the different types of stone. Quartzite was worked mainly in a manner which resulted in the production of choppers and chopping tools. As noted above, flint and quartz were preferred for retouched flake tools. Moreover, flakes were obtained by somewhat different methods from these very dissimilar materials.

Overall, quartzite appears to have been the least extensively exploited raw material. It was employed less frequently than flint and quartz (Table 5.2), and artifacts made of quartzite were not reduced or shaped to any great extent. Flakes and blanks of quartzite were not retouched as often as those made of either flint or quartz. In fact, of the three common raw materials, quartzite is the only one for which there are more unretouched flakes and fragments than retouched flake tools (Table 5.2). Cores of quartzite also appear less extensively consumed than cores of other raw materials. The great majority of quartzite cores are simply "tested pebbles", cobbles with a small number of unsystematic removals. The relatively casual exploitation is reflected in the small number of scars and large amounts of residual cortex found on quartzite cores (Table 5.7). In contrast, cores of flint and quartz were much more extensively worked. Not only are the consistently smaller than quartzite cores, they tend preserve significantly smaller amounts of cortex and they possess more removal scars as well (Table 5.7)

Manufacture Technology: Quartzite

Although quartzite accounts for only about 14% of the total lithic assemblage from Yarımburgaz Cave, fully 75% of the "core tools" are made of this raw material (Table 5.2). It appears that the predominant mode of working quartzite at Yarımburgaz resulted in the production of choppers and chopping tools. Most flakes of this raw material have forms typical of byproducts from chopper manufacture. On average, quartzite flakes have fewer than two dorsal scars, and nearly 1/3 of them have no dorsal scars at all. More than 84% of flakes preserve cortex on their dorsal surfaces, and most preserve a considerable amount: the median amount of dorsal cortex coverage for quartzite flakes is 75%. On flakes with dorsal scars, the scars most commonly originate

Table 5.7. Median flake scar number and mean cortex coverage per core.

	flint	quartz	quartzite
median flake scars			
chopper/chopping tools excluded	8	7	2
chopper/chopping tools included	8	7	4
median cortex %			
chopper/chopping tools excluded	32.90%	30.40%	77.90%
chopper/chopping tools included	35.40%	34.90%	75.50%

at the proximal (platform) end of the flake, and the majority of all quartzite flakes and flake tools have cortical butts (Table 5.8). The existence of a variety of platform and dorsal scar patterns, especially multidirectional and bipolar, demonstrate that not all quartzite flakes were produced from choppers and chopping tools. Nonetheless, it is reasonable to infer that a great many were the results of such manufacture.

Referring to choppers and chopping tools as "core tools" implies that these items were manufactured intentionally to be used as tools. Although many of the Yarımburgaz choppers appear suitable for use as heavy-duty cutting tools, we should be cautious about making such a functional inference. In discussing much older Oldowan assemblages, many authors have suggested that "choppers" are really informal cores used to produce flakes rather than serving as implements themselves (e.g., Toth 1985). However, at Yarımburgaz some evidence does appear to support the notion that these items were in fact shaped to create a "heavy duty" working edge, and not simply as a sources of flakes. First, there is no reason to believe that a technique yielding "chopper-like" cores was the only practical means for working quartzite: a variety of other methods were apparently used to produce flakes of quartzite, if only occasionally. Thus, we are left with the question of why this particular method might have been employed predominantly if the only goal were to make usable flakes. Second, in contrast to Oldowan "chopper cores", many (though

Table 5.8. Platforms and dorsal scar origins, quartzite flakes and tool blanks.

			dorsal scar origins					
					1 side +			
platform	**no scars**	**proximal**	**bipolar**	**lateral**	**proximal**	**centrip.**	**indet**	**%**
missing	8	2	1	4	0	0	5	20.60%
cortical	13	21	7	4	6	1	5	63.90%
plain	5	1	5	1	1	0	1	14.40%
dihedral	0	1	0	0	0	1	0	3.50%
faceted	2	0	0	0	0	0	0	3.50%
collapsed	2	0	0	0	0	0	0	3.50%
Total	30	25	13	9	7	2	11	

Note: Row percentages calculated using only specimens with "legible" dorsal scar patterns. Column percentages calculated using only specimens with identifiable platforms.

not all) quartzite choppers from Yarımburgaz have relatively acute edge angles. On the other hand, quartzite flakes were retouched quite frequently, though less often than flakes of other raw materials. Thus, even if the choppers and chopping tools could have served as tools themselves, it is quite obvious that they were also exploited as sources of flakes.

Manufacture Technology: Quartz

As Table 5.2 shows, quartz was frequently employed for producing flake tools in the Yarımburgaz Cave assemblage. Quartz flakes show evidence of retouch or secondary modification nearly as often as do flakes of flint. In fact, it is difficult to estimate the true frequency of retouch and utilization for this raw material. The variety of quartz used at Yarımburgaz varies from translucent to nearly transparent, and it is difficult to recognize subtle modification of the edge due to either use or intentional modification. Quartz was also used occasionally to make choppers and chopping tools, but not nearly as often as quartzite.

Despite the rather small sample, quartz exhibits the widest variety of core forms of any of the three principal raw materials at Yarımburgaz. No single core form accounts for more than 30% of the total. The most abundant cores are "amorphous" specimens, with two or more platforms in a variety of orientations (Table 5.6). The second most abundant form of core consists of bipolar cores, pieces with splintered or wedge-shaped platforms of detachment that appear to have been worked by hammer-and-anvil technique.

Like the cores, the forms of quartz flakes are highly varied. However, a large proportion of quartz flakes and blanks possess attributes consistent with the use of bipolar, hammer-and-anvil percussion. More than 40% of flakes have no recognizable platforms (Table 5.9), a common occurrence when bipolar percussion is used. In marked contrast to other raw materials, the great majority of remaining specimens have either plain (single facet) or crushed/collapsed platforms. Because of the difficulty of discerning flake scar directions on translucent materials like quartz, more than one third of specimens have dorsal scar patterns classified as "indeterminate". Of those specimens with "legible" scars, however, more than 60% have scars originating either at the proximal end or both the proximal and distal ends (Table 5.9), also consistent with hammer-and-anvil technique.

Table 5.9. Platforms and dorsal scar origins, quartz flakes and tool blanks.

			dorsal scar origins					
					1 side +			
platform	**no scars**	**proximal**	**bipolar**	**lateral**	**proximal**	**centrip.**	**indet**	**%**
missing	12	4	10	2	0	2	36	
cortical	2	2	7	6	1	1	5	29.10%
plain	1	16	10	2	2	4	12	54.70%
dihedral	0	0	1	0	1	1	1	4.70%
collapsed	0	3	4	0	0	0	4	12.70%
Total	15	25	32	10	4	8	58	

Note: Row percentages calculated using only specimens with "legible" dorsal scar patterns. Column percentages calculated using only specimens with identifiable platforms.

Bipolar, hammer-and-anvil percussion has been a favored technique for working quartz pebbles in a variety of cultural contexts and periods. It may be especially appropriate for working quartz because it is less likely to be disrupted by the material's natural tendency to break along flat planes. However, the range of other core forms and flake morphologies seen in the quartz artifacts from Yarımburgaz Cave may well reflect the unpredictable working properties of the raw material itself. Vein quartz like that used at the site yields flakes with sharp and very durable edges, but it also tends to exhibit unpredictable fracture as cores break apart preferentially along boundaries between individual macro-crystals. A core of quartz may thus begin with one form, but an unanticipated fracture may necessitate changing the strategy of exploitation after subsequent removals. While such "surprises" may occur with even the most homogeneous of raw materials, they are especially common in working vein quartz. Thus, the diversity of core forms in this material may simply be the outcome of a series of decisions made by tool makers in response to chance events.

Manufacture technology: Flint

Choppers and chopping tools were infrequently produced from flint, although the few specimens recovered are quite "typical" in form. Flint was most frequently used for the manufacture of retouched flake tools (Table 5.2). It appears that flint flakes and tool blanks were often produced using a distinctive method of core preparation and flake detachment. Although flint cores exhibit a variety of forms, the majority (62.7%) show evidence of centripetal patterns of flake detachment (Table 5.6). These centripetal cores are relatively flat, and the faces of detachment are either flat or slightly convex. As such, they are quite distinct from the steeply-angled, "conical" or "biconical" types of discoid core described from many Middle and Lower Paleolithic sites (e.g., Isaac 1977:175; Slimak 2003; Terradas 2003). They are not classical centripetial Levallois cores either, in that platforms were never extensively prepared and there is no evidence either of "preferential" removals or predetermining flakes. The Yarımburgaz cores also differ markedly from both classic discoid and Levallois cores in that striking platforms were normally cortical over most or all of the core's perimeter.

The predominantly centripetal exploitation of flint cores is reflected in dorsal scar patterns on flint flakes and tool blanks, in which multidirectional centripetal and orthogonal (one side, one end) scar patterns predominate (Table 5.10). However, the forms of flakes and tool blanks do not conform to the products of better known Middle Paleolithic methods of centripetal core exploitation. As discussed above, there are no true Levallois flakes, and very few flakes with faceted platforms, in the Yarımburgaz assemblage. Pseudo-Levallois points and similar flat, triangular *dejete* flakes typically produced from Mousterian discoid cores (Boëda 1993; Brezillon 1968:95) are also essentially absent from the collection. In fact, the flint flakes and tool blanks from Yarımburgaz show an unusual mixture of attributes for well-documented methods of Lower and Middle Paleolithic core reduction. Given the high frequencies of centripetal and multi-directional dorsal scar patterns, one might expect to find little dorsal cortex, and this is indeed the case: while almost half of flint pieces preserve traces of cortex, the median coverage of the dorsal surface is only 5.0%. However, platforms are somewhat less typical. Most well documented variants of discoid method involve at least some adjustment of platform angles during reduction (see Boëda 1993; Terradas 2003), which typically results in high frequencies of either plain or dihedral butts. In contrast, more than half of the flint flakes and tool blanks from Yarımburgaz were struck from completely unprepared, cortical surfaces. Where the platform

Table 5.10. Platforms and dorsal scar origins, flint flakes and tool blanks.

			dorsal scar origins					
platform	no scars	proximal	bipolar	lateral	1 side + proximal	centrip.	indet	%
missing	13	18	10	8	21	3	77	
cortical	6	73	36	9	68	60	68	54.00%
plain	4	53	19	11	42	30	40	33.60%
dihedral	0	9	2	0	9	4	7	5.30%
faceted	0	4	0	1	2	2	3	2.00%
collapsed/	0	5	7	2	8	1	7	5.10%
Total	23	162	75	31	150	100	202	

Note: Row percentages calculated using only specimens with "legible" dorsal scar patterns. Column percentages calculated using only specimens with identifiable platforms.

type can be identified, 54% of all flint flakes and tool blanks, and about 51% of specimens with multidirectional or centripetal scar patterns, have cortical butts (Table 5.10). A number of illustrative specimens are shown in Figure 5.15 (also Fig. 5.1, n. 4, 6, 10, 11).

From the attributes of flakes, it is clear that the variety of centripetal flake production used at Yarımburgaz involved little systematic shaping or adjustment of the platform. The reconstructed sequence of operations (Figure 5.16) begins with the splitting of a flint pebble into two hemispherical or relatively flat sections (e.g., Fig. 5.11, n. 1; Fig. 5.12, n. 1). Flakes were subsequently detached from the newly created surface by blows directed at the cortical margins of the core. If detached early in the sequence, the resulting flakes would be relatively flat, while pieces struck off later in the sequence may be more wedge-shaped in saggital section. The presence of single-facet, dihedral, or even faceted butts on some flakes suggest that platforms were prepared on occasion. Some residual cores also bear traces of at least partial platform modification (e.g., Fig, 5.12, n. 1, 2; Fig. 5.13, n. 3, 4), showing that platforms were sometimes reshaped, perhaps late in the reduction of a core.

The mode of flake manufacture used on flint at Yarımburgaz Cave yields blanks rather different than those found in most other Middle and Upper Paleolithic assemblages. However, it is not unique. A very similar method of blank production is represented at open air sites on the terraces of the Tarn River in France (Tavoso 1978), as well as at the southern French Mousterian sites of Coudelous and Mauran, where it is referred to as the "unifacial discoid" method (Jaubert 1994; Jaubert and Farizy 1995). The so-called "Lahuti technique" described at late Middle Pleistocene sites in central Asia (Ranov et al. 1995) is also quite similar.

In order to illustrate the peculiarities of the technique used to work flint at Yarımburgaz, as well as its similarity to techniques documented on Tarn river terrace sites, we have adapted A. Tavoso's system for the classification of flakes and tool blanks, modifying it slightly to fit with the categories used in this study. In the following graphs, flakes and blanks are classified according to the distribution of dorsal cortex and the type of butt. In each graph, four patterns of dorsal cortex distribution are represented. In order of presentation, these include: (1, 5) 75-100% cortex; (2, 6) a strip of cortex or cortical "back" along one margin; (3, 7) an isolated patch of cortex on the distal end or in the center of the flake; and, (4, 8) no cortex. These four categories are

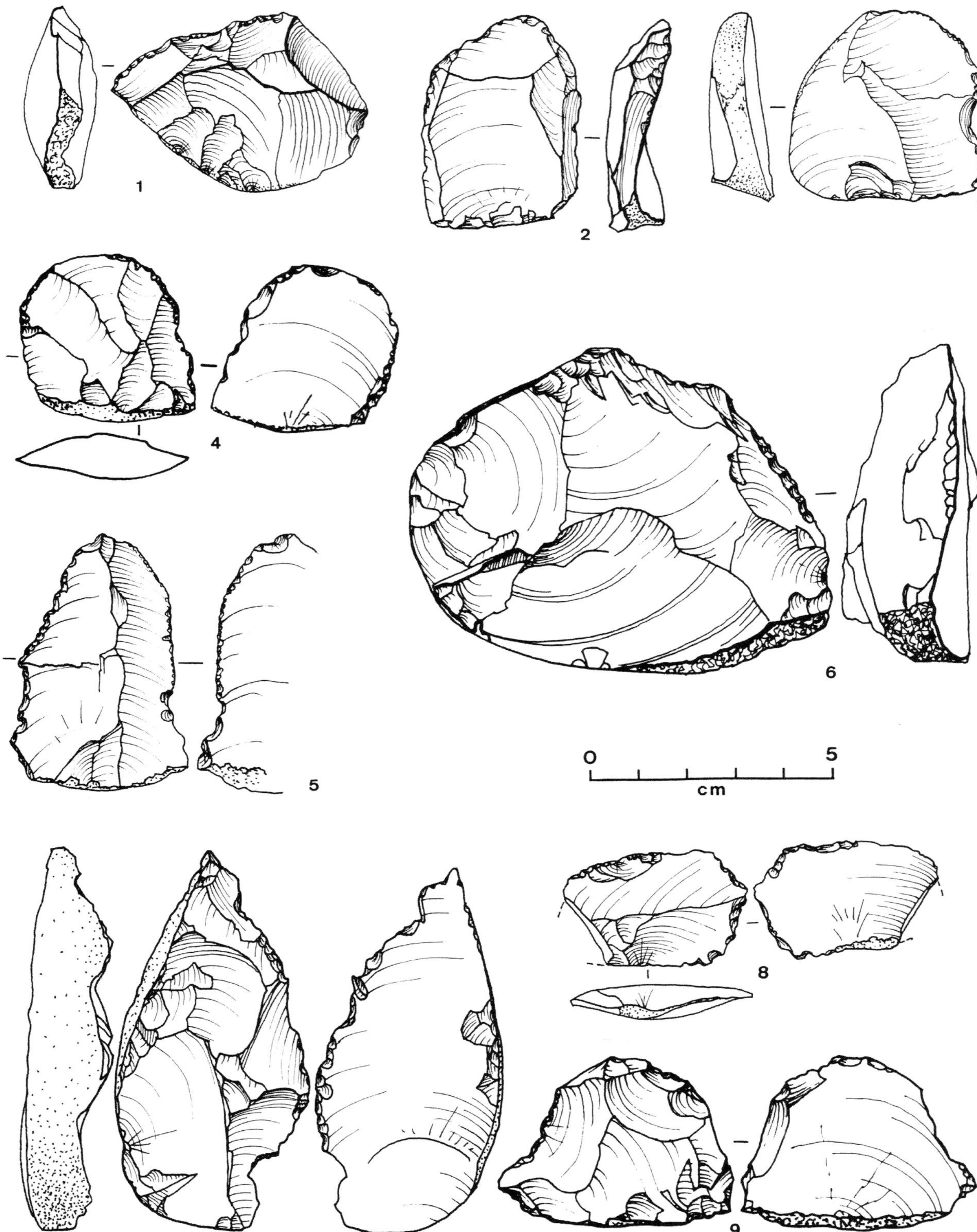

Figure 5.15. Flakes and tools with cortical butts and multi-directional dorsal scars.

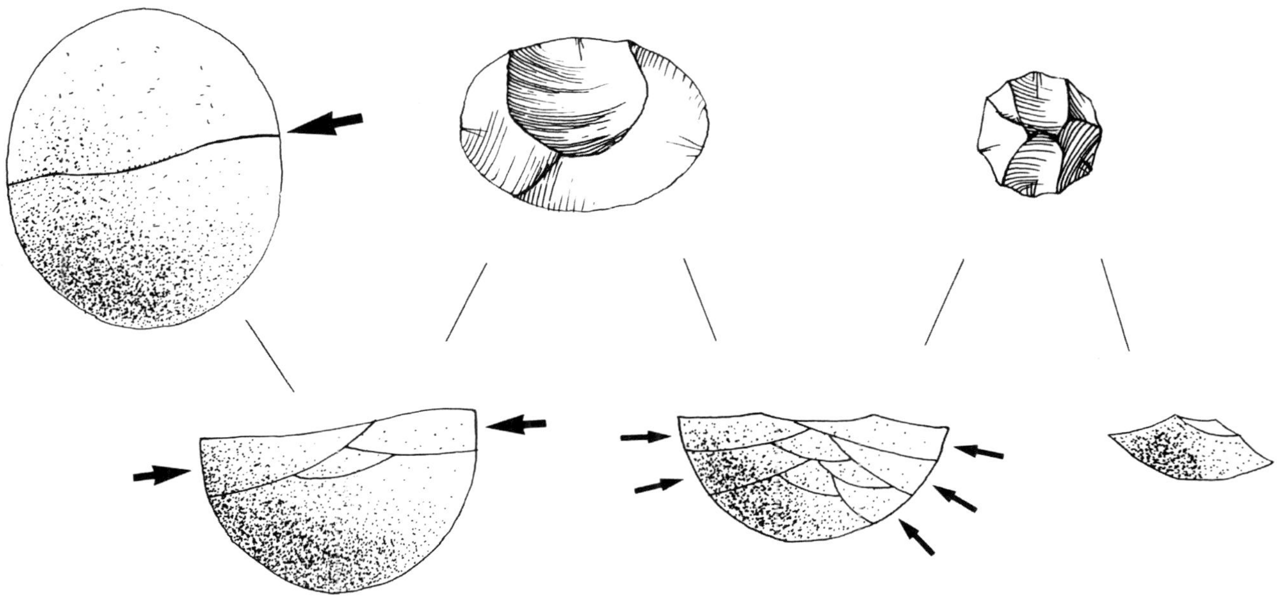

Figure 5.16. Schematic illustration of flake production technology used to work flint at Yarımburgaz.

repeated for each major platform type, cortical (1-4) and plain (single facet) (5-8). Because of their scarcity in the Yarımburgaz assemblage, other platform types are not included.

Figure 5.17 compares the three principal raw materials from Yarımburgaz, and illustrates the marked differences in flake production. For flint (Figure 5.17a), the largest single group consists of flakes and blanks with cortical butts and no dorsal cortex. This particular combination of attributes is quite scarce among the other two materials. Quartzite is dominated by flakes with cortical platforms and a large proportion of cortex covering (Figure 5.17c), while quartz artifacts tend to have plain, single faceted platforms and little or no dorsal cortex (Figure 5.17b).

Figure 5.18 compares the Yarımburgaz flint assemblage with Mousterian assemblages from Riparo Mochi, an Upper Pleistocene rockshelter located near the Balzi Rossi, in Italian Liguria (Kuhn 2004; de Lumley 1969). The Mochi assemblages are included to show how strongly the Yarımburgaz material differs from the products of common Middle Paleolithic flake production systems. The assemblage from layers 30-35 at Riparo Mochi (Figure 5.18b) is dominated by classic Mousterian discoid core technology and contains numerous pseudo-Levallois points and similar flake forms. In layers 50-55 at Mochi (Figure 5.18c), the production of elongated flakes and blades from unidirectional, predominantly Levallois cores complements the discoid method. Bar graphs for the two Mousterian assemblages have markedly different profiles from Yarımburgaz flint flakes and tool blanks. In the Middle Paleolithic cases, cortical platforms are very rare, and the most common flake types have plain platforms and little or no cortex coverage. Flakes and blanks with dihedral and/or faceted platforms have not been included. As these are rare at Yarımburgaz, but relatively common at Riparo Mochi, their incorporation would magnify differences between the various assemblages.

Figure 5.19 illustrates the strong resemblance between the Yarımburgaz flint artifacts and those from some open-air French sites on the terraces of the Tarn River. The topmost graph again represents flint artifacts from Yarımburgaz, but here the lower graphs (Figure 5.19b, 5.19c) are derived from two open-air archaeological sites, Tecou and Le Bouy, situated on a Middle

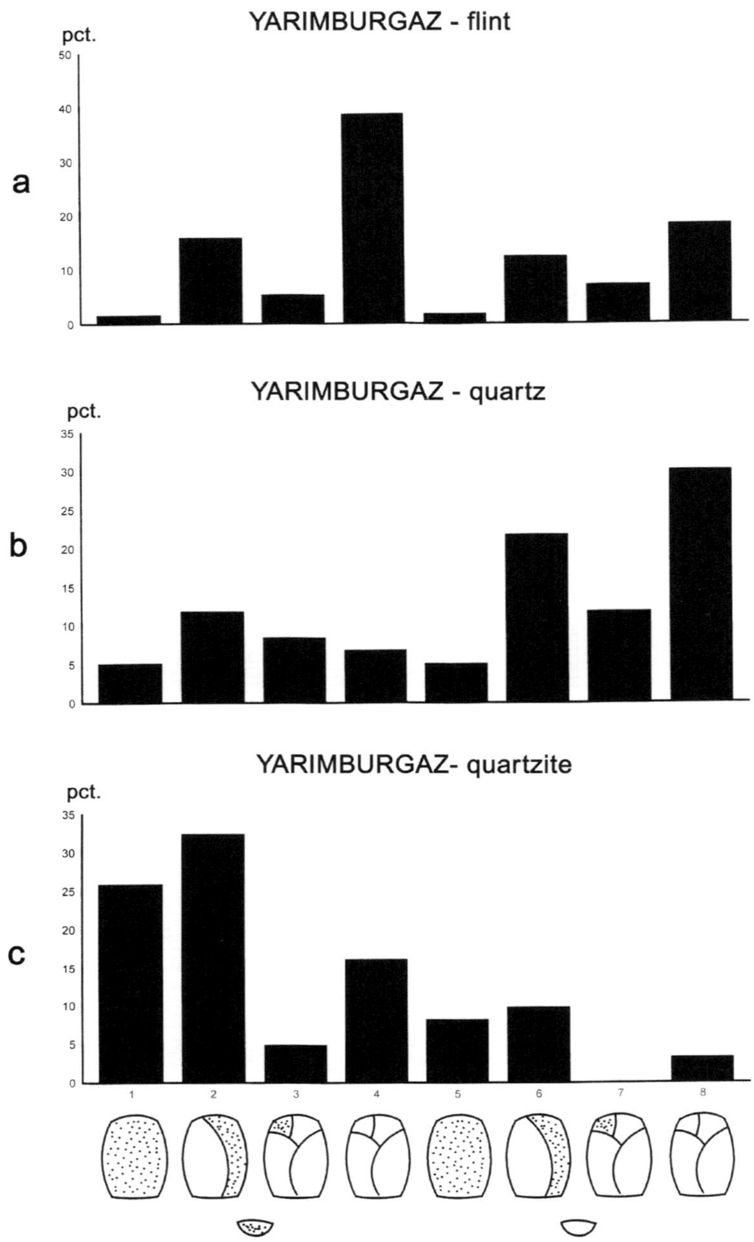

Figure 5.17. Comparisons of cortex distribution by two platform types, Yarımburgaz raw materials.

Pleistocene terrace of the Tarn River, as reported by A. Tavoso (1978). These are two of a series of localities from the same region that have yielded assemblages with similar technological profiles. The similarities among this set of bar charts are remarkable, especially in light of the marked dissimilarities described in the previous graphs. Most notable is the fact that flakes with cortical platforms but little or no cortex on the dorsal surface make up the largest single category in all cases.

In the Tarn River sites, these distinctive flake assemblages are associated with residual cores called *epanneles* or "truncated cobbles" (Tavoso 1978), which appear to be a logical residue of the methods of core exploitation illustrated in Figure 5.16. While a few similar specimens are

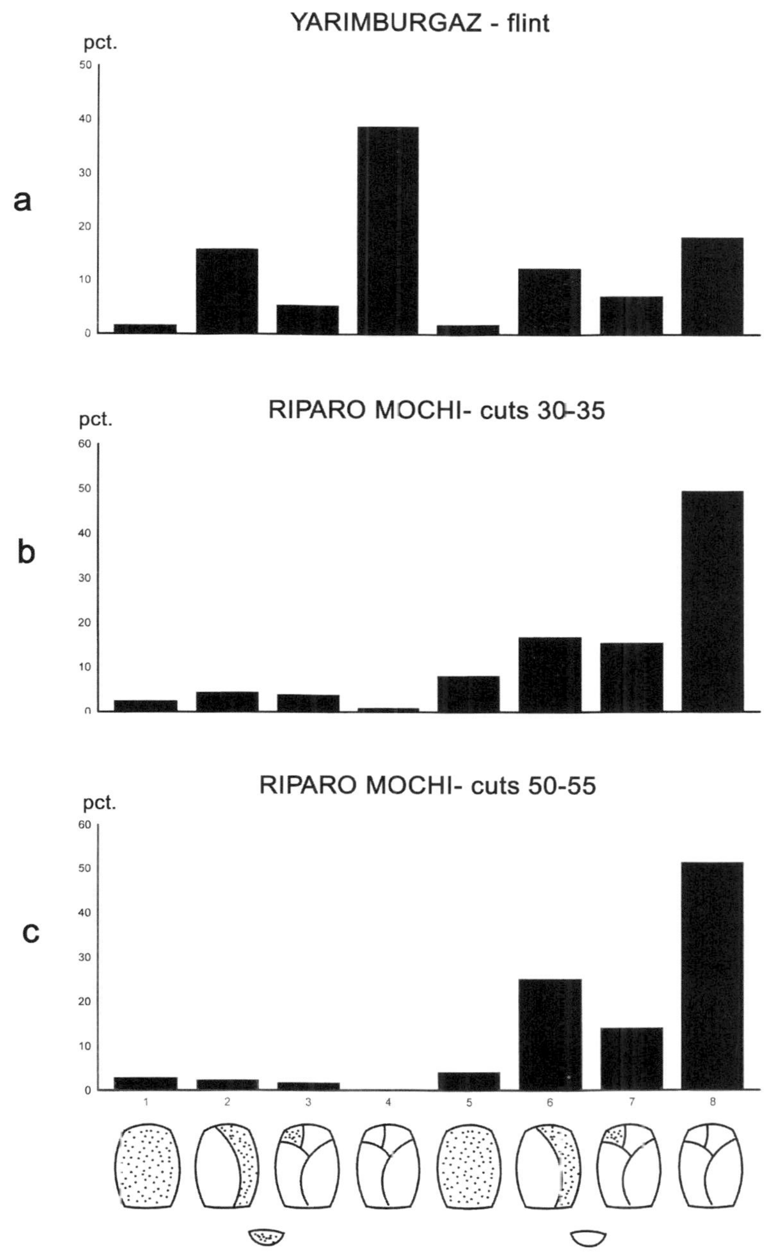

Figure 5.18. Comparisons of cortex distribution by two platform types, Yarımburgaz flint and Riparo Mochi (Mousterian).

found at Yarımburgaz, they are not especially common. Instead, it appears that as cores were reduced, they often took on a more classic radial/disc form (e.g., Fig. 5.11, n. 1, 2, 3, 5; Fig. 5.2, n. 2), or else were transformed into other core forms. Figure 5.13 (no. 6) shows an example of such a transformation. This core was apparently exploited for a time in a centripetal manner, but at a certain point the strategy of exploitation shifted so that the former face of detachment was used as a striking platform.

The comparison with the Tarn River sites serves primarily to demonstrate that the method of flake production used at Yarımburgaz, while distinctive, is not absolutely unique. These locales are quite far apart, and perhaps chronologically distant as well, and there is no particular reason

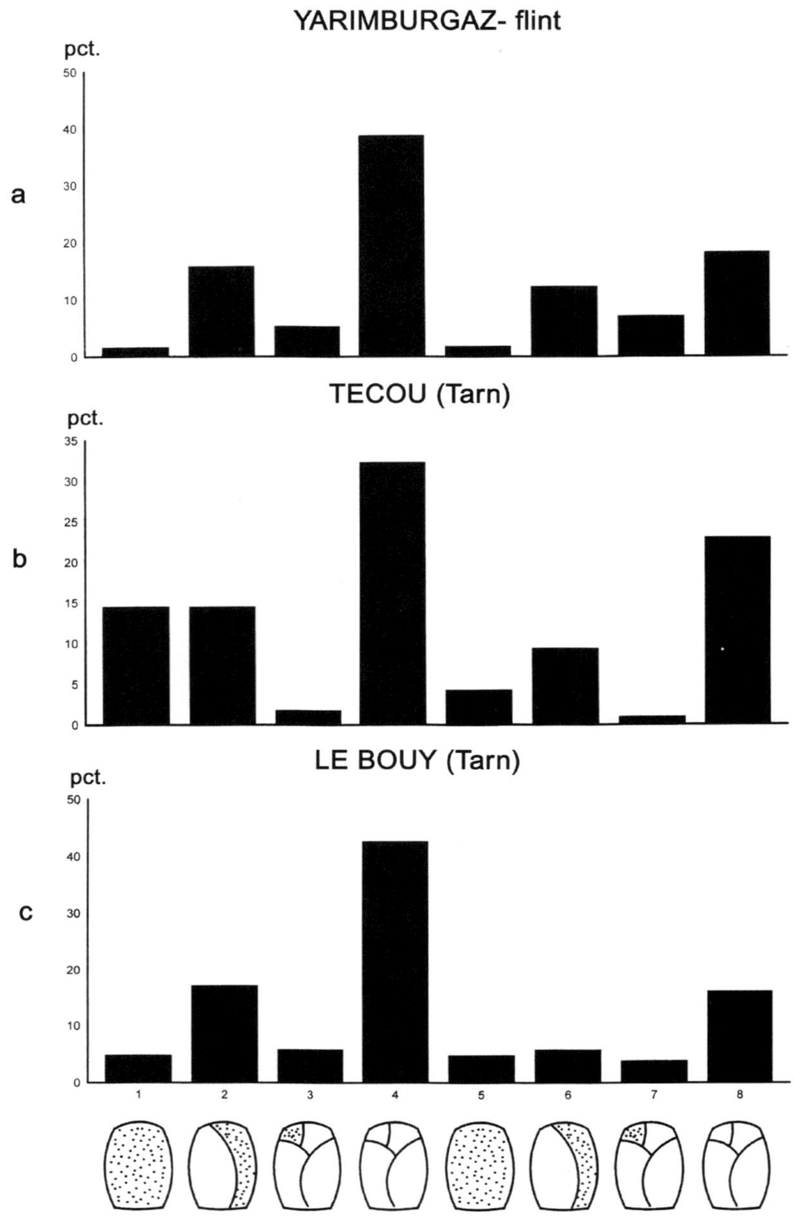

Figure 5.19. Comparisons of cortex distribution by two platform types, Yarımburgaz flint and Tarn river terrace sites.

to expect that similarities in methods of core reduction demonstrate any direct cultural linkage. However, the comparison also demonstrates that the use of this particular variety of centripetal core exploitation was not in some way determined by the type of raw material employed. At Yarımburgaz, the "unifacial discoid" method was employed primarily to work flint, whereas in the Tarn River sites the dominant stone is quartz. The technological convergence may reflect a response to similar pebble shapes, or to requirements for similar kinds of tool blanks.

Context of the Yarımburgaz lithic assemblages

The technology of the Yarımburgaz Cave lithic assemblages was clearly oriented around the production of flake tools, often on thick blanks. A variety of technological procedures were employed for producing flakes, including bipolar, hammer-on-anvil technique and a distinctive variety of discoid flaking. Choppers and chopping tools are comparatively infrequent and many may well have served as cores. Flake tools frequently received heavy, invasive retouch, and some appear to have been resharpened. Levallois and bifacial shaping are effectively lacking from the Yarımburgaz assemblages. Given the size of the collection (>1600 clearly artifactual specimens), the absence of evidence for bifacial technology is not an artifact of sampling in smaller assemblages (cf., Villa 2001; White 2000: 16-20).

There is limited comparative context for the assemblage from Yarımburgaz. It is currently the only well documented, geologically *in situ* Middle Pleistocene assemblage from western Turkey (western Anatolia and Thrace), although flake-tool and chopper assemblages without bifaces and Levallois have been collected from surface sites in the region (Dinçer and Slimak 2007). Comparable, though not identical industries have been recovered from geologically sealed Middle Pleistocene-age deposits at several sites in central Turkey. The lower levels of Karain Cave near Antalya in south-central Anatolia contain industries variously termed Clactonian and "proto-Charentian", with many heavily-modified tools made on thick flakes (Otte et al. 1995; Yalçınkaya et al. 1993). These layers are undated, but have been attributed to the later Middle Pleistocene on stratigraphic and paleontological grounds. However, the Karain assemblages appear to contain a much higher proportion of sidescrapers than Yarımburgaz, and the techniques of blank production may differ as well. Kaletepe Dersi 3, in central Anatolia, is a multi-component stratified open-air site with both Middle and Lower Paleolithic components, including the only geologically sealed Acheulean deposits with handaxes and cleavers excavated to date within Turkey. Materials from levels 2' and 3 at Kaletepe Deresi 3 are later Middle Pleistocene in age, being stratified beneath tephra dated to about 160ka. Both assemblages have been described as "Clactonoid," in terms of flake production (Slimak et al. 2007, 2008). However, the collections from these particular levels are small, and, owing to the fact that Kaletepe is situated within a source of raw materials (obsidian, rhyolite, and andesite), they contain few retouched pieces.

Comparisons with southern and central Europe and western and central Asia are also relevant. Due to the predominance of heavily retouched flake tools with steep irregular, denticulate edges along with well-made choppers and chopping tools, as well as the absence of both true bifaces and Levallois technology, the assemblage from Yarımburgaz Cave resembles Tayacian and allied industries from southern Europe (e.g., Ferrari et al. 1991; de Lumley 1976a, 1976b; Rolland 1985), and the chopper and small tool industries of eastern and central Europe (Svoboda 1987, 1989). Middle Pleistocene industries composed of heavily modified flake tools and choppers are also found in central Asia (Clark 1993: 163; Davis et al. 1980; Ranov et al. 1995). In the Caucasus, Acheulean assemblages are predominant (Ljubin and Bosinski 1995). Dmanisi, the earliest site in the larger region and possibly the earliest site outside Africa, also yields assemblages of stone tools lacking in handaxes and Levallois (de Lumley et al. 2005). However, Dmanisi is so much older than Yarımburgaz that technological comparisons have little value.

Early "core and flake tool" assemblages are also known from southwest Asia (e.g., Copeland 2003; Kirkbride et al. 1983; Solecki 1968), although they are less common than Acheulean assemblages (Bar-Yosef 1994). Representative collections come from the Dursunlu quarry in central Turkey (Güleç et al. 2009), Bizat Ruhama in Israel (Burdukiewicz and Ronen 2000; Zaidner 2003) and layer G at Tabun (Garrod and Bate 1937; Howell 1959). Here again most of these are much older than Yarımburgaz. Dursunlu and Bizat Ruhama both predate the Brunhes/Matuyama boundary. A single TL date of around 415 ka has been attributed to Garrod's layer G at Tabun (equivalent to Jelinek's Bed 90) (Mercier et al. 2000). However, the nature of the industry associated with the dates remains in contention (compare Jelinek 1982; Ronen and Tsatskin 1995).

The absence of Levallois method and bifacial *faconnage* from the Yarımburgaz assemblage certainly is significant in terms of the technological diversity and cultural geography of the later Middle Pleistocene, topics that will be examined in detail in the concluding chapter of this volume. However, these features alone are not justification enough to assign the Yarımburgaz assemblage to a "cultural" unit such as Tayacian. Defining a category such as Tayacian based mainly on its negative characters produces an overly general, "catch all" category (Laville et al. 1980). After all, assemblages from the Plio-Pleistocene pre-Oldowan to the Quin Mousterian, *sensu stricto* may also lack Levallois and bifaces, yet a class of assemblages that contained such diverse technologies representing such a vast span of time would be essentially meaningless. Detailed analyses of *chaînes opératoires,* not only methods of blank production and *faconnage* but also techniques of retouch and artifact rejuvenation, may be useful for drawing hypothetical links between different local traditions of artifact making that lack more spectacular technological features. In fact, the Yarımburgaz assemblage does not resemble technologically the narrower description of Levantine Tayacian assemblages recently published by Copeland (2003).

One of the most striking features of the Yarımburgaz assemblage is the sharply differentiated exploitation of the three main raw materials. Some of the associations between technology and raw material—most notably the frequent association between quartz and bipolar, hammer-and-anvil percussion—may reflect the working properties of specific types of stone. On the other hand, there is no obvious reason why choppers should be made of quartzite and not flint. Similarly, the distinctive centripetal method of flake production that was used mainly on flint at Yarımburgaz was apparently employed successfully to work quartz or quartzite at sites in France (Tavoso 1978).

In fact, Middle and Lower Pleistocene assemblages from throughout the world often exhibit such differentiated use of raw materials (e.g., Bar-Yosef and Goren-Inbar 1993; Clark 1968; Leakey 1975; Martinez 1998; Miller-Antonio et al. 2004; Svoboda 1989; Villa 1983; Zaidner 2003). The rigid association between specific raw materials and specific *chaînes opératoires* may even be a defining characteristic of the Lower Paleolithic (Goren-Inbar and Belfer-Cohen 1998). Significantly, the associations between raw materials and tool forms are consistent within sites but not between them, suggesting that the physical properties of different types of stone were not absolute determining factors. At the lower Pleistocene site of 'Ubeidiya, for instance, choppers and chopping tools were made almost exclusively on flint (Bar-Yosef and Goren-Inbar 1993), while at Yarımburgaz they almost never were. Perhaps the only general constant is that flint and other fine-grained raw materials usually seem to have been preferred for making small tools (e.g., Martinez 1988: 20). However, this could also be due the fact that use-damage and retouch tend to be more evident on fine-grained rocks.

The behavioral basis for these locally rigid but globally variable patterns of raw material selection remains an intriguing question. One possibility is that they reflect a series of more or less opportunistic choices from among a limited range of technological options made in response to the sizes and shapes of pebbles. The methods of flake production found in many early assemblages—including Yarımburgaz—seem to have been oriented toward the production of flakes with minimal core shaping or preparation. While these technologies were certainly systematic, they were also quite opportunistic. The fact that assemblages often contain a variety of raw materials shows that Middle Pleistocene toolmakers tended to take advantage of many types of workable stone available to them. In a situation where flakes were produced rapidly and with little shaping of the core, the forms of the pebbles or nodules would have had a strong influence on both the trajectories of reduction and the attributes of the final products. In Middle Pleistocene technologies, the relationships between pebble shapes and the forms of final products were probably not as direct as some suggest for the Oldowan (e.g., Schick and Toth 1993:130; but cf. Delagnes and Roche 2005). Nonetheless, pebbles of different forms would have provided different kinds of opportunities, different "paths of least resistance" in the manufacture of flakes. If this is indeed the case, then associations between raw material and technological procedures may be largely fortuitous, reflecting instead variation in the shapes of pebbles of different kinds of stone.

On the other hand, the characteristics of raw materials as different as flint and quartz should be considered for more than their effects on efficient blank production. Makers of fairly simple, unstandardized flake technologies may control the functional properties of tools through selection of raw materials for properties such as hardness, toughness, and the microscopic topologies of flake edges. Recent findings suggest that even the earliest Pliocene toolmakers preferentially selected stones for their functional properties as well as workability (Braun et al. 2008; Stout et al. 2005), so it is reasonable to expect that later Middle Pleistocene hominins exploited these same properties. Experimental work and broad inter-assemblage comparative studies should help to clarify the significance of this general tendency.

Another striking aspect of the Yarımburgaz assemblage is the fact that the intensity of raw material exploitation seems to be decoupled from raw material availability. Flake tools show frequent and often heavy retouch. The steep, highly invasive retouch on many of the artifacts suggests that they were extensively modified and used, perhaps even resharpened. This kind of intensive use of tools is often thought to reflect a scarcity of raw materials or to occur as the repeated reuse and/or long-term transport of implements (Dibble 1995; Holdoway et al. 1996; Kuhn 1991). However, aside from this intensive modification of flake tools, there is no strong evidence that pebble raw materials were especially scarce in the vicinity of Yarımburgaz Cave. In fact, methods of core reduction at Yarımburgaz involved little systematic preparation and shaping, and many cores were worked in at best a summary fashion (fully one third of them have fewer than four negatives). On the other hand, some quartz and flint cores are small and extensively modified, and many flakes of both materials were retouched. Yet quartz and flint pebbles are known to occur naturally in sediments at the site.

The heavy modification of many artifacts at Yarımburgaz could be a product of the activities in which tools were used. Tasks like woodworking produce heavy attrition to tool edges and could result in extensive resharpening, even where raw materials are plentiful. This hypothesis will probably remain untested. Because of the extensive chemical modification of the flint artifacts, it is very unlikely that microwear studies would prove successful. However, the fact that such intense exploitation of raw materials occurred in the absence of evidence for raw material shortage is noteworthy.

In evaluating raw material economy at Yarımburgaz, it is important to recall that "assemblages" are units defined by the archaeologists based on *post hoc* geological and archaeological criteria. Any assemblage of artifacts almost certainly represents an accumulation of many independent behavioral events, each involving the creation, use and discard of a few artifacts. Because the Yarımburgaz record a time-transgressive palimpsest, the behavior of hominins must have varied over the period over which the assemblage accumulated as a whole. A range of circumstances affected the immediate advantages of reducing existing artifacts as opposed to collecting more stone, such that hominins using the cave sometimes used raw materials quite casually, and other times exploited tools and cores almost to the point of exhaustion. This kind of fine-grained behavioral dynamic cannot always be captured in analytical units as coarse as assemblages based on geologically-defined layers.

Analysis of variability in artifact density and intensity of raw material exploitation within Yarımburgaz Cave can help to clarify whether the unusual features of the assemblage are a consequence of lumping together smaller depositional units with differing characteristics. At many Middle and Upper Paleolithic sites in Europe and the Near East, there is a strong relationship between the density of stone artifacts within deposits and the composition of assemblages (Clark 2008; Kuhn 2004; Meignen et al. 2005; Riel-Salvatore and Barton 2004; Riel-Salvatore 2007). Across a range of cases and contexts artifact density is negatively correlated with retouch frequency. High-density deposits exhibit low frequencies of retouch and, often, high frequencies of cores, whereas low density deposits are marked by higher frequencies of retouch and fewer cores. This robust tendency is thought to index variable intensity of human occupations. Low density deposits represent debris left by shorter, more episodic occupations. Under these conditions, deposits are dominated by transported artifacts brought to the cave and abandoned during brief stays, and there is comparatively little *in situ* production. Higher density deposits represent more prolonged, intensive occupations. Hominins spending more time in and around particular sites are more likely to produce artifacts locally, generating quantities of waste material and lower frequencies of retouched pieces. Of course, few assemblages contain debris just one kind of occupation, so the significant fact is the correlations between variables, and not the variables, actual values.

The excavation techniques used at Yarımburgaz make it impossible to reconstruct the precise volume of particular excavation or sedimentological units. However, it is possible to estimate volumes by adding up the number of squares in which artifacts were found within a particular trench, and multiplying it by the range of depths over which artifacts occur. Dividing the total number of artifacts by this pseudo-volume provides an estimate of the artifact density for a particular stratigraphic unit in a specific trench. Confining analysis to trenches and levels with more than eight specimens, it was possible to produce density estimates for 115 sedimentary units across 6 trenches.

Table 5.11. Correlations between technological indices and estimated artifact density.

	r value	df	p
retouch index 1	0.08	14	>>0.10
log retouch index 1	0.103	14	>>0.10
retouch index 2	-0.192	14	>>0.10
log retouch index 2	-0.157	14	>>0.10
tools+flakes/cores	-0.085	13	>>0.10

Retouch index 1 = number of retouched pieces/(large flakes+retouched pieces); retouch index 2 = number of retouched pieces/(large flakes +retouched pieces+debris).

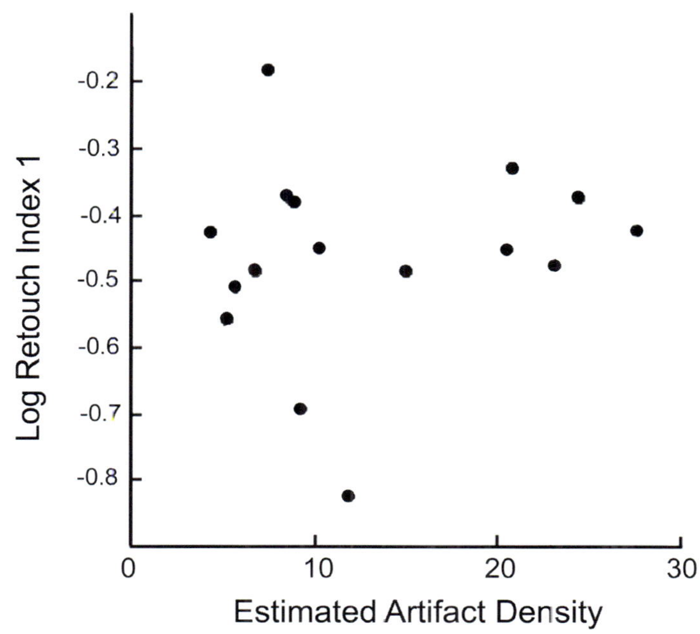

Figure 5.20. Plot of logged retouch index against estimated artifact density.

Correlations (Pearson's *r*) between density and various technological indices are shown in Table 5.11. In many studies the relationship between the retouch index and density is curvilinear so both raw and logged figures are shown. The Yarımburgaz results demonstrate clearly that there is no relationship between estimated artifact density and frequency of retouch across sedimentary units and excavation trenches in this cave. The correlations are very low, and none approaches statistical significance at the standard 0.05 α level. Figure 5.20 shows a sample plot of a logged retouch index against estimated density, confirming they lack of any clear linear trend. The number of blanks per core also shows no relationship with the estimated density of artifacts.

These results confirm that the unusual combination of heavy and frequent retouch with light reduction of cores and near-exclusive use of local raw materials at Yarımburgaz is not due to mixing material debris from radically different kinds of occupation. As has also been observed for the pre-Mousterian Middle Pleistocene assemblages from Tabun Cave (Clark 2008), there is no clear relationship between artifact density and intensity of raw material exploitation. Although there is appreciable variation in both the density of stone tool finds and the frequency of retouched tools across various proveniences, they do not show any correlative relationship. This points toward a rather different mode of land use and technological provisioning than is typical for the later Middle and Upper Paleolithic. We will return to this question in the concluding chapter of the monograph.

ACKNOWLEDGMENTS

In early 1991, F. Clark Howell called Mary Stiner and I out of the blue with an invitation to come to Istanbul to work on the material from the recently-completed excavations at Yarımburgaz Cave. This was the beginning of a long, happy and productive period of research on the Paleolithic in Turkey. I am forever grateful to Clark for this, and for his good advice and support. Güven Arsebük helped us in so many ways to get established and to understand the site and its context. M. Özbaşaran, S. Gülçur, N. Balkan-Atlı, O. Tanındı, M. Özdoğan and others provided us with crucial advice and insights on the site and the collections, in the process forming friendships and scientific collaborations that have lasted to this day. Conversations with O. Bar-Yosef, A. Clark, S. McPherron, L. Meignen, L. Slimak, and M. Stiner have helped sharpen many of the ideas presented in this chapter. Substantial portions of data and basic analyses presented here have appeared previously in Kuhn et al. (1996) and Kuhn (2003).

REFERENCES CITED

Ashton, N., 2007. Flakes, cores, flexibility and obsession: situational behaviour in the British Lower Palaeolithic, in: McPherron, S., (Ed.), *Tools versus Cores: Alternative Approaches to Stone Tool Analysis*. Cambridge Scholars Publishing, Newcastle, pp. 1-16.

Bar-Yosef, O., 1994. The Lower Paleolithic of the Near East. *Journal of World Prehistory* 8, 211-265.

Bar-Yosef, O., Goren-Inbar, N., 1993. *The Lithic Assemblages of 'Ubeidiya, a Lower Paleolithic Site in the Jordan Valley*. QEDEM 34, Monographs of the Institute of Archaeology, The Hebrew University. Hebrew University, Jerusalem.

Boëda, E., 1993. Le débitage discoïdes et le débitage Levallois récurrent centrripète. *B.S.P.F.* 90, 392-404.

Bourguingnon, L., 1996. La conception de débitage Quina. Quaternaria *Nova* 6, 149-166.

Braun, D.R., Plummer, T., Ditchfield, P., Ferraro, J.V., Maina, D., Bishop, L.C., Potts, R., 2008. Oldowan behavior and raw material transport: perspectives from the Kanjera Formation. *Journal of Archaeological Science* 36, 1605-1614.

Brézillon, M., 1968. *La Dénomination des Objets de Pierre Taillée*. 4th supplément à Gallia Préhistoire. Éditions du CNRS, Paris.

Burdukiewicz, J.M., Ronen, A., 2000. Ruhama in the northern Negev Desert. A new microlithic site of the Lower Palaeolithic in Israel. *Praehistoria Thuringica* 5, 3-5.

Clark, A.E., 2008. Changes in Occupation Intensity during the Lower and Middle Paleolithic at Tabun Cave, Israel. Unpublished M.A. thesis, Department of Anthropology, University of Arizona.

Clark, J.D., 1968. The middle Acheulean occupation site at Latamne, northern Syria (2nd paper). Further excavations (1965): general results, definitions and interpretation. *Quaternaria* 10, 1-72.

Clark, J.D., 1993. African and Asian perspectives on the origins of modern humans, in: Aitken, M., Stringer, C., Mellars, P., (Eds.) *The Origin of Modern Humans and the Impact of Chronometric Dating*. Princeton University Press, Princeton, NJ, pp. 148-178.

Copeland, L., 2003. The Tayacian of the Cordon Littoral, Ras Beirut (Lebanon) and its relations with other Tayacian sites in the Levant. *Paléorient* 29/2, 87-107.

Davis, R., Ranov, V., Dodonov, A., 1980. Early man in Soviet Central Asia. *Scientific American* 243, 130-137.

Delagnes, A., Roche, H., 2005. Late Pliocene hominid knapping skills: the case of Lokalalei 2C, West Turkana, Kenya. *Journal of Human Evolution* 48, 435-472.

Dibble, H.L., 1995. Raw material availability, intensity of utilization, and Middle Paleolithic assemblage variability, in: Dibble, H.L., Lenoir, M., (Eds.), *The Middle Paleolithic Site of Combe-Capelle Bas (France)*. The University Museum, University of Pennsylvania, Philadelphia, pp. 289-315.

Dibble, H., McPherron, S., 2006. The missing Mousterian. *Current Anthropology* 47, 777-803.

Dinçer, B., Slimak, L., 2007. The Paleolithic of Turkish Thrace: synthesis and recent results. *TÜBA-AR* 10, 49-61.

Ferrari, M., Peretto, C., Vianello, F., 1991. Aspetti tecno-tipologici e distribuzione areale dell'industria litica del II settore di Isernia la Pineta (Molise, Italia), in: Peretto, C., (Ed.), *Isernia la Pineta: Nuovi Contributi Scientifici*. Istituto Regionale per gli Studi Storici del Molise, Isernia, pp. 49-78.

Garrod, D.A.E., Bate, D., 1937. *The Stone Age of Mt. Carmel*. Vol. 1. Clarendon Press, Oxford.

Goren-Inbar, N., Belfer-Cohen, A., 1998. The technological abilities of the Levantine Mousterian: cultural and mental capacities, in Akazawa, T., Akoi, K., Bar-Yosef, O., (Eds.), *Neandertals and Modern Humans in Western Asia*. Plenum Press, New York, pp. 205-221.

Güleç, E., White, T., Kuhn, S., Özer, I, Sağır, M., Yılmaz, H., Howell, F.C., 2009. The Lower Pleistocene Lithic Assemblage from Dursunlu (Konya), central Anatolia, Turkey. *Antiquity* 83, 11-22.

Holdoway, S., McPherron, S., Roth, B., 1996. Notched tool reuse and raw material availability in French Middle Paleolithic sites. *American Antiquity* 61, 377-387.

Hovers, E., 2007. The Many Faces of Cores-On-Flakes: A Perspective from the Levantine Mousterian, in: McPherron, S., (Ed.), *Tools versus Cores: Alternative Approaches to Stone Tool Analysis*. Cambridge Scholars Publishing, Newcastle, pp. 42-74.

Howell, F.C., 1959. Upper Pleistocene stratigraphy and early man in the Levant. Proceedings of the *American Philosophical Society* 103, 1-65.

Isaac, G. Ll., 1977. *Olorgesailie: Archaeological Studies of a Middle Pleistocene Lake Basin in Kenya*. University of Chicago Press, Chicago.

Jaubert, J., 1994. Les industries lithiques, in: Farizy, C., David, F., Jaubert, J., (Eds.) *Hommes et Bisons du Paléolithique Moyen à Mauran (Haute Garonne). XXX supplément à Gallia Préhistoire*. Éditions du CNRS, Paris.

Jaubert, J., Farizy, C., 1995. Levallois debitage: Exclusivity, absence or coexistence with other operative schemes in the Garonne Basin, Southwestern France, in: Dibble, H., Bar-Yosef, O., (Eds.), *The Definition and Interpretation of Levallois Technology*. Mongraphs in World Archaeology, 23. Prehistory Press, Madison, WI, pp. 227-248.

Jelinek, A.J., 1982. The Tabun cave and Paleolithic man in the Levant. *Science* 216, 1368-1376.

Kirkbride, D., de Saint-Mathurin, S., Copeland, L., 1983. Results, tentative interpretation and suggested chronology, in: Roe, D.A., (Ed.), *Adlun in the Stone Age*. BAR International Series 159 (II). British Archaeological Reports, Oxford, pp. 415-432.

Kuhn, S., 1991. Unpacking reduction: lithic raw material economy in the Mousterian of west-central Italy. *Journal of Anthropological Archaeology* 10, 76-106.

Kuhn, S., 2003. Flexibility and variation in the Lower Paleolithic: a view from Yarımburgaz Cave, in: Özbaşaran, M., Tanındı, O., Boratav, A., (Eds.), *Homo Amatus: Archaeological Essays in Honor of Güven Arsebük*. Ege Yayınları, Istanbul, pp. 149-158.

Kuhn, S., 2004. Middle Paleolithic Assemblage Formation at Riparo Mochi, in: Johnson, A., (Ed.), *Processual Archaeology*. Praeger, London, pp. 31-60.

Kuhn, S., Arsebük, G., Howell, F.C., 1996. The Middle Pleistocene lithic assemblage from Yarımburgaz Cave, Turkey. *Palèorient* 22/1: 31-49.

Laville, H., Rigard, J.-P., Sackett, J., 1980. *Rocksheiters of the Perigord*. Academic Press, New York.

Leakey, M.D., 1975. Cultural patterns in the Olduvai sequence, in: Butzer, K.W., Isaac, G.L., (Eds.), *After the Australopithecines*. Mouton Press, The Hague, p.p. 477-494.

Ljubin, V.P., Bosinski. G., 1995. The earliest occupation of the Caucasus region, in Roebroeks, W., van Kolfschoten, T., (Eds.), *The Earliest Occupation of Europe*. University of Leiden, Leiden, pp. 207-254.

Lumley-Woodyear, H. de, 1969. *Le Paleolithique inferieur et moyen di midi Mediterraneen dans son cadre Geologique*. Tome 1, Ligurie-Provence. 5th supplément à Gallia Prehistoire. Éditions du CNRS, Paris.

Lumley, H. de, 1976a. Les civilisations du Paléolithique inférieur en Provence, in: deLumley, H., (Ed.), *La Préhistoire Français*. Tome 1: Les Civilisations Paléolithiques et Mésolithiques de la France. Éditions du CNRS, Paris, pp. 819-851.

Lumley, H. de, 1976b. Les civilisations du Paléolithique inférieur en Languedoc méditerranéen et en Rousillon, in: deLumley, H., (Ed.), *La Préhistoire Français*. Tome 1: Les Civilisations Paléolithiques et Mésolithiques de la France. Éditions du CNRS, Paris, pp. 852-874.

Lumley, H. de, Nioradzé, M., Barsky, D., Cauche, D., Celiberti, V., Nioradzé, G., Notter, O., Zvania, D., Lordkipanidze, D., 2005. Les industries lithiques préoldowayennes du début du Pléistocène inférieur du site de Dmanissi en Géorgie. *l'Anthropologie* 109, 1-182.

Martinez, M.M., 1998. Differential raw material use in the Middle Pleistocene of Spain: Evidence from Sierra de Atapuerca, Torralba, Ambrona and Aridos. *Cambridge Archaeological Journal* 8, 15-28.

Meignen, L., Bar-Yosef, O., Speth, J.D., Stiner, M.C., 2005. Middle Paleolithic settlement patterns in the Levant, in: Hovers, E., Kuhn, S., (Eds.), *In Transitions before the Transition: Evolution and Stability in the Middle Paleolithic and Middle Stone Age*. Plenum Press, New York, pp. 149-170.

Mercier, N., Valladas, H., Froget, L., Joron, J.-L., Ronen, A., 2001. Datation par thermoluminescence de la base du gisment Paléolithique de Tabun (Mt. Carmel, Israël). *Comptes Rendues Academie di Sciences*, Paris 330, 731-738.

Miller-Antonio, S., Schepartz, L.A., Karkanas, P., Hou Y., Huang, W., Bekken, D., 2004. Lithic Raw Material Use at the Late Middle Pleistocene Site of Panxian Dadong. *Asian Perspectives* 43, 314-332.

Otte, M., Yalçınkaya, I., Kozlowski, J., Bar-Yosef, O., Taşkıran, H., Noiret, P., 1995. Évolution technique au Paléolithique ancien de Karain. *L'Anthropologie* 90, 529-561.

Ranov, V.A., Carbonell, E., Rodriguez, X.P., 1995. Kuldara: Earliest human occupation in central Asia in its Afro-Asian context. *Current Anthropology* 36, 336-346.

Riel-Salvatore, J., 2007. The Uluzzian and the Middle-Upper Paleolithic Transition in Southern Italy. Unpublished Ph.D. dissertation, Arizona State University, Tempe, AZ.

Riel-Salvatore, J., Barton, C.M., 2004. Late Pleistocene technology, economic behavior, and land-use dynamics in southern Italy. *American Antiquity* 69, 273-290.

Rolland, N., 1985. Recent findings from La Micoque and other sites in south-western and Mediterranean France: their bearing on the 'Tayacian' problem and Middle Paleolithic emergence, in: Bailey, G., Callow, P., (Eds.), *Stone Age Prehistory*. Studies in Memory of Charles McBurney. Cambridge University Press, Cambridge, pp. 121-151.

Ronen, A., Tsatskin, A., 1995. New interpretation of the oldest part of thew Tabun sequence, Mt. Carmel, Israel, in: Ulrich, H., (Ed.), *Man and Environment in the Levant*. ERAUL 62, Liège, pp. 247-263.

Schick, K., Toth, N., 1993. *Making Silent Stones Speak*. Simon and Schuster, New York.

Slimak, L., 2003. Les débitages Discoïdes moustériens : évaluation d'un concept technologique, in: Peresani, M., (Ed.) *Discoid Lithic Technology: Advances and Implications*. BAR International Series 1120. British Archeological Reports, Oxford, pp. 33-65.

Slimak, L., Kuhn, S., Balkan-Atlı, N., Binder, D., and Dinçer, B., 2007. Kaletepe Deresi 3: de 'l'Acheuleen au Mousterien en Anatolie centrale. *Anatolia Antiqua/Eski Anadolu* 15, 257-273.

Slimak, L., Kuhn, S., Roche, H., Mouralis, D., Buitenhuis, H., Balkan-Atlı, N., Binder, D., Kuzucuoğlu, C., Guillou, H., 2008. Kaletepe Deresi 3 (Turkey): archaeological evidence for early human settlement in Central Anatolia. *Journal of Human Evolution* 54, 99-111.

Solecki, R., 1968. The Shemsi industry, a Tayacian-related industry at Yabrud, Syria. preliminary report, in: *La Préhistoire*. Problèmes et Tendances. Éditions du CNRS, Paris, pp. 401-410.

Stout, D., Quade, J., Semaw, S., Rogers, M.J., Levin, N.E., 2005. Raw material selectivity of the earliest stone toolmakers at Gona, Afar, Ethiopia. *Journal of Human Evolution* 48, 365-380.

Svoboda, J., 1987. Lithic industries of the Arago, Véresszöllös, and Bilzingsleben hominids: comparison and evolutionary interpretation. *Current Anthropology* 28, 219-227.

Svoboda, J., 1989. Middle Pleistocene adaptations in central Europe. *Journal of World Prehistory* 3, 33-70.

Tavoso, A., 1978. *Le Paléolithique Inférieur et Moyen du Haut-Languedoc*. Études Quaternaires, Mémoire N. 5, Laboratoire de Paléontologie Humaine et de Préhistoire, Université de Provence. Institut de Paléontologie Humaine, Paris.

Terradas, X., 2003. Discoid flaking method: conception and technological variability, in: Peresani, M., (Ed.), *Discoid Lithic Technology: Advances and Implications*. BAR International Series 1120. British Archaeological Reports, Oxford, pp. 19-30.

Toth, N., 1985. The Oldowan reassessed: A close look at early stone artifacts. *Journal of Archaeological Science* 12, 101-120.

Villa, P., 1983. Terra Amata and the Middle Pleistocene Archaeological Record of southern France. University of California Publications in *Anthropology*, Vol. 13. University of California Press, Berkeley.

Villa, P., 2001. Early Italy and the colonization of Western Europe, in: Bar-Yosef, O., Straus, L., (Eds.), *Out of Africa in the Pleistocene. Quaternary International* 75 (special issue), 113-130.

White, M., 2000. The Clactonian question: On the interpretation of core-and-flake assemblages in the British Lower Paleolithic. *Journal of World Prehistory* 14, 1-63.

Yalçınkaya, I., Otte, M., Bar-Yosef, O., Kozlowski, J., Léotard, J.-M., Taşkıran, H., 1993. The excavations at Karain Cave, southwestern Turkey: an interim report, in: Olszewski, D., Dibble, H., (Eds.), *The Paleolithic Prehistory of the Zagros-Taurus*. University Museum Monograph 83. The University of Pennsylvania, Philadelphia, p.p. 101-118.

Zaidner, Y., 2003. The use of raw material at the Lower Palaeolithic site of Bizat Ruhama, Israel, in: Burdukiewicz, J.M., Ronen, A., (Eds.), *Lower Palaeolithic Small Tools in Europe and the Levant*.

CHAPTER 6

A TAPHONOMIC STUDY OF BONE ASSEMBLAGE FORMATION AND ARTIFACT-BEAR ASSOCIATIONS IN YARIMBURGAZ CAVE

Mary C. Stiner

INTRODUCTION

Karstic landscapes have preserved more than their share of Pleistocene faunas, of which hominin fossils and cultural records are only one part. Bear skeletal remains frequently are found in Paleolithic caves, particularly Eurasian sites whose deposits date to the Middle and Late Pleistocene geologic periods. Thanks in particular to the works of Björn Kurtén (1958, 1976; Kurtén and Poulainos 1981), many investigators now recognize that co-occurrences of human artifacts and bear remains in cave deposits may arise from alternating occupations of natural shelters by these species spaced over decades, centuries or millennia. Water and gravity may also promote stratigraphic associations among historically unrelated materials. Of course, Pleistocene humans also hunted bears from time to time.

The story of human-bear interactions is important to our general understanding of Pleistocene hominin ecology. Converging patterns of omnivory would seem reason enough for Pleistocene humans and bears to have stayed out of each other's way. Yet the remains of cave bears (*U. deningeri, U. spelaeus*) or brown bears are found in nearly every Middle and Upper Paleolithic site with preserved bone—sometimes many bear bones, but more usually a few. In fact, bear remains turn up in Paleolithic cave sites more consistently than the remains of almost any other large carnivore, at least prior to 20,000 years ago, when greater inter-specific exclusion among cave residents is apparent (Gamble 1986; Stiner 1994). Human exploitation of bears by humans in later periods is relatively well known, if infrequent overall (Bárta 1989; Rogers 1981: 69; Stiner 1990, 1994).

Yarımburgaz represents a particularly early example of hominid-bear associations. It is one of very few early Paleolithic cave sites documented in the area between western Asia and central Europe. Part of an extensive karst system in Eocene limestone, the cave extends up to 600 m into the limestone bedrock and consists of two parallel chambers with separate south-facing entrances (Upper and Lower Caves, see Chapter 1, this volume). Deposits within 70 m of the lower entrance are rich in both cave bear remains and Paleolithic artifacts that date to the middle Pleistocene (Blackwell et al., this volume). Taphonomic analyses of the large mammal remains were undertaken by the author in 1992 (Stiner et al. 1996) to better understand the circumstances of bone assemblage formation and the agencies responsible for the accumulations. Other aspects of the macromammal faunas, including isotopic and mortality data for bears, are presented elsewhere (Stiner 1998; Stiner et al. 1998). Most of the bones from Yarımburgaz Cave are in an extraordinary state of macroscopic preservation, possibly due to an alkaline sedimentary environment (Figure 6.1); some of the flint artifacts from the same deposits have not fared nearly as well. A few bones display unusual forms of chemically induced break-down but this kind of

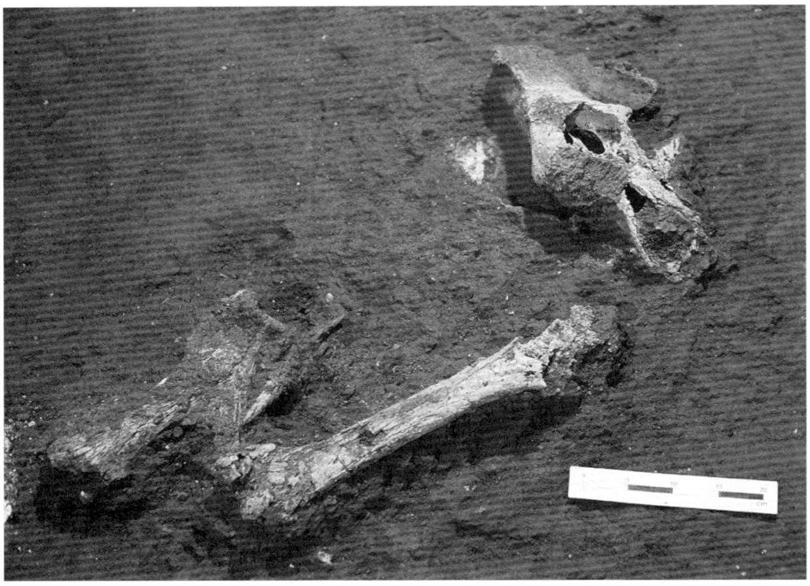

Figure 6.1.
Cave bear bone scatter from Trench P (1990), level 2, plan 9 showing some typical variation in bone condition.

Figure 6.2.
Excavated level in Yarımburgaz Cave (Trench Y, 1988, level 1, plan 7a) showing a large scatter of cave bear bones, Paleolithic artifacts and loose limestone rocks.

Figure 6.3.
Nearly complete bear cranium (detail) from Trench Y (1988), level 1, plan 7a.

damage is highly localized in the deposits. Most of the faunal remains could be identified to genus if not species because fragmentation is limited. Only the material from the 1988, 1989 and 1990 excavations at Yarımburgaz Cave is the focus of consideration here. Many other bones and artifacts were recovered during earlier excavation campaigns, particularly in the Upper Cave, but those materials date to later periods and underwent considerable post-depositional mixing (Özdoğan and Koyunlu 1986).

FRAMING THE PROBLEM AND ANALYTICAL METHODS

There are close spatial associations of bear bones and lithic artifacts throughout most of the area excavated during 1988 through 1990 (Figures 6.2-6.5). One of the most obvious questions about the Yarımburgaz Cave assemblage is why these materials are associated in the middle Pleistocene layers. Unfortunately there seldom is a straightforward explanation for the spatial associations of Paleolithic stone artifacts and carnivore bones in natural shelters (Brugal and Jaubert 1991; Gamble 1986; Jenkinson 1984; Kurtén 1976; Lindly 1988; Stiner 1994; Straus 1982). Slow or variable rates of sediment build-up are typical in caves, potentially allowing materials of different ages to accumulate in the same geological strata.

Cave bears constitute 93% of all macromammal specimens recovered (Table 6.1) from the middle Pleistocene layers of the Lower Cave, interspersed with about 1700 Paleolithic artifacts (see Kuhn et al. 1996; Kuhn, this volume). Rare brown bear specimens were also encountered in the collection (Figure 6.6). Ungulate and non-ursid carnivore bones are represented in low frequencies—4% and 3% of identifiable specimens respectively—and their distributions seem no less restricted within the cave. The Yarımburgaz collection does contain ungulate bones modified by hominins, but they are exceedingly rare. It is unlikely that the mammal bones were carried into the cave by water or gravity, although water flowed through the karst during earlier periods (Farrand and McMahon, this volume). The scant presence of other carnivore species, such as wolves, lions, spotted hyenas, and caracals, is an important taphonomic signal in its own right. Some of these predators are known to gather prey bones at den and rest sites (Ewer 1973; Fentress and Ryon 1982; Fox 1984; Mech 1970; for hyenas, see Hill 1980; Sutcliffe 1970).

Table 6.1. Summary of macromammal NISP and stone artifact counts from the 1988, 1989, and 1990 excavations of the lower chamber of Yarımburgaz Cave.

a. Macromammal NISP (bones and teeth):

Mammal group	Tooth NISP	Bone NISP	Composite NISP	Composite %
Herbivores	42	109	151	4
Bears	761	3159	3920	93
Other carnivores	33	79	109	3
Total	836	3344	4180	100

b. Paleolithic artifact counts:

Total lithic artifacts	1674
Modified (retouched) lithic artifacts	602

Note: Composite NISP refers to sum of tooth (articulated and isolated) NISP and bone NISP.

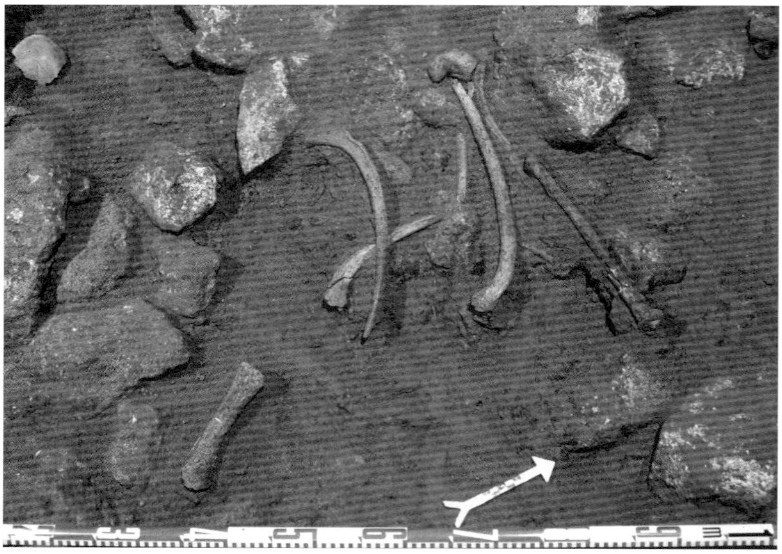

Figure 6.4.
Scattered whole ribs and other long bones of cave bears in Trench Y (1988), level 1a.

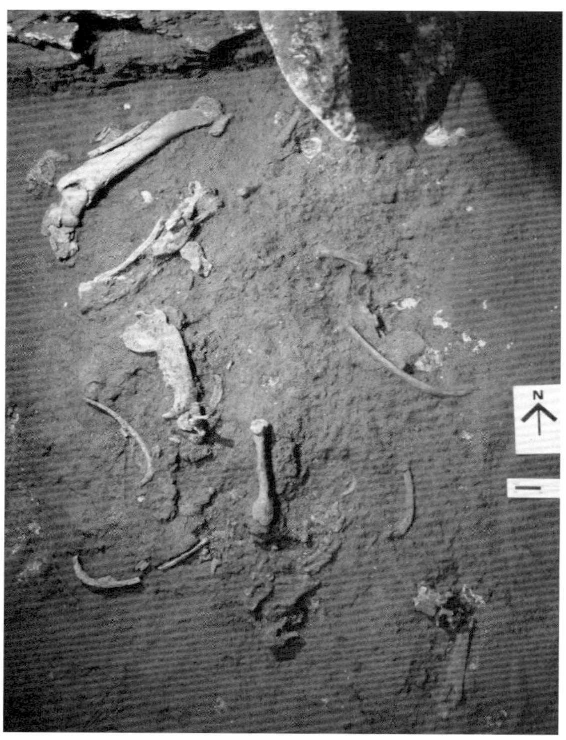

Figure 6.5.
Concentration of disassembled cave bear bones in the northeast corner of Trench P (1990).

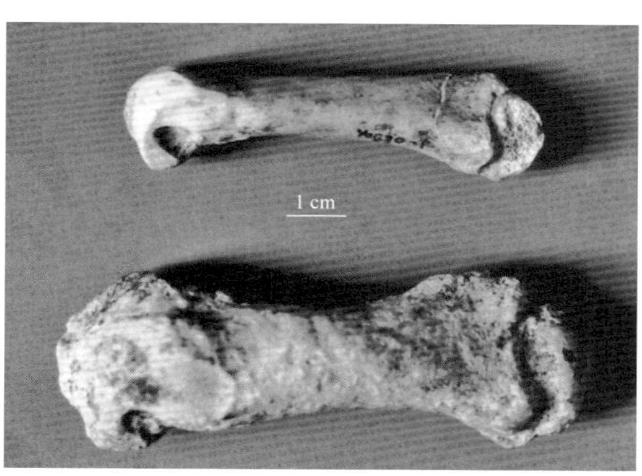

Figure 6.6.
Gracile and robust bear fifth metatarsals indicating the presence of at least two *Ursus* species in the Yarımburgaz Cave deposits.

As shown below, the influence of large canids on the vertebrate remains Yarımburgaz Cave was periodic but dramatic.

Whereas only 4% of all bones are from ungulates, the array of species represented by the 151 pieces is great, including horse (*Equus caballus*), roe deer (*Capreolus* aff. *süssenbornensis*), fallow deer (*Dama* sp.), red deer (*Cervus elaphus*), giant deer (*Megaloceros* sp.), aurochs (cf. *Bos primigenius*), bison (*Bison* cf. *priscus*), wild pig (*Sus scrofa*), wild goat and/or ibex (*Capra* aff. *aegagrus*, *C.* cf. *ibex*), possibly gazelle (*Gazella* sp.), and an undetermined species of megafauna (Howell, this volume). The array of carnivores other than bears is also diverse, given the small number of specimens (NISP = 109), and includes large and small cats of the genera *Panthera* (*P. leo*, possibly also *P. pardus*) and *Felis* (*F. caracal* and *F. sylvestris*), as well as spotted hyena (*Crocuta crocuta*), wolf (*Canis lupus*), fox (*Vulpes* spp.), possibly dhole (*Cuon*) and jackal (*Canis aureus?*), and a small mustelid. The Middle Pleistocene levels of Yarımburgaz Cave contain many species, but the number of taxa represented is not significantly different across excavation units and trenches. The number of genera (N-genera) represented in each excavation unit is explained principally by differences in assemblage size (N = 23, $r = 0.739$, $p < 0.01$). Removing bears from the comparison greatly strengthens the relation between N-genera and NISP (N = 23, $r = 0.904$, $p < 0.001$).

One hypothesis for the association between artifacts and ursid remains at Yarımburgaz Cave is that the bear deaths resulted from non-violent causes in the context of hibernation, so that bears' presence in the cave was not linked in time to human activities there. Alternatively, the bears may have been hunted by humans and other predators, implying that the spatial proximity of bear bones and stone artifact reflects a temporal and causal link between them. To address these hypotheses, one first needs to know (1) whether cave bears were hibernators, and, if so, (2) whether bears would collect bones of prey in dens, and (3) the likelihood that bears would use caves for either purpose. The rich literature on modern brown and black bears (*U. americanus*) is used to build testable predictions about how bears may contribute to faunal assemblage formation (Stiner et al. 1996, 1998). *Ursus deningeri*, the most common bear in the middle Pleistocene layers (Tsoukala, this volume) is extinct and its adaptations poorly known, but some of the behavioral tendencies documented for modern *Ursus* help establish what was possible in the past.

Tests of the hypotheses will rely on archaeological data such as patterns of bone damage, skeletal element representation, element completeness, and mortality patterns (age structure) of the fossil population based on tooth eruption and wear. The potential bone-collecting and bone-modifying roles of the other large carnivores must also be evaluated. It is possible that some or all of these animals were periodic residents of the Lower Cave. The origins of the herbivore remains are additional points of interest: they almost certainly represent food items of predators and therefore could have been collected by hominins or members of the order Carnivora.

Conventional zooarchaeological counting units were used to construct a faunal inventory by provenience, taxonomic affiliation, and skeletal anatomy (see Appendixes 6.1-6.5). NISP refers to the number of identified specimens in fragmented or whole condition, following Grayson (1984; also Lyman 1994). Teeth and bones were recorded separately, following Stiner (1994). Species representation and bone damage patterns are examined primarily in terms of NISP, and the taphonomic observations focus on evidence of burning, tool marks, gnawing damage, fracture forms, and weathering damage. The minimum number of individual animals (MNI) represented by teeth versus cranial bone is used to evaluate the extent of in situ bone destruction caused by various attritional factors for the bears (following Stiner 1994: 99-103). The question of in situ destruction is explored further through comparisons of numbers of various skeletal elements (minimum number of elements, or MNE) and "portion" representation. Contrasting body part profiles for ungulates, bears, and other carnivores illuminate possible differences in the circumstances of accumulation among these major taxonomic groups. Spatial associations of artifacts and the bones of bears, ungulates and non-ursid carnivores are evaluated by comparing relative abundances across provenience units. Finally, mortality patterns and body part representation are used to evaluate the hypothesis that the bears of Yarımburgaz Cave represent hibernation-related deaths. The mortality analysis uses tooth eruption and wear data based on scheme developed by the author for *Ursus* bears (Stiner 1998).

BEHAVIORAL MODELS IN THE ABSENCE OF LIVING ANALOGS

Cave bears were widely distributed in Eurasia during the Middle and Late Pleistocene. They possessed a specialized, expanded dentition with powerful chewing muscles for crushing and grinding functions, implying a heavy reliance on tough or fibrous plant foods (Koby 1940; Kurten 1976: 11-26; Stiner et al. 1998). Isotope studies have yielded highly varied results on cave bear diet (Bocherens et al. 1994; Hilderbrand et al. 1996; Richards et al. 2008; Stiner et al. 1998). Some controversy stems from complications that taphonomic and morphologic evidence pose to the isotope results, but a parsimonious interpretation is that cave bear habits varied in response to environmental circumstances (Mattson 1998; Stiner et al. 1998). These animals were opportunistic feeders that ate meat when it was available, but strong selection for a specialized chewing apparatus argues for a fundamental importance of plant foods throughout much of their evolutionary history. The oft-noted prevalence of cave bear skeletons in the Pleistocene sediments of Eurasian caves suggests that cave bears were hibernators (Kurtén 1958, 1976) which is consistent a diet that included much plant food.

Although distinct species, modern black and brown bears have much in common biologically, including hibernation behavior and their nutritional and reproductive habits (Clevenger 1990, 1991; Clevenger and Purroy 1991; Clevenger et al. 1987, 1988; Garshelis and Pelton 1980; Hellgren et al. 1990; Johnson and Pelton 1980; Murie 1985; Nelson et al. 1980; Rogers 1981, 1987; Tassi 1983.) Basic similarities among modern members of the genus *Ursus* permit a few generalizations about the relations between bear diet and hibernation behavior, hibernation-related mortality, and the conditions that govern den site choice. The metabolic and reproductive aspects of hibernation in modern bears are thought to be intrinsic (Hellgren et al. 1990; Johnson and Pelton 1980; McNamee 1984: 253-257; Watts et al. 1987; Watts and Jonkel 1988). We may assume that these qualities are unlikely to have been profoundly different in Middle and Late Pleistocene cave bear populations. The mortality that normally accompanies hibernation in bears therefore should be comparable as well.

Do bears den in caves?

Even if one accepts that cave bears were hibernators, we are left to explain their presence in caves. The abilities of modern bears to make their own dens are well known, and excavated dens are the norm in many modern study areas (e.g., Hayes and Pelton 1994; Kolenosky and Strathearn 1987; Mack 1990; Miller 1990; Rogers 1981; Schwartz et al. 1987; Smith et al. 1994).

Wildlife studies from Europe and North America show, however, that modern bears willingly hibernate in natural caves and fissures where these conveniences are present (Clevenger and Purroy 1991: 113-123; Hayes and Pelton 1994; Murie 1985: 133-135). The frequent presence of bear remains in Pleistocene cave sediments does not necessarily mean that bears preferred caves over earthen dens in the past. Limestone caves and fissures are merely better preservation environments for the bones of deceased bears accumulating over hundreds or thousands of years.

Do bears collect bones of prey in their dens?

Carnivores that rely on natural or excavated shelters seem to divide into two behavioral categories with respect to bone accumulation—bone-collectors and non-collectors. Many of the canids and hyaenids deliberately gather bones at den and rest sites (Ewer 1973; Fentress and Ryon 1982; Kruuk 1972; Mech 1970) and may amass great quantities of prey bones in some cases (Hill 1980; Binford 1981). Hominins did essentially the same at residential sites during the Pleistocene, sometimes in caves. The habits of bears are quite different, however. Wildlife accounts show that black and brown bears do not normally carry food of any sort to dens, and they consume little food while preparing their winter beds (McNamee 1984; Rogers 1987: 23). Bears often mound piles of inedible vegetation in hibernation dens (Clevenger 1991; Manville 1987; McNamee 1984: 252-253; Murie 1985: 133-135; Rogers 1987: 20-22; Smith et al. 1994), but bones other than those of unlucky bears generally are very rare or not found at all.

Bears are secretive about the locations of their winter dens (Hayes and Pelton 1994; Kolenosky and Strathearn 1987; Mack 1990), because, despite their great size, sleeping bears are vulnerable to attack (Kurten 1976; Rogers 1987: 53; Ross et al. 1988; Tietje et al. 1986). Predation on denning bears by wolves, humans, and other bears is well known (LeCount 1987; Mattson et al. 1992). Food debris in and around dens would betray the location of a hibernating bear to predators, and presumably conflicts with the bear's need to remain hidden. An exception to this generalization is reported by Rogers (1987: 23), in which a lactating female black bear found a deer carcass near her den in spring and dragged it into the den. Such reports are extremely rare. Thus it is reasonable to generalize that the quantities of non-bear bones that might normally accumulate in bear dens are nominal, if perceptible at all, from a paleontological point of view.

What sorts of bear body part profiles and patterns of bone damage occur in hibernation-death situations?

While bears normally do not collect ungulate bones in hibernation dens, bears are very likely to contribute their own remains to the deposits of the shelters they occupy. The hibernation death scenario implies that bears perished inside the den or nearby in the cave system, so that complete bear skeletons therefore should be present in the deposits. The bones may not stay in articulation, however, because surviving bears return and renovate bedding areas each year (see references above); the scale of this kind of disturbance hinges on the frequency with which a site is re-used. Bears may hibernate in natural caves and fissures in regions where these conveniences are plentiful (Clevenger and Purroy 1991: 113-123; see also Murie 1985: 133-5) or excavate their own (e.g., Rogers 1981).

Mortality rates in modern bears are high toward the end of the hibernation period (Garshelis and Pelton1980; Kurtén 1976; McNamee 1984; Rogers 1981, 1987; see also Stuart 1982: 85).

Starvation appears to be the most common cause of hibernation-related deaths in most modern study areas, especially for cubs and yearlings (Rogers 1987). Bears generally do not die in their sleep. Rather, they awaken because their energy stores are depleted and may, as a last resort, make short forays in search of nearby food. It is in situations such as this that bears sometimes collapse and die in the vicinity of their dens. In addition to non-violent causes of death, bears may be attacked in their hibernation dens by other predators including humans. It is most feasible for other predators to hunt these large animals while the bear is dormant. Cannibalism within and between bear species may also be commonplace during lean years or where population densities are high (Mattson et al. 1992; Ross et al. 1988; Tietje et al. 1986).

Whereas hibernating bears are nearly odorless, a dead bear is not, and therefore is likely to attract scavengers. It would be difficult from a taphonomic perspective to distinguish skeletons resulting from attacks on live, dormant individuals from those that were scavenged after death. Attacks may be encouraged by the distress of a starving, wakeful bear, but it must also be recognized that scavengers would be attracted by the odor of a dead bear in the same place. In either case, much gnawing damage by marauding carnivores is expected on bear remains in cave deposits.

Which bear age groups are commonly affected by hibernation-related mortality?

Hibernation by bears is both a means for supporting exceptionally altricial infants and enduring seasonal scarcity of food plants, invertebrates, and small vertebrates (Clevenger et al. 1992; Ewer 1973; Garshelis and Pelton 1980; McNamee 1984; Murie 1985; Rogers 1981, 1987; Tassi 1983). Rather than turning to large prey when plants and invertebrates are no longer available, the bear possesses the metabolic option to drastically reduce its need for energy (e.g., Hellgren et al. 1990; Johnson and Pelton 1980; Nelson et al. 1980; Watts and Jonkel 1988; Watts et al. 1987). Both sexes may hibernate if the population is highly omnivorous. The polar bear (*Ursus maritimus*) is a pure carnivore and unaffected by plant productivity cycles. Hence, only pregnant females (and cubs) of this species normally hibernate.

Death in dens is driven principally by starvation, so the success of the hibernation strategy hinges upon food availability during the previous season, especially autumn. Deaths are especially common toward the close of the hibernation season, when bears are using dens as short-term foraging hubs. Natural, nonviolent mortality under dire conditions may in turn lead to a gradual, steady rain of bear bones into the sediments of shelters frequented across generations. Because so many deaths occur from starvation and disease, and because young are harbored by female bears in hibernation dens, the juvenile and old adult age groups are preferentially affected by hibernation-related mortality.

Summary of expectations for paleontological assemblages

If the bears in Yarımburgaz Cave represent hibernation deaths, at least four expectations about the distribution, content, and condition of their remains may be forwarded:

Expectation I. Because modern bears normally do not collect food inside hibernation dens, the relative abundances of bear bones and those of other species in cave sediments may not show strong positive spatial correlations to one another, especially with regard to horizontal units. It may not be valid to expect a strongly negative relation between the frequencies of bear bones and other mammal remains either, because (a) the geological units recognized by excavators may have formed over long periods of time, and/or (b) the bones may become jumbled and re-concentrated subsequently by, among other things, more bed-seeking bears. The expectation therefore is one of spatial independence among bear bones, artifacts, and other mammal bones.

Expectation II. Evidence of intra- and inter-specific violence or scavenging is likely to be found on at least some individuals in a large hibernation death assemblage. Adult male bears and wolves are the most common perpetrators in modern settings. On the other hand, non-violent bear deaths represent scavenging opportunities that few predators could ignore. Considerable gnawing damage on bear carcasses in dens is likely in either case.

Expectation III. Hibernation deaths in caves should result in relatively complete body part representation for bears, unlike the situation for large mammal parts carried into the cave by hominins and other predators. Bear bones may ultimately fall out of articulation, however, because bears typically modify bedding areas by digging and lateral raking with powerful claws. Disturbance of skeletal order in carcasses on or buried just below a cave floor is especially likely if the den area is used again and again. Selective deletion of bear body parts by predators is also possible, but the expectation of complete skeletons, qualified only in terms of density-mediated destruction of bone tissue, is a useful baseline with which to begin the comparisons.

Expectation IV. Starvation, disease, senescence, and occasional cannibalism all may contribute to bear deaths in hibernation dens. A U-shaped age structure (a.k.a. "concave" or "attritional" when plotted in histogram format) therefore should result over many generations of cave use (Andrews and Turner 1992; Kurtén 1958: 4-5; Stiner 1994: 318, 1998). This type of mortality pattern characteristically is composed of many juveniles and a smaller complement of old adults; it is nearly devoid of prime-aged adults. The juvenile fraction may be exaggerated further by high infant mortality in the den, although the relatively small litter sizes that are characteristic of bears (Ewer 1973: 306) will suppress this tendency relative to most other den-using carnivore species (Stiner 1994: 316-331).

If we find that bear skeletons in a cave are due to hibernation deaths—places to which bears should not also carry food—then the presence and condition of other macromammal remains in cave deposits need to be explained by other means. In the case of Yarımburgaz, "other" macromammal remains are primarily from ungulates and non-ursid carnivores (see Table 6.1 and Appendixes 6.1-6.2). The condition of ungulate bones should, at the very least, differ from that of the bears. Body part profiles for these taxonomic categories should represent another point of difference, because ungulate remains may have been selectively transported to the cave from procurement spots outside; such a contrast assumes, however, that the effects of *in situ* bone decomposition can be controlled. As for spatial distributions of bears, ungulates, non-ursid carnivores, and stone artifacts, it is clear from the outset that no drastic differences exist, although diverse taphonomic histories are suspected. The vertical distributions of these materials relative to one another depend foremost on sedimentation rates. Horizontal distributions may more consistently expose behavioral differences among the various cave occupants under conditions of slow sediment formation (see, for example, Gargett 1996; Stiner 1994: 150-153).

BONE DAMAGE AND SPATIAL DISTRIBUTIONS OF MATERIALS IN YARIMBURGAZ CAVE

The taphonomic history of the Yarımburgaz faunas is explored three ways in this section. First, we examine how much bone material was lost *in situ* through decomposition of fragile skeletal tissues or damage from consumers. Second, we focus on patterns of visible bone damage, including fractures, the forms and frequencies of marks made by stone tools, and gnawing damage from carnivores and rodents. All excavation units are lumped in this analytical step to avoid potentially serious aggregation errors. Finally, we search for possible differences in the spatial distributions of stone artifacts and ungulate, bear, and non-ursid carnivore remains across excavation units in the cave. The relative abundances of these materials are compared using a Pearson's (r) correlation matrix. The animal remains are divided into only three categories—bears, ungulates, and non-ursid carnivores—on the grounds that they could have found their ways into the cave through distinct pathways.

Bone preservation as a function of skeletal tissue density

The question of *in situ* bone loss focuses on how much skeletal material may have been destroyed after reaching the cave through processes such as decomposition and gnawing. The quality of macroscopic preservation for bones recovered from the Middle Pleistocene deposits ranges from very good to extraordinary. Substantial molecular transformations almost certainly have taken place, however, judging from the advanced state of fossilization. The great majority of bones and all teeth nonetheless retain even the finest natural textures, along with traces of damage inflicted when the tissues were fresh.

The degree of in situ bone loss can be evaluated in gross fashion by comparing the minimum number of individual animals (MNI) derived for two distinct classes of skeletal tissues—tooth and bone—across excavation units. Here the comparison is confined to the skull, because the head is the only transportable unit of the body that is composed of both very hard and very fragile skeletal tissue types. If teeth were deposited in the site, chances are that cranial bones were once there too. Cross-element comparisons, such as between cranial and post-cranial elements, potentially conflate the problem of *in situ* attrition with that of differential transport (see Lyman 1994), something to be avoided in this investigation. The comparison is further confined to bear remains, because they greatly outnumber those of all other mammals and therefore provide the only sample large enough to provide meaningful results (but see skeletal portion data below).

If no *in situ* bone attrition has occurred, MNI counts based on teeth and those based strictly on cranial bone for differently sized assemblages should be approximately equal. Under excellent preservation conditions, a regression plot of tooth-based and bone-based MNIs should yield a slope of 1. The model (Stiner 1994: 100-101) assumes that teeth are denser than any type of bone tissue and therefore have a better chance of being preserved. A regression slope that is significantly less than 1 (with tooth-based MNI forming the x-axis) therefore indicates that some bone attrition has occurred, the severity of which increases as the slope decreases.

A regression of bear MNI values for 18 horizontal units in the cave yields a slope of 0.285 ($n = 18$, y-intercept $= 0.431$, $r = 0.782$, $p < 0.001$). Only the assemblage from square P2 diverges significantly from the regression line; oddly, the MNI based on cranial bone slightly exceeds MNI based on teeth in this assemblage. With the P2 assemblage removed, all of the other cases

Table 6.2. Comparison of bear MNE and MNI counts for small dense bones, long limb bones, and cheek teeth of Yarımburgaz bears from all proveniences combined in the lower cave.

Skeletal element	MNE	MNI
Small bones:		
Scapholunate	33	17
Pisiform	29	15
Patella	22	11
Astragalus	40	20
Calcaneum	24	12
Limb bones:		
Scapula	26	13
Humerus	31	16
Radius	26	13
Ulna	28	14
Femur	22	11
Tibia	29	15
Fibula	30	15
Cheek teeth:		
P^4	28	14
M^1	37	19
M^2	30	15
P^4	38	19
M^1	59	30
M^2	74	37
M^3	64	32

Note: MNI here calculated simply by dividing MNE by 2, with right and left sides combined and value rounded to the next highest whole number.

conform well to the regression line (n = 17, r = 0.881, $p < 0.001$). Clearly, some bone loss relative to teeth did occur for the bear remains in Yarımburgaz Cave. The MNI values for bears based on teeth are consistently higher than those based on cranial bone landmarks, a conclusion that probably also loosely applies to other, scantier mammal remains.

This point is explored further through a comparison of minimum numbers of skeletal elements (MNE) and MNI (MNE/2) for various small dense and large limb bones of bears (Table 6.2). Here the data for all proveniences are pooled, and right and left sides are summed. The small dense elements are two types of carpals, the patella, and two types of tarsals. Some of the counts of small bones exceed those for large limbs, but the MNIs are broadly analogous overall; for example, the MNI for bear humeri is 16 using this counting scheme, whereas the highest tarsal MNI is 20, both of which are much lower than the MNIs based on bear lower molars (e.g., M_2 MNI = 37).

Because limb elements usually have heterogeneous structures, it is also possible to compare the representation of softer and harder components in an assemblage. Table 6.3 quantifies skeletally unique distal, diaphyseal, and proximal portions of the major limb elements of bears. Many of the bear bone specimens represent complete or nearly complete skeletal elements, and of those broken, shafts are no more likely to be preserved than one or both limb ends (epiphyses). With the exception of the scapula and fibula, no distinct bias toward one or the other of bear limb bone ends is apparent.

Bone damage caused by hominins, carnivores and rodents

Figure 6.7. Possible cut marks and rodent gnawing damage on a large ungulate rib fragment (Yar 90, specimen no. 303); detailed view of rodent damage below.

Figure 6.8. Specimen with chemical pitting and related damage (rare), in addition to rodent gnawing traces and possible cut marks.

The Yarımburgaz faunas contain only small quantities of herbivore bones. Among them are only 2 probable and 2 possible cases of cut marks from Paleolithic stone tools (Figures 6.7 and 6.8). The marks are relatively small, ranging between 0.5 and 2.0 cm in length, and they occur on ungulate rib and limb shaft fragments. None of these specimens is also gnawed by carnivores. None of the bear or non-ursid carnivore bones shows signs of tool marks. A few specimens were instead scarred by excavators' tools.

Although absence or scarcity of cut marks does not, in itself, preclude the possibility that hominins also collected the unmarked herbivore bones, other types of bone damage weigh against this possibility. There is no evidence of burning whatsoever, nor are any of the breakage patterns exclusively attributable to hominins. Cone fractures (*sensu* Binford 1978, among others) occur on 16 cortical limb bone specimens, mostly of ungulates. Both hominins and large carnivores are physically capable of generating large cone fractures, although hominins tend to produce them in higher frequencies (i.e., usually well in excess of 6% of bone NISP, see Stiner 1994: 106, 130-132). The frequency of cone fractures in the Yarımburgaz faunas is quite low (<<1%), which is more consistent with carnivore agencies than with stone-wielding hominins. Cone fracture diameters vary between 1.0 and 6.0 cm, and eight of these co-occur with gnawing damage from large carnivores.

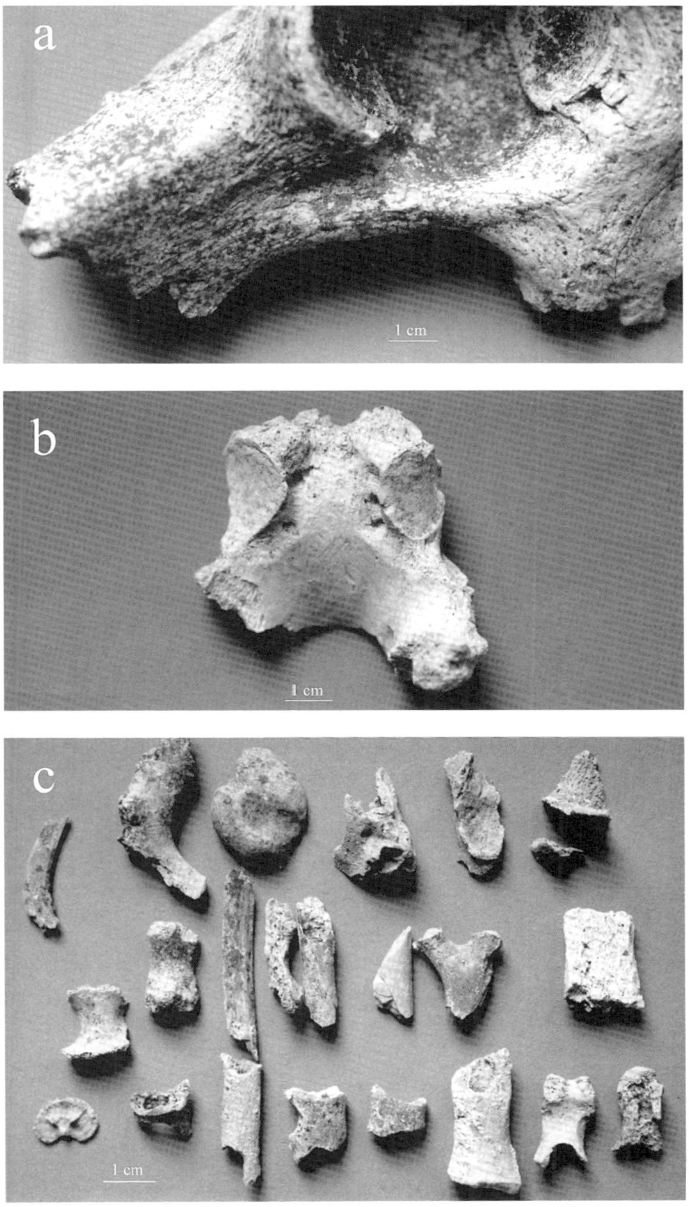

Figure 6.9. Damage to bear remains: (a) carnivore gnaw marks and salivary rounding on cave bear innominate, probably from large canids; (b) arch of canid tooth punctures on bear thoracic vertebra, probably by wolf; (c) infant bear, rabbit, deer antler, and other bone fragments scarred by digestive acids as they passed through the gut of a large canid.

In contrast to the scarcity of tool marks, damage caused by carnivore teeth is common on the surfaces of bones from Yarımburgaz cave (Table 6.4). Most of the tooth marks are from wolf-sized or larger carnivores, judging from the diameters and shapes of tooth punctures and general chewing patterns (cf. Binford 1981; Haynes 1983; Stiner 1994). Some tooth puncture diameters help narrow the probable range of carnivore agents, providing that the punctures in bone are deep and the distinct shapes and alignments of teeth are expressed. Some 59 measurable deep punctures were recorded, mostly in trabecular (spongy) tissue. Nearly all of the puncture diameters fall between 0.3 and 0.5 cm, and most are attributable to carnivores of intermediate body size. The deep imprints of canid teeth, including premolars and lower first molars, are unmistakable in some instances (Figures 6.9a-b and 6.10b-c).

Four young bear bones display exceptionally large punctures, ranging between 0.7 and 0.8 cm in diameter, the arrangements of which best match partial anterior dental arcades of adult bears. Figure 6.11 illustrates the condition of infant bear crania apparently punctured by the canine teeth of adult bears. However, evidence of cannibalism or scavenging by bears on other bears is much less common in the Yarımburgaz faunas than damage caused by non-ursid carnivores.

Some infant bear bones digested by larger carnivores were defecated inside Yarımburgaz Cave (Figure 6.9c); these remains tend to co-occur with hare bones, also marred by digestive acids. The largest concentration of fecal bone was found in the stratigraphically uncorrelated square U, which also contained the largest sample of microfaunal remains (see von Koeningswald et al., this volume). Other instances, presumably representing once-coherent fecal pellets, occur in units U0, S2, S4, and T4. These digested bones could not have been passed by cats or hyenas,

Table 6.3. Summary of "portion" counts (subset of NISP) for major limb elements of Yarımburgaz bears from all proveniences combined in the lower cave.

Portion type	Scapula	Humerus	Radius	Ulna	Femur	Tibia	Fibula
Complete	7	3	11	10	7	12	8
Proximal	1	7	11	15	5	10	2
Distal	13	8	16	10	6	5	16
Diaphysis	8	15	3	?	6	6	5

(COMPLETE) complete or nearly complete specimen; (PROXIMAL) fragment that includes some or all of the proximal epiphysis; (DISTAL) includes some or all of the distal epiphysis; (DIAPHYSIS) fragment in which some or all of the limb shaft cross-section is intact; (?) no data, short cross-sections could have been mistaken for rib fragments.

Note: Differential bone density distributions in bear skeletons are not known. On the whole, shafts are no more likely to be preserved than epiphyses in fragmented material, and, with the exception of scapula and fibula, no distinct bias toward one or other of the ends (epiphyses) is apparent. Bears (massive plantigrade) seem to be robust all over. Also, portion counts presented here are not appropriate for direct calculation of MNE, because portion categories partly overlap (with respect to unique landmarks needed to count and control for counting same element twice or more). See instead Appendix 2 for bone MNE data.

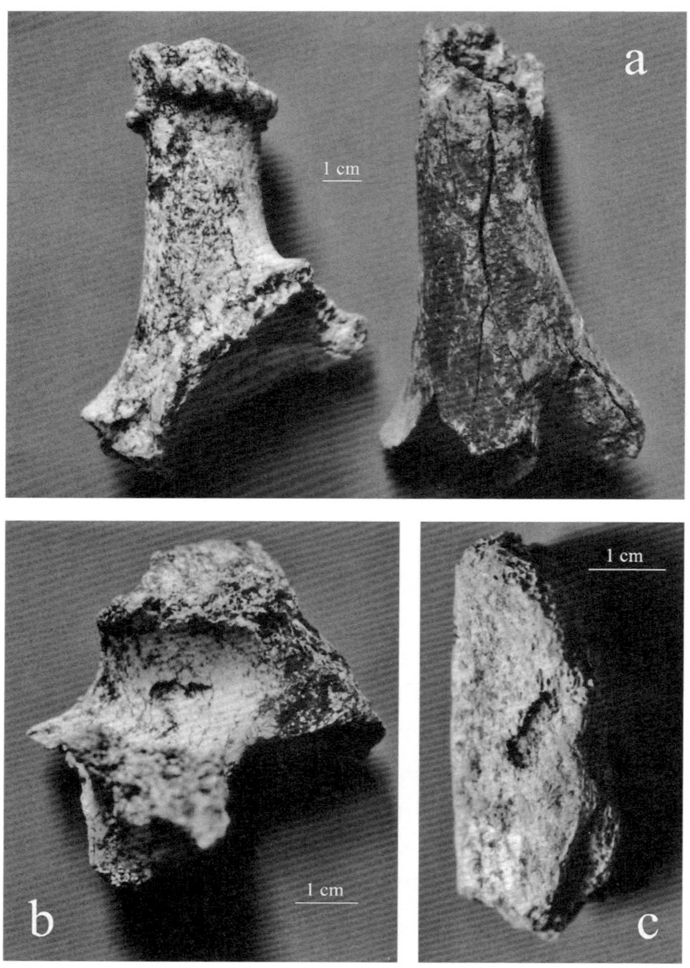

Figure 6.10. Examples of canid and other carnivore gnawing damage traces: (a) deer antler base (unshed) and humerus gnawed by large carnivores, probably canids but possibly hyena; (b-c) characteristic canid cheek tooth impressions in bear vertebra fragments.

as the degree of etching by gut acids is too subtle (see Schmitt and Juell 1994; Stiner 1994: 168-171). The general pattern of breakage, the average fragment size (mode = 1.4 cm, range = 0.6 to 5.2 cm, NISP sampled = 173), and the quality of etching damage in the fecal bone samples are most typical of large canids, although this interpretation suffers from the lack of comparative data on modern bear scats. Etched bones (Figure 6.9c), rounded bone tablets (Figure 6.12), and gnawed antlers and horns (Figure 6.10a) tend to cluster in the units named above.

Carnivore gnawing damage is twice as frequent on ungulate (23%) and non-ursid carnivore bones (18%) as it is on bear bones (10%) (Table 6.4). Moreover, gnawing damage is much more common on ungulate bones than are marks from hominins' tools (roughly 2%). The gnawing frequencies on ungulate and non-ursid carnivore bones compare favorably with those observed in Pleistocene wolf and spotted hyena dens of Italy, wherein gnawing damage on bones varies greatly but averages around

Table 6.4. Frequencies of carnivore and rodent gnawing damage on ungulate, bear, and other carnivore remains in the lower chamber of Yarımburgaz Cave, expressed at a percentage of bone NISP.

	Ungulates	Bears	Other Carnivores
Bone NISP (N observations)	117	3129	67
Percent carnivore gnawed	23%	10%	18%
Percent rodent gnawed	6%	11%	1%

a Primarily canids and bears, but possibly also *Panthera*, *Felis* or *Vulpes* in some instances

roughly 22% and 30% for wolf and hyena dens respectively (Stiner 1994: 110, 121). The lower frequencies of gnawing on bear remains distinguish them from all other large mammals in the cave.

Large canids (mainly but not exclusively wolves) appear to have played especially significant roles in collecting and modifying ungulate remains in the cave. These carnivores either transported many or most of the ungulate bones to the cave, or created scavenging opportunities for hominins, the spoils of which retained marks made by the primary feeders. The fact that ungulate and non-ursid carnivore remains occur at comparable frequencies in the deposits argues against the latter interpretation. This fact, along with the nature of damage to the ungulate bones, instead suggests that canids (and possibly hyenas at times) used the cave as a den or rest site, to which they carried food.

Damage from gnawing rodents is more prevalent on the remains of bears than on any other large mammal (Table 6.4). Based on the width of the tooth grooves the rodents were vole-sized

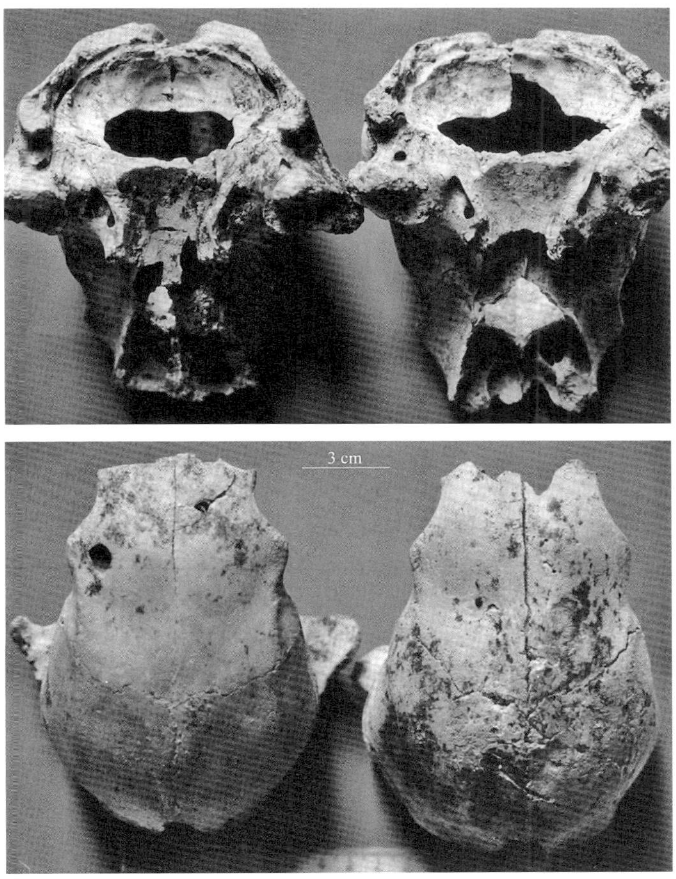

Figure 6.11. Examples of cannibalism by adult cave bears: twin cub crania with (a) opened brain cases and anterior faces destroyed; and (b) a large puncture, probably from an adult bear canine, in the left frontal of one cub cranium.

Figure 6.12.
Three compact bone fragments rounded by salivary and/or gut enzymes of a large carnivore (wolf, hyena or large cat).

Figure 6.13. Partial frontal bone fragment of a young cave bear punctured by the incisors of a large adult bear.

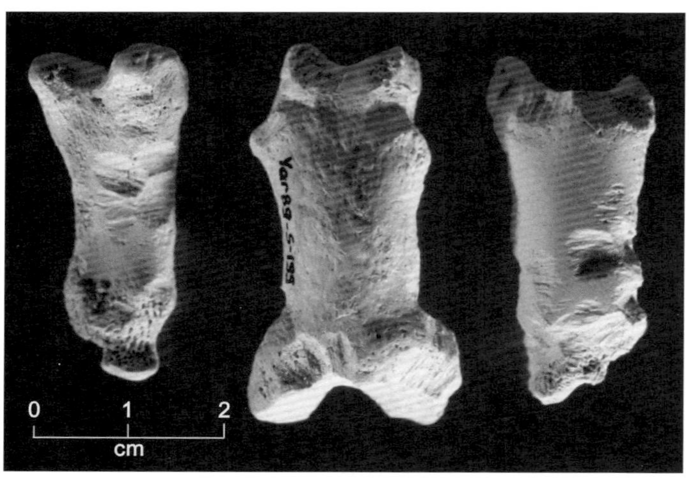

Figure 6.14. Cave bear phalanges extensively gnawed by small rodents.

species, which are quite common in the microfaunal samples (see Koenigswald et al., this volume). Rodent gnawing is most common on bear phalanges and metapodials in units S 0-3 (10%), U 0 (67%), U 0b (40%), and U 1 (65%), and the damage is especially severe on bear phalanges (Figure 6.14). Whereas rodents had little impact on other bear parts, rodents frequently attacked the exposed phalanges of the paws and worked inward (proximally). Distal phalanges show the greatest damage: MNE counts for first through third phalanges should be equal under good preservation conditions but in fact decline markedly below the first phalanx element (MNE = 384, counts for the second and third are 213 and 204, respectively). Because the terminal phalanges of bears tend to be quite large, size-dependent bone recovery techniques by archaeologists cannot explain the discrepancy. Rodents may have been attracted to the bear bones whilst they were still rich in minerals and fat, since the marrow cavities of the bear phalanges often were breached.

The frequencies of carnivore and rodent gnawing show essentially inverse distributions across taxa. The high incidence of rodent gnawing on bear bones in particular may reflect differing conditions of accumulation for bear carcasses and other large mammals, reinforcing the possibility that the causes of ungulate and non-ursid carnivore bone accumulations are largely independent of those responsible for the bear bone accumulations. Because rodents would be sensitive to foot traffic, digging by carnivores, and the risk of being eaten, the intensity of rodent activity should be inversely proportional to the intensity of predator activity within the cave.

Weathering data also suggest differences in the circumstances of accumulation among ungulates, bears, and non-ursid carnivores. Consistent with situations in other Mediterranean caves (Stiner 1994), bones affected by atmospheric weathering (Behrensmeyer 1978; Lyman 1994) are comparatively rare in the Yarımburgaz faunas overall (1% of total mammal bone NISP), and the damage tends to be mild when it occurs. But weathering damage is found only on bear bones, indicating that bear carcasses lay exposed on the cave floor for longer periods than the remains of any other large mammal species. Rare types of damage—as atmospheric weathering damage

Table 6.5. Frequencies with which medullary cavities of major limb bones[1] of bears, non-ursid carnivores, and ungulates from all proveniences combined in Yarımburgaz Cave were breached by gnawing carnivores or hammer-wielding hominids, presumably to extract marrow.

Mammal group	Total MNE	MNE complete	MNE fragmented	Percent with opened medulla
Bears	108	33	75	69%
Non-ursid carnivores	6	2	4	67%
Ungulates	11	1	10	91%

1 Major limb bones considered are humerus, radius, femur, and tibia, because these were judged to possess substantial medullary cavities potentially rich in marrow and therefore worthy of extraction efforts by large mammalian consumers.

Note: The frequency with which medullae were opened is calculated as follows: Total MNE count for the four limb bone elements considered subtracted by the MNE count for complete or nearly complete specimens. This yields a derived MNE count for fragmented specimens, which is then converted to a percentage of total MNE. Counts for the four major limb elements are combined due to small sample sizes for ungulates and non-ursid carnivores.

certainly is in this case—may only be apparent in relatively large samples, but ungulate and other carnivore remains are not so few in Yarımburgaz Cave as to minimize chances of finding this kind of damage.

If any of the large mammal groups—bears, non-ursid carnivores, and ungulate—were food to another large-bodied species, there should be evidence of marrow extraction. A comparison of the condition of the humerus, radius, femur, and tibia (Table 6.5), elements that possess substantial medullary cavities, shows that 69% of bear long bones and 67% of long bones of other carnivores were opened, whereas the limb bones of ungulates were breached 91% of the time. The crudity of this comparison notwithstanding, it is clear that the major limb bones of ungulates were more consistently opened by all consumers. The fate of carnivore remains of any sort was more variable, with the medullary cavities of bones left whole or nearly whole approximately 30% of the time. Although any mammal can be food to another, bears and non-ursid carnivores appear to represent a different economic, and taphonomic, category than do the ungulates in Yarımburgaz Cave.

Spatial associations among mammalian groups and stone artifacts

Bear bones are widespread among the trenches and levels in the Lower Cave of Yarımburgaz, and they generally co-occur with stone artifacts. Very few mammal bone specimens in the Lower Cave were found in articulation (see Figures 6.2, 6.4-6.5). The only significant exceptions are some bear foot bones (Figures 6.15 and 6.16). The generally close spatial associations between bones and stone artifacts in the sediments could be explained by slow sediment accumulation, mechanical mixing or as part of a single, contemporary process. Only the horizontal distribution of material in Yarımburgaz Cave holds much promise for evaluating the nature of the spatial associations, and even in this regard may not be justified in expecting clean separation among material classes.

Table 6.6 compares lithic artifact and faunal abundances (NISP) across excavation units, using a Pearson's correlation matrix. The large mammal remains are grouped into three broadly defined taxonomic categories as before: bears, non-ursid carnivores, and ungulates. Stone artifacts represent the fourth class of material, and they the only material unequivocally traceable

ACKNOWLEDGMENTS

I am very grateful to G. Arsebük, F. Clark Howell, M. Özbaşaran, M. Özdoğan, S. Gülçur, O. Tanındı, N. Balkan-Atlı, K. Atlı, and others for making our work a stimulating and enjoyable experience. Our study was facilitated in 1992 by the Section of Prehistory of the University of Istanbul, and by accommodations provided by the American Research Institute in Turkey (ARIT, Istanbul) and much practical assistance from Dr. A. Greenwood (Director). The data collection phase of this research was supported by a grant from the Wenner-Gren Foundation. Information and critiques provided by S. Kuhn, R. L. Lyman, P. Martin, M. Pelton, A. Clevenger, T. Fuller, and anonymous reviewers have greatly improved this manuscript. Substantial portions of data and basic analyses presented here have appeared previously in Stiner et al. (1996), Stiner (1998), and Stiner et al. (1998).

REFERENCES CITED

Andrews, P., Turner, A., 1992. Life and death of the Westbury bears. *Annalidi Zoologica Fennici* 28, 139-149.

Bárta, J., 1989. Hunting of brown bears in the Mesolithic: evidence from the Medvedia Cave near Ruzin in Slovakia, in: Bonsall, C., (Ed.), *The Mesolithic in Europe*. John Donald, Edinburgh, pp 456-460.

Behrensmeyer, A.K., 1978. Taphonomic and ecologic information from bone weathering. *Paleobiology* 4, 150-162.

Binford, L.R., 1978. *Nunamiut Ethnoarchaeology*. Academic Press, New York.

Binford, L.R., 1981. *Bones: Ancient Men and Modern Myths*. Academic Press, New York.

Binford, L.R., Bertram, J., 1977. Bone frequencies and attritional processes, in: Binford, L.R., (Ed.), *For Theory Building in Archaeology*. Academic Press, New York, pp. 77-156.

Bocherens, H., Fizet, M., Mariotti, A., 1994. Diet, physiology, and ecology of fossil mammals as inferred from stable carbon and nitrogen isotope biochemistry: implications for Pleistocene bears. *Palaeogeography, Palaeoclimatology and Palaeoecology* 107, 213-225.

Brain, C.K, 1981. *The Hunters or the Hunted?* University of Chicago Press, Chicago.

Brugal, J.-P., Jaubert, J., 1991. Les gisements paleontologiques Pleistocenes a indices de frequentation humaine: un nouveau type de comportement de predation? *Paléo* 3, 15-41.

Bunnell, F.L., Tait, D.E.N., 1981. Population dynamics of bears—implications, in: Fowler, C.W., Smith, T.D., (Eds.), *Dynamics of Large Mammal Populations*. John Wiley and Sons, New York, pp. 75-98.

Caughley, G., 1977. *Analysis of Vertebrate Populations*. John Wiley and Sons, London.

Clevenger, A., 1990. Biologia del Oso Pardo (Ursus arctos) en la Cordillera Cantabrica. Memoria de Tesis Doctoral, Universidad de Leon, León, Spain.

Clevenger, A., 1991. The phantom bear of the Spanish Sierras. *Wildlife Conservation* 94, 34-45.

Clevenger, A., Purroy, F.J., 1991. *Ecologia del Oso Pardo en Espana*. Monographias del Museo Nacional de Ciencias Naturales, Madrid, Spain.

Clevenger, A., Purroy, F.J., Pelton. M.R., 1988. Movement and activity patterns of a European brown bear in the Cantabrian Mountains, Spain. *International Conference on Bear Research and Management* 8, 205-211.

Clevenger, A., Purroy, F.J., Pelton. M.R., 1992. Food habits of brown bears (Ursus arctos) in the Cantabrian Moutains, Spain. *Journal of Mammalogy* 73, 415-421.

Clevenger, A., Purroy, F.J., Saenz de Buruaga, M., 1987. Status of the brown bear in the Cantabrian Mountains, Spain. *International Conference on Bear Research and Management* 7, 1-8.

Dittrich, L., 1960. Milchgebißentwicklung und Zahnwechsel beim Braunbären (Ursus arctos L.) und anderen Ursiden (Dissertation). *Morphologisches Jahrbuch* 101, 1-141.

Eberhardt, L.L., Knight, R.R., Blanchard, B.M., 1986. Monitoring grizzly bear population trends. *Journal of Wildlife Management* 50, 613-618.

Ewer, R.F., 1973. *The Carnivores*. Weidenfeld and Nicholson, London.

Fentress, J.C., Ryon, J., 1982. A long-term study of distributed pup feeding in captive wolves, in: Harrington, F.H., Paquet, P.C., (Eds.), *Wolves of the World: Perspectives of Behavior, Ecology, and Conservation*. Noyes Publications, Park Ridge, NJ, pp. 238-260.

Fox, M.W., 1984. *The Whistling Hunters: Field Studies of the Asiatic Wild Dog (Cuon alpinus)*. SUNY Press, Albany.

Gamble, C., 1986. *The Palaeolithic settlement of Europe*. Cambridge University Press, Cambridge.

Gargett, R.H., 1996. *Cave Bears and Modern Human Origins: The Spatial Taphonomy of Pod Hradem Cave, Czech Republic*. University Press of America, Lanham, MD.

Garshelis, D.L., 1994. Density-dependent population regulation of black bears, in: Taylor, M.K., (Ed.), Density-Dependent Population Regulation in Black, Brown, and Polar Bears. *International Conference on Bear Research and Management* 3, 3-14.

Garshelis, D.L., Pelton, M.R., 1980. Activity of black bears in the Great Smokey Mountains National Park. *Journal of Mammalogy* 61, 819.

Glenn, L.P., Lentfer, J.W., Faro, J.B., Miller, L.H., 1976. Reproductive biology of female brown bears (Ursus arctos), McNeil River, Alaska. *International Conference on Bear Research and Management* 3, 381-390.

Grayson, D.K., 1984. *Quantitative Zooarchaeology*. Academic Press, Orlando.

Hayes, S.G., Pelton, M.R., 1994. Habitat characteristics of female black bear dens in northwestern Arkansas. *International Conference on Bear Research and Management* 9, 411-418.

Haynes, G., 1983. A guide for differentiating mammalian carnivore taxa responsible for gnaw damage in herbivore limb bones. *Paleobiology* 9, 164-172.

Hellgren, E.C., Vaughan, M.R., Kirkpatrick, R.L., Scanlon, P.R., 1990. Serial changes in metabolic correlates of hibernation in female black bears. *Journal of Mammalogy* 71, 291-300.

Hilderbrand, G.V., Farley, S.D., Robbins, C.T., Hanley, T.A., Titus, K., Servheen, C., 1996. Use of stable isotopes to determine diets of living and extinct bears. *Canadian Journal of Zoology* 74, 2080-2088.

Hill, A.P., 1980. Hyaena provisioning of juvenile offspring at the den. *Mammalia* 44, 594-595.

Horwitz, L.K., Smith, P., 1988. The effects of striped hyaena activity on human remains. *Journal of Archaeological Science* 15, 471-481.

Jenkinson, R.D.S., 1984. *Creswell Crags, Late Pleistocene Sites in the East Midlands*. BAR British Series 122. British Archaeological Reports, Oxford.

Johnson, K.G., Pelton, M.R., 1980. Environmental relationships and the denning period of black bears in Tennessee. *Journal of Mammalogy* 61, 653-660.

Judd, S.L., Knight, R.R., Blanchard, B.M., 1986. Denning grizzly bears in the Yellowstone National Park area. *International Conference on Bear Research and Management* 6, 111-117.

Koby, F.E., 1940. Les usures séniles des canines d'Ursus spelaeus et al préhistoire. *Verhandlungen der Naturforschenden Gesellschaft* LI, 76-95.

Kolenosky, G.B., Strathearn, D.M., 1987. Winter denning of black bears in east-central Ontario. *International Conference on Bear Research and Management* 7, 305-316.

Kruuk, H., 1972. *The Spotted Hyaena*. University of Chicago Press, Chicago.

Kuhn, S.L., Arsebük, G., Howell, F.C., 1996. A Middle Pleistocene lithic assemblage from Yarımburgaz Cave, Turkey. *Paléorient* 22/1, 31-49.

Kurtén, B., 1958. Life and death of the Pleistocene cave bear: a study in paleoecology. *Acta Zoologica Fennica* 95, 4-59.

Kurtén, B., 1976. *The Cave Bear Story*. Columbia University Press, New York.

Kurtén, B., Poulainos, A.-N., 1981. Fossil carnivora of Petralona Cave: status as of 1980. *Anthropos* 8, 1-56.

LeCount, A.L., 1987. Causes of black bear cub mortality. *International Conference on Bear Research and Management* 7, 75-82.

Lindly, J.M., 1988. Hominid and carnivore activity at Middle and Upper Paleolithic cave sites in eastern Spain. *Munibe* 40, 45-70.

Lyman, R.L., 1984. Bone density and differential survivorship of fossil classes. *Journal of Anthropological Archaeology* 3, 259-299.

Lyman, R.L., 1987. On the analysis of vertebrate mortality profiles: sample size, mortality type, and hunting pressure. *American Antiquity* 52, 125-142.

Lyman, R.L., 1994. *Vertebrate Taphonomy*. Cambridge University Press, New York.

Mack, J.A., 1990. Black bear dens in the Beartooth Face, south-central Montana. *International Conference on Bear Research and Management* 8, 273-277.

Manville, A.M., 1987. Den selection and use by black bears in Michigan's northern Lower Peninsula. *International Conference on Bear Research and Management* 7, 317-322.

Mattson, D.J., 1998. Diet and morphology of extant and recently extinct northern bears. International *Conference on Bear Research and Management* 10, 479-476.

Mattson, D.J., Knight, R.R., Blanchard, B.M., 1992. Cannibalism and predation on black bears by grizzly bears in the Yellowstone ecosystem 1975-1990. *Journal of Mammalogy* 73, 422-425.

McCullough, D.R., 1981. Population dynamics of Yellowstone grizzly bear, in: Fowler, C.W., Smith, T.D., (Eds.), *Dynamics of Large Mammal Populations*. John Wiley and Sons, New York, pp. 173-196.

McNamee, T., 1984. *The Grizzly Bear*. Alfred A. Knopf, New York.

Mech, L.D., 1970. *The Wolf: The Ecology and Behavior of an Endangered Species*. Natural History Press, Garden City, NJ.

Miller, S.D., 1990. Denning ecology of brown bears in southcentral Alaska and comparisons with a sympatric black bear population. *International Conference on Bear Research and Management* 8, 279-287.

Murie, A., 1985. *The Grizzlies of Mount McKinley*. University of Washington Press, Seattle.

Nelson, R.A., Folk, G.E, Pfeiffer, E.W., Craighead, J.J., Jonkel, C.J., Wellik, D.L, 1980. Behavior, biochemistry, and hibernation in black, grizzly, and polar bears. *International Conference on Bear Research and Management* 4, 284-290.

Özdoğan, M., Koyunlu, A., 1986. Yarımburgaz Mağarası: 1986 Yılı Çalışmalarının İlk Sonuçları ve Bazı Gözlemler, *Arkeoloji ve Sanat* (Istanbul) 32/33, 4-14.

Reynolds, H.V., Curatolo, J.A., Quimby, R., 1976. Denning ecology of grizzly bears in northeastern Alaska. *International Conference on Bear Research and Management* 3, 403-409.

Richards, M.P., Pacher, M., Stiller, M., Quilès, J., Hofreiter, M., Constantin, S., Zilhão, J., Trinkaus, E., 2008. Isotopic evidence for omnivory among European cave bears: Late Pleistocene *Ursus spelaeus* from the Peştera cu Oase, Romania. *Proceedings of the National Aacademy of Science* 105, 600-604.

Rogers, L.L., 1981. A bear in its lair. *Natural History* 90, 64-70.

Rogers, L.L., 1987. Effects of food supply and kinship on social behavior, movements, and population growth of black bears in northeastern Minnesota. *Wildlife Monographs*, no. 97.

Ross, P.I., Hornbeck, G.E., Horejsi, B.L., 1988. Late denning black bears killed by grizzly bear. *Journal of Mammalogy* 69, 818-820.

Schmitt, D.N., Juell, K.E., 1994. Toward the identification of coyote scatological faunal accumulations in archaeological contexts. *Journal of Archaeological Science* 21, 249-262.

Schwartz, C.C., Miller, S.D., Franzmann, A.W., 1987. Denning ecology of three black bear populations in Alaska. *International Conference on Bear Research and Management* 7, 281-291.

Smith, M.E., J.L. Hechtel, Follmann, F.H., 1994. Black bear denning ecology in interior Alaska. *International Conference on Bear Research and Management* 9, 513-522.

Stiner, M.C., 1990. The use of mortality patterns in archaeological studies of hominid predatory adaptations. *Journal of Anthropological Archaeology* 9, 305-351.

Stiner, M.C., 1991. Food procurement and transport by human and non-human predators. *Journal of Archaeological Science* 18, 455-482.

Stiner, M.C., 1994. *Honor among Thieves: A Zooarchaeological Study of Neandertal Ecology*. Princeton University Press, Princeton, NJ.

Stiner, M.C., 1998. Mortality analysis of Pleistocene bears and its paleoanthropological relevance. *Journal of Human Evolution* 34, 303-326.

Stiner, M.C., 2004. A comparison of photon densitometry and computed tomography standards of bone density in ungulate body part profiles. *Journal of Taphonomy* 2(3), 117-145.

Stiner, M.C., Achyuthan, H., Arsebük, G., Howell, F. C., Josephson, S., Juell, K., Pigati, J., Quade, J., 1998. Reconstructing cave bear paleoecology from skeletons: a cross-disciplinary study of Middle Pleistocene bears from Yarımburgaz Cave, Turkey. *Paleobiology* 24, 74-98.

Stiner, M.C., Arsebük, G., Howell, F.C., 1996. Cave bears and Paleolithic artifacts in Yarımburgaz Cave, Turkey: dissecting a palimpsest. *Geoarchaeology* 11, 279-327.

Straus, L.G., 1982. Carnivores and cave sites in Cantabrian Spain. *Journal of Anthropological Research* 38, 75-96.

Stuart, A.J., 1982. *Pleistocene Vertebrates in the British Isles.* Longman, London.

Sutcliffe, A., 1970. Spotted hyaena: Crusher, gnawer, digestor, and collector of bones. *Nature* (London) 227, 1110-1113.

Tassi, F., 1983. L'orso bruno Marsicano. Panda 12 (Estratto), Parco Nazionale d'Abruzzo, Italia.

Tietje, W.D., Pelchat, B.O, Ruff, R., 1986. Cannibalism of denned black bears. *Journal of Mammalogy* 67, 762-766.

Torres, T.J., 1988. *Osos (Mammalia, Carnviora, Ursidae) del Pleistoceno de la península Ibérica.* Publicaciones Especiales de Boletín Geologica y Minero (reimpresión del I-VI, T. XCIX 1988).

Watts, P.D., Jonkel, C., 1988. Energetic cost of winter dormancy in grizzly bear. *Journal of Wildlife Management* 52, 654-656.

Watts, P.D., Øritsland, N.A., Hurst, R.J., 1987. Standard metabolic rate of polar bears under simulated denning conditions. *Physiological Zoology* 60, 687-691.

Wiszniowska, T., 1982. Carnivora, in: Kozlowski, J.K. (Ed.), *Excavation in the Bacho Kiro Cave (Bulgaria), Final Report.* Pannstwowe Wy Dawnictwo Naukowe, Warsaw, pp. 52-55.

STINER CHAPTER 6 - APPENDIXES

Appendix 6.1. Tooth NISP by macromammal taxon and excavation unit in the Lower Chamber of Yarımburgaz Cave.

Taxon TEETH	P1	P2	R0	R1	R2	R6	S1	S4	T1	T2	T3	T4	U1	Y0	Y1	Y1a	Y1b	Y5	Z3	Total
Equus												1								1
Capreolus													3				1			4
Cervidae	2		1	3			2				1	2	1		1					13
Dama	9																			9
Cervus				4								1			1					6
Bos/Bison											1		1							2
Sus							1													1
Hippopotamus			1?				1?													2?
Capra				1									2						1	4
Indet. Carnivore	1			1								1								3
Crocuta									1?			1?								2?
Felis				1							1		1							3
Canis/Cuon	1											1	6							8
Vulpes	1						1			1?		1	4		2					10
Panthera	2			1				32				1								4
Ursus	90	32	1	281	8	1	3		14	28	20	139	23	18	7	51	13			761
Mustela				1								1						1		3
Total by unit	106	32	3	293	8	1	8	32	15	29	23	149	41	18	11	51	14	1	1	836

Note: NISP is the number of identifiable specimens. Data represent provenienced material only. Note that some very rare species are missing from this sample, because they were removed for taxonomic study (see Howell, this volume).

Appendix 6.2. Bone NISP by macromammal taxon and excavation unit in the Lower Chamber of Yarımburgaz Cave.

Taxon	P1	P2	R0	R1	R2-3	R5-6	S0-3	S4	T0-1	T2	T3	T4	U0	U0b	U1	U<5	Y0	Y0b	Y1	Y1a	Y1b	Y1c	Y4	Y5	Z2	Z3	Z4	Z5	Total
Large mamm.[a]	29	14	1	102		3	6	29	31	116	17	119	5	4	13			4		29			26						548
Med. mamm.[b]				1								2			1	1													5
Small mamm.[c]										3	1	1			2	1			1										9
Equus	1			2				1		3		1																	8
Capreolus	1									3		3			2														9
Cervidae	3			2				4		4	2	3	2	3															23
Dama	3													1															4
Cervus	4			9			1	3		1		2		2						1									23
Megaloceros								1?												1									2
Bos/Bison														2										1					3
Sus													1?	1						1									3
Capra																										1			1
Small ungul.										1			1																2
Medium ungul.	1										1	1																	3
Large ungul.	2	1		1						9	2	1								1									17
Megafauna	1	4													1														6
Indet. Carniv.	2			1											2					1									6
Crocuta																													
Felis	1			1	1			1				3	1						2										10
Canis	1	1		4				3				2			1		1												13
Cuon	1?			3?																									4
Vulpes								2							3				2										7
Panthera	4	2		4				1	3	1?		10									1								26
Ursus	279	199	1	550	20	3	94	156	48	164	41	381	267	6	167		11	4	9	182	18	3	4		1	1	1	1	2611
Mustela													1																1
Total by unit	333	221	2	680	21	6	101	201	82	305	64	529	278	19	192	2	12	8	14	216	19	3	30	1	1	2	1	1	3344

Note: [a] mostly bear remains; [b] mostly ungulates; [c] mostly other carnivores. Small ungulate remains may include *Capreolus*, *Sus*, and *Gazella*; medium ungulate remains may include *Dama*, *Cervus*, and *Megaloceros*; large ungulate remains may include *Equus*, *Bos*, and *Bison*. NISP is the number of identifiable specimens. Data represent provenienced material only. Note that some very rare species are missing from this sample, because they were removed for taxonomic study (see Howell, this volume).

Appendix 6.3. Bear bone MNE by skeletal element and excavation unit in the Lower Chamber of Yarımburgaz Cave.

Bone element	P1	P2	R0	R1	R2	R3	R5	R6	S0	S1	S2	S3	S4	T0	T1	T2	T3	T4	U0	U0b	U1	Y0	Y0b	Y1	Y1a	Y1b	Y1c	Y4	Z2	Z3	Z4	Z5	Total
Hyoid	1			3									2			2		5							1								14
Half cranium, L or R	3	10		~4	1								2	1	2	2	4	6			1	2			2			2					42
Mandible, L or R	4	4	1	9									4	1	1	2	1	9				2			2	2							42
Atlas	3			~5									1				2	2							3								16
Axis	3	1		~4									2			1		1							1	1							14
Cervical vertebra	5	1		8		1					4		5		1	7	1	1	1				2		2	1				1			41
Thoracic vertebra	9	4		14	1						1		6		2	3		13				3			10	1							67
Rib	11	10		34	2		1		3	2	1		~24		13	~40	5	64	27	1	17			5	~40	3		~11				1	315
Lumbar vertebra	6	2		11	2								6			1		6							6	1							41
Sacral vertebra	2	1		3	1											1	1																9
Innominate, L or R	~3	~2		11							1		1			3		8											1				30
Caudal vertebra	2	5		8					1				4			1		3	2		3					1							30
Sternal segment	2	1		3									1					7	1		2												17
Scapula	3	1		~11									1			4	~1	3						1							1		26
Humerus	3	3		~11									2		2	3	1	4			1	1											31
Radius	2	2		1	1	1							1	1	~1	3	1	6		1					5								26
Ulna	1	1		10	1					1						4		4			1	1			4								28
Metacarpal	13	13		37	2					3	1		5		2	12	4	20	25	1	9		1	1	12		1						162
Femur	1	3		~6									1		1	3	1	6															22
Tibia	4	2		6	1						1?		1		1	1		7			1				4								29
Patella	3	1		11						1							1		1		1				3								22
Astragalus	8	4		11						3			3					3			2				5	1							40
Calcaneum	1	2		6						2			1				1	4	2		1		1		2	1							24
Fibula	5	4		9					1							1	1	4	1		2				2								30
Metatarsal	29	13		31					1	8			9			10	2	18	15	1	5				7		1						150
1st phalanx	53	24		61	2			1	4	11	2	1	11		6	26	4	36	78	1	44	2			16	1		1					385
2nd phalanx	24	26		32						6	1		9		4	9	4	23	43		27				5	1		1					215
3rd phalanx	26	16		34	1					7			7		3	14	1	28	27		25				12	2	1	1					205
Sesamoid	4	18		4						12						2	1	4	3		6				2								56
Indet. metapodial	15	4	1	15	1			2		1			4		3	~4	1	9	16		3				4								83
Carpal/tarsal	15	18		57	1					11			6		2	19	4	34	24	1	14				18	1		1					226
Indet. vertebra	2	1		~45						2	1		~11		~4	~9	~2	~20		1	1		~2	~3	1			1					106
Os penis	1			1						1						1		2															6
Total by unit	267	197	2	516	17	2	1	3	10	71	13	1	130	3	48	188	44	360	266	7	166	11	6	10	169	17	3	18	1	1	1	1	2550

Note: MNE is the minimum number of each type of skeletal element and is based on the most common portion or unique anatomical "landmark"; counts are mutually exclusive across categories; some MNE values could only be approximated (~). Cranial MNE is based exclusively on bone tissues (not teeth) and represents a combination of left and right halves. Metapodials are counted using the proximal end, and carpal/tarsal MNE excludes the astragalus and calcaneum. Data represent provenienced material only.

Appendix 6.4. Ungulate (all species) bone MNE by skeletal element and excavation unit in the Lower Chamber of Yarımburgaz Cave.

Bone element	P1	P2	R1	S0	S4	T2	T3	T4	U0	U1	U<5	Y1a	Y5	Z3	Total
Horn core												1			1
Antler			2		2	~1	~1								6
Half cranium w/antler					1										1
Half cranium, L or R	2		1		1					1					5
Mandible, L or R	1													1	2
Axis	1														1
Cervical vertebra								~1							1
Rib	1					5	1								7
Lumbar vertebra			1												1
Innominate, L or R		1	1												2
Scapula						1				1					2
Humerus					1	1		2				1			5
Radius						1	1	1							3
Metacarpal			1			1		1							3
Femur					~1										1
Tibia			1				1								2
Astragalus	1		1	1	2			2							7
Calcaneum			2					1	1						4
Metatarsal	1		1			1			2	1					6
1st phalanx	4		1			3	1		1	4	~1	1			16
2nd phalanx			2			1		1		2		1	1		8
3rd phalanx								1							1
Sesamoid						2									2
Indet. metapodial					1					2					3
Carpal/tarsal	1							1							2
Total by unit	12	1	14	1	9	17	5	11	4	11	1	4	1	1	92

Note: Ungulate species are summed together in this appendix table because too few of each taxon are represented to warrant species by species analyses of body part representation. MNE is the minimum number of each type of skeletal element, based on the most common portion or unique anatomical "landmark"; counts are mutually exclusive across categories; some MNE values could only be approximated (~). Cranial MNE is based exclusively on bone tissues (not teeth), and is computed in terms of right (R) and left (L) halves. Metapodials are counted using the proximal end, and carpal/tarsal MNE excludes the astragalus and calcaneum. Data represent provenienced material only.

Appendix 6.5. Non-ursid carnivore* bone MNE by skeletal element and excavation unit in the Lower Chamber of Yarımburgaz Cave.

Bone element	P1	P2	R1-2	S4	T1-2	T4	U0	U1	Y0	Y1-1b	Total
Half cranium, L or R			1								1
Mandible, L or R								1			1
Cervical vertebra								1			1
Thoracic vertebra			1			1					2
Lumbar vertebra	1		1							1	3
Scapula								1			1
Humerus							1	1			2
Radius	1		1								2
Ulna			1		1					2	4
All carpals				1		4				1	6
Metacarpal	3		1					1			5
Femur				1							1
Tibia							1				1
Astragalus						1					1
Calcaneum		1	1								2
All other tarsals					1	1					2
Fibula		1	1		1	1				1	5
Metatarsal	2		2			4			1	2	10
1st phalanx		1	1	3							5
2nd phalanx				1		3					4
3rd phalanx	1			1							2
Sesamoid	2										2
Indet. metapodial			2		1	1		1			5
Total by unit	10	3	12	7	4	16	2	6	1	7	68

Note: (*) Mostly lion and secondly wolf, followed by other carnivore taxa. MNE is the minimum number of each type of skeletal element, based on the most common portion or unique anatomical "landmark"; counts are mutually exclusive across categories. Cranial MNE is based exclusively on bone tissues (not teeth), and is computed in terms of right (R) and left (L) halves. Metapodials are counted using the proximal end, and carpal/tarsal MNE excludes the astragalus and calcaneum. Data represent provenienced material only.

CHAPTER 7

MIDDLE PLEISTOCENE BEARS FROM YARIMBURGAZ CAVE (TURKEY)

Evangelia Tsoukala

INTRODUCTION

The bear material from Yarımburgaz Cave, along with other faunal specimens, were recovered during three field seasons of extensive excavations (1988-1990) of the lower chamber, carried out under the direction of Prof. Güven Arsebük and the late Prof. F. Clark Howell (Arsebük 1993, 1996; Arsebük and Özbaşaran 1994, 2001; Arsebük et al. 1995; Howell and Arsebük 1989). The material is stored in the Prehistory Section of the Faculty of Letters, Istanbul University, Turkey. The bear remains represent animals of all ages, as commonly occurs in caves. The presence of milk teeth and of recently-erupted permanent teeth suggests that many bears used the cave as a den during Middle Pleistocene. Obviously the cave has been alternately used by bears and humans (Stiner et al. 1996, 1998) over seasonal or longer intervals. This study is based on the extensive and detailed report submitted by the author to F. C. Howell and G. Arsebük, in 1994 (Arsebük et al. 1995; Tsoukala 2003).

The paleontology of the bears is the focus of this presentation. The Middle Pleistocene bear collection is treated as a single time-averaged sample for the most part. The objective is to isolate distinct species within this sample, and to examine for the possibility of intergrading forms as opposed to palimpsest accumulations of distinct species that lived in different periods. Of particular concern is the taxonomic status of the cave bears in the collection.

Material

The representation of the various ursid bones and teeth (n = 2.567) by trenches is shown in Figs 7.1-7.4. The following abbreviations are used in the figures and discussions: (I) Incisor, (C) Canine, (P) Premolar, (M) Molar, (Cr) Cranium, (Mx) Maxilla, (Md) Mandible, (At) Atlas, (Ax) Axis or Epistropheus, (Sc) Scapula, (Pe) Pelvis, (Hu) Humerus, (Fe) Femur, (Pa) Patella, (Ra) Radius, (Ul) Ulna, (Ti) Tibia, (Fi) Fibula, (As) Astragalus, (Ca) Calcaneus, (Sl) Scapholunar, (Pi) Pisiform, (Py) Pyramidal, (Tm) Trapezium, (Td) Trapezoid, (Ma) Magnum, (Un) Unciform, (Cu) Cuboid, (Sd) Scaphoid, (T1) Cuneiform 1, (T2) Cuneiform 2, (T3) Cuneiform 3, (Ses) Sesamoideum, (Mp) Metapodial, (Mc) Metacarpal, (Mt) Metatarsal, (Ph) Phalanx, (V) Vertebra, (Vc) cervical, (Vd) thoracal, (Vl) lumbar, (Vs) sacrum, (Vcd) caudal, (Coc) Coccyx, (Hy) Hyoid, (Pen) Penis bone. General aspectual terms are as follows: (s) superior, (i) inferior (on Fig. 7.1), (s) left, (d) right, (int) internal, (ext) external, (a.c.) accessory cuspid, (art, a.s.) articular surface, (cond) condylus, (inc. sem.) *incisura semilunaris*, (pro. ang.) *processus angularis*, (sym.) symphysis, (Id) Infradental, and for dental elements, (pr) protocone, (hy) hypocone, (pa) paracone, (me) metacone, (pr^d) protoconid, (hy^d) hypoconid, (pa^d) paraconid, (me^d) metaconid, (en^d) entoconid, (cg) cingulum, (a.c.) accessory cuspid, (pas) parastyle, (mes) metastyle, (mcl) metaconulus, (tr^d) trigonid, (tl^d) talonid, (tl) talon, (m.f.) middle field, (cons) constriction. Measurement orientations are (L) length, (B) breadth, and (H) height. Measurements for various elements are presented in Appendices 7.1-7.9.

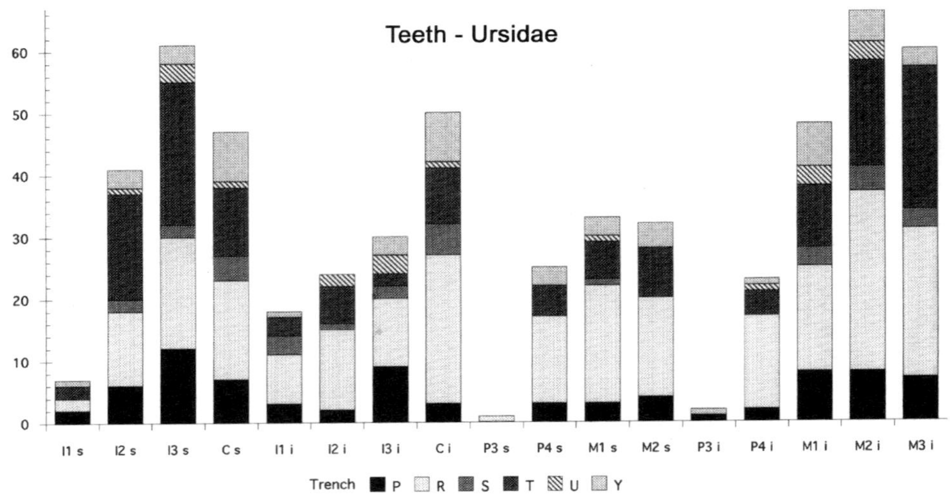

Figure 7.1. Yarımburgaz ursids: Teeth (n = 568)

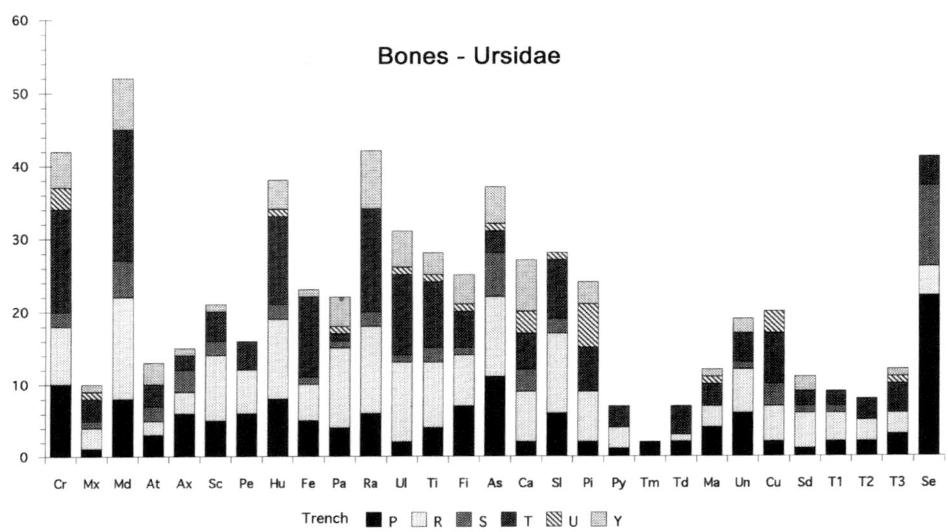

Figure 7.2. Yarımburgaz ursids: Bones (n = 642)

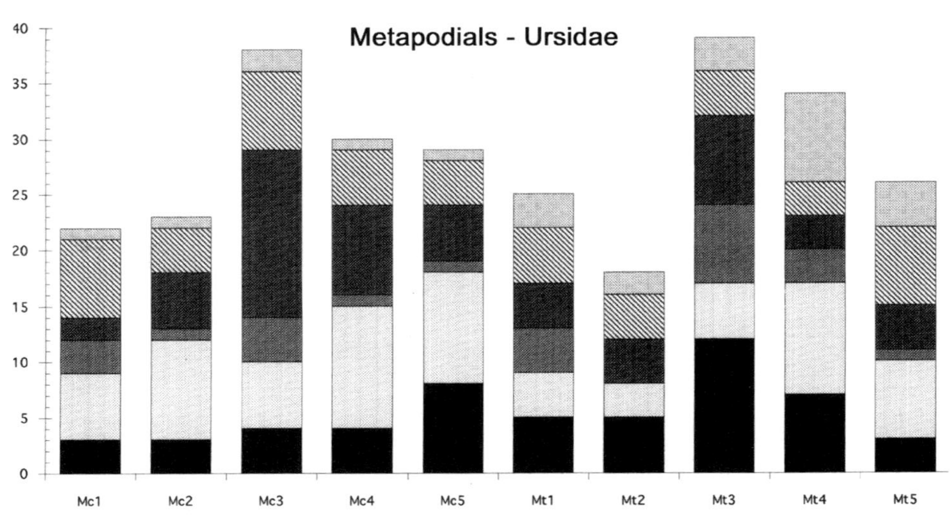

Figure 7.3. Yarımburgaz ursids: Metapodials (n = 284)

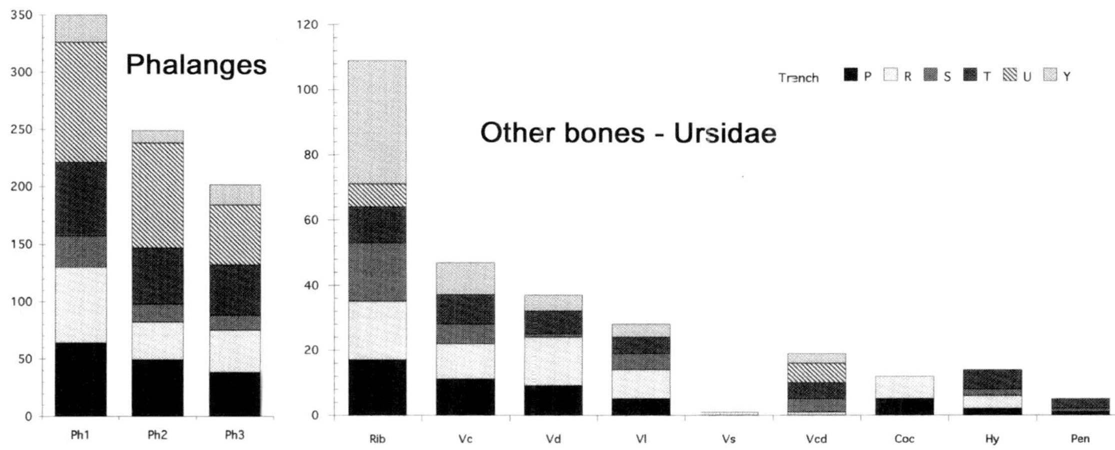

Figure 7.4. Yarımburgaz ursids: Other bones (n = 801) - Phalanges (n = 272)

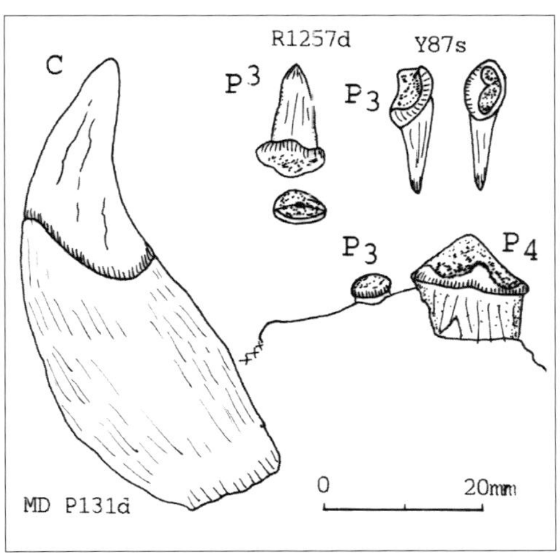

Figure 7.5. *Ursus arctos* YAR: Mandible fr. with C_i, P_3, P_4 *in situ* YAR P131d. Isolated teeth: P^3 YAR R1257d and P_3 YAR Y89s

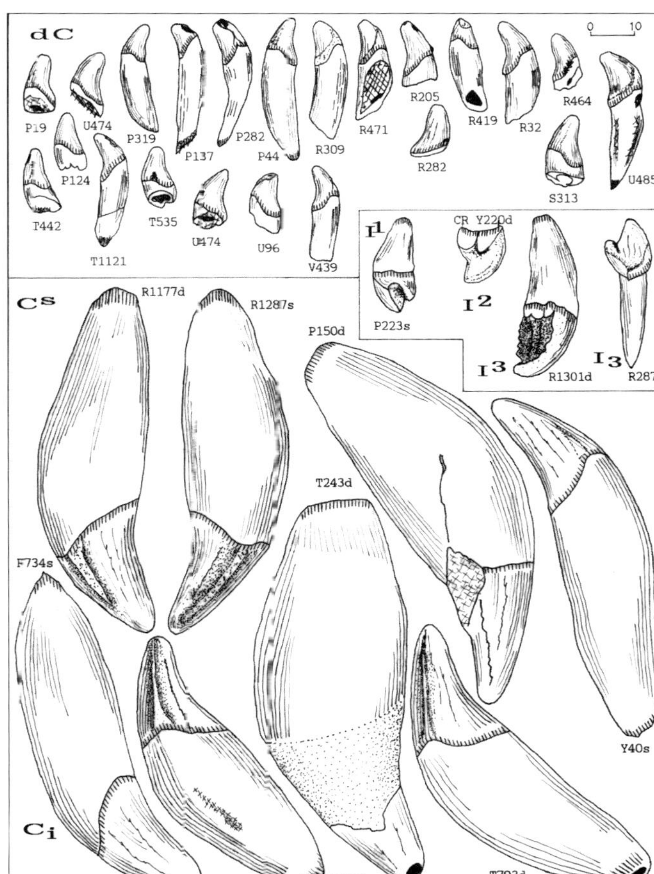

Figure 7.6. Yarımburgaz ursids. Upper: dC-milk canines, Middle: upper and lower incisors, Lower: upper and lower canines

PALEONTOLOGY

Taxonomy

Order: CARNIVORA BOWDICH, 1821

Superfamily: Canoidea SIMPSON 1931a-Arctoidea FLOWER, 1869

Family: Ursidae GRAY, 1825

Genus: ***Ursus*** LINNAEUS, 1825

U. deningeri v. REICHENAU, 1906 - ***U. spelaeus*** ROSENMÜLLER, 1794 and ***U.*** cf. ***arctos*** L.

Description - Morphological Analysis

Cranium–Maxilla: Although many cranial fragments—mainly parietal, temporal, frontal and orbital bones—were recovered, they all belong to juveniles, and no complete adult cranium is present in the sample. There is one complete cranium (Y220, Plate 7.1) of a sub-adult with the full cheek dentition (Fig. 7.9): the right second incisor and the canines of this specimen were still in the course of eruption at the time of death. There is also an incomplete cranium of a juvenile (Y447). Large maxillary elements are also uncommon, and very few are preserved with teeth still in place (R1059, Plate 7.2a).

Mandible: Mandibles are common in the Yarımburgaz ursid sample. The degree of fragmentation varies a great deal, but some of the mandibles retain full dentitions (Plates 7.2-7.7). The mandibles exhibit much variation in morphological and metrical features, some of which may be due to sexual dimorphism. The mandibles are mostly long and have brachygnathic to hypsognathic forms. They are slender to robust, with a small and slender to very robust (broad and high) condylus and symphysis. One speleoid[1] type mandible (T426, Plate 7.7f) with a full but worn dentition is exceptionally robust, with a very well developed *processus coronoideus*, and it is very hypsognathic: this specimen probably represents a male individual. The arctoid[2] form of mandible is represented by fragment P131. The C_i, P_3, P_4 are present but not so well preserved, and the alveoli of the M_1 and M_2 are also present (Fig. 7.5, Plate 7.2d). The corpus of this specimen is slender and brachygnathic.

The teeth in the bear assemblages are mostly well preserved. About half of them are unworn or lightly worn, while very few are extensively worn. They present a great variation both in morphotype and in their metrical proportions. Among the milk teeth, the deciduous canines (dC) all have a flattened appearance laterally, and they are straight or lightly curved (Fig. 7.6). The lower deciduous canines are usually shorter and thicker at the base of the root than the upper ones (Ehrenberg 1964; Erdbrink 1953). The majority of the 24 deciduous canines are crowns, some of with a partial root and signs of osteoclasts. The crowns are curved with a distinct basal cingulum.

Incisors: **I^1** - There are few first upper incisors, and these are lightly worn for the most part. One (P223, Fig. 7.6) is rather short, though unworn, with palatinal crest and tubercle, while the rest are longer. **I^2** - The second upper incisors are also lightly worn. They vary from long and slender (P49, R405) to short and robust (T371) with intermediate types present. There are two with well-developed palatinal cuspids and tubercular cingulum (R745, R855). Some of the specimens have open roots (T1276, T1036, T384, T829). The speleoid form (R487) is very robust, while the arctoid

1 Speleoid character is that, which is attributed to the *Ursus spelaeus.*

2 Arctoid character is that, which is attributed to the *Ursus arctos* (Erdbrink 1953; Mottl 1964; Rode 1935; Zapfe 1946]

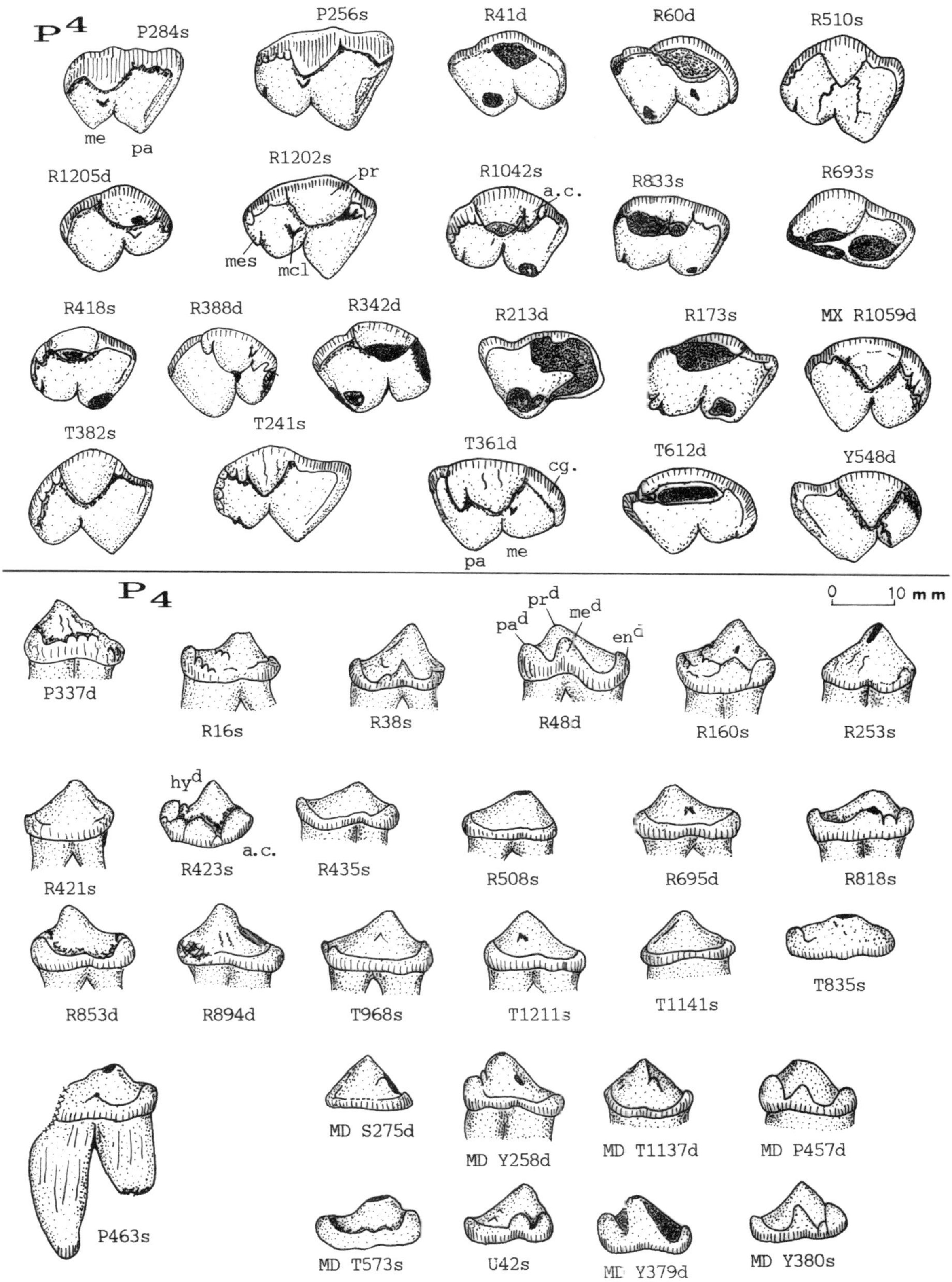

Figure 7.7. Yarımburgaz ursids. Upper: upper, lower: fourth lower premolars.

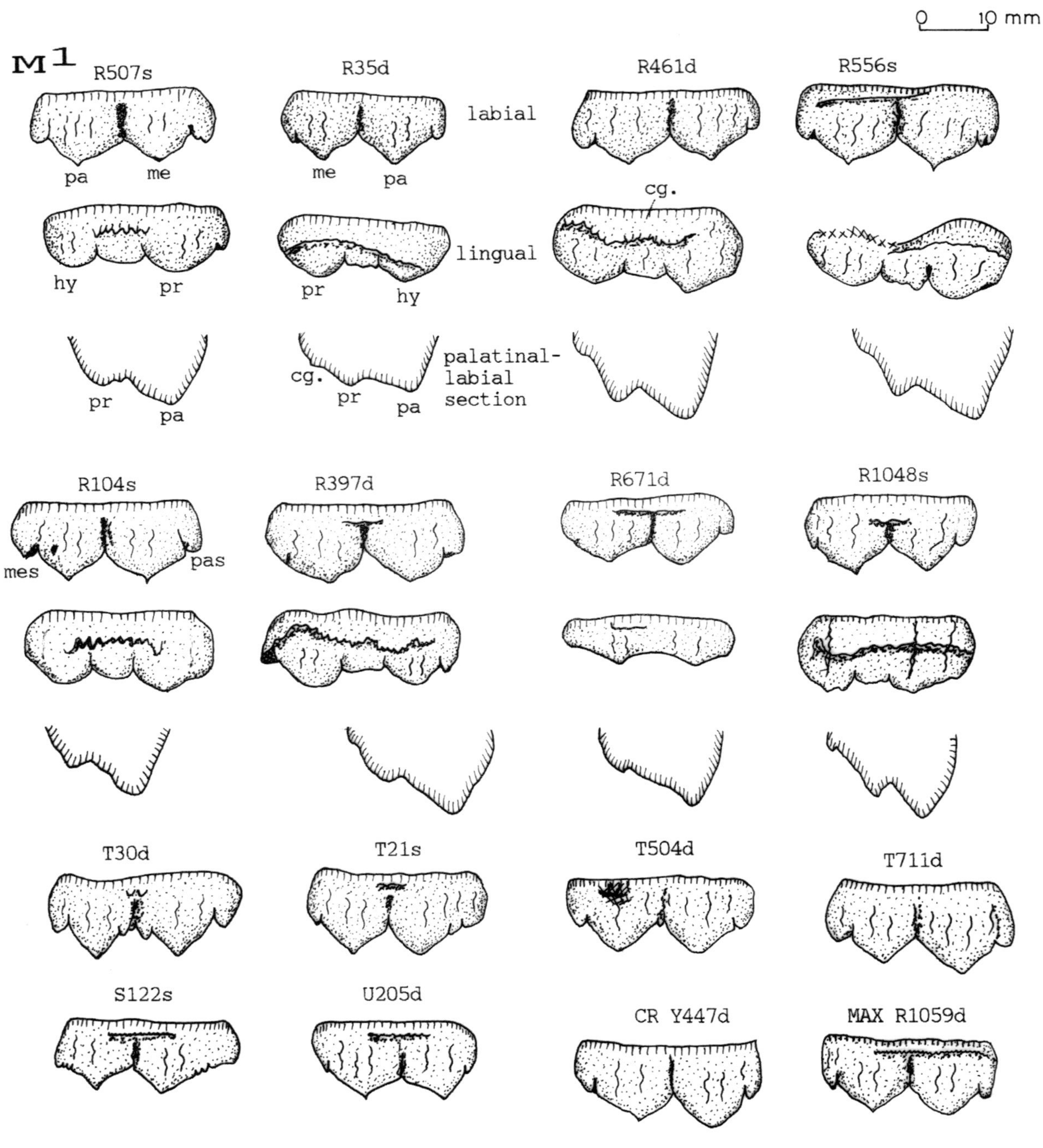

Figure 7.8. Yarımburgaz ursids: first upper molars

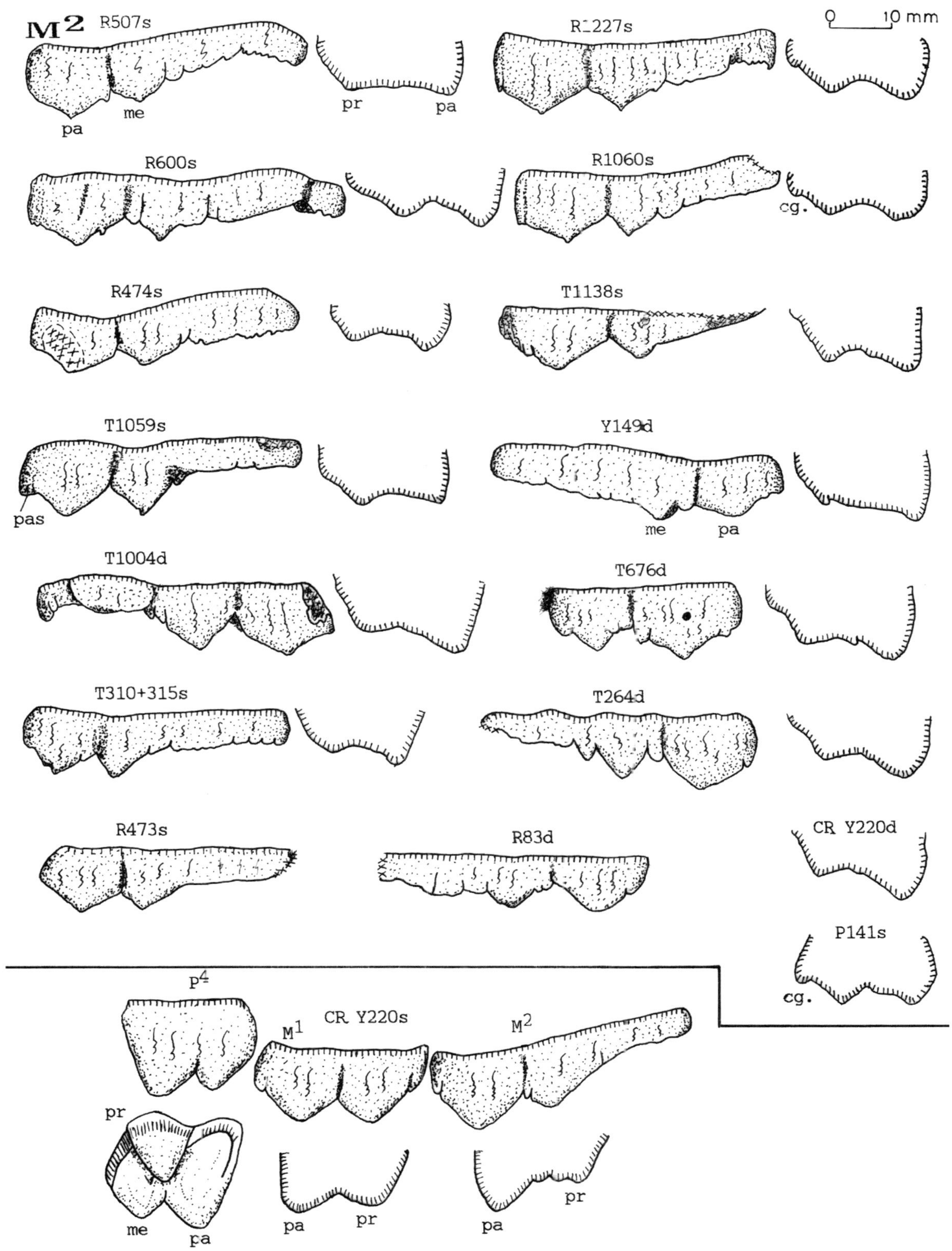

Figure 7.9. Yarımburgaz ursids: second upper molars, lower left tooth-row of the skull Y220

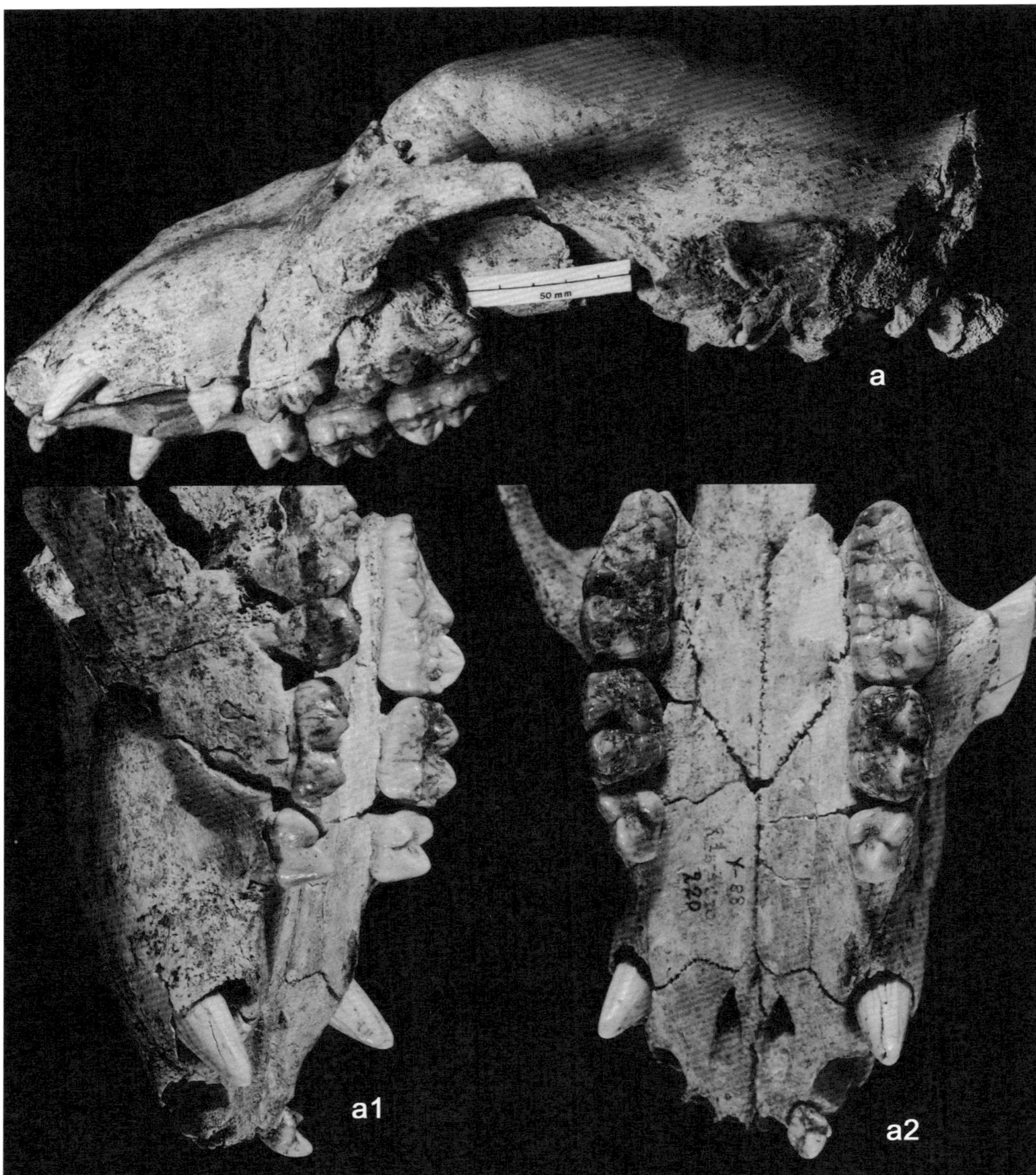

Plate 7.1. *U. deningeri* YAR
a) Skull Y220 of subadult with full cheek dentition and right I^2, a_1) occlusal, a_2) labial.

form (P31) is very slender. **I^3** - The third upper incisors are lightly worn, with either long or short and robust roots. There is also a strong palatinal cingulum between pronounced crests. The speleoid specimens are very robust and have high crowns. **I_1** - Some specimens of first lower incisors are relatively slender, while the speleoid ones (P50, T399) are very robust with a distinct labial protuberance. **I_2** - The second lower incisors are either germs or possess open roots, and they are either unworn or lightly worn. Only one (P784) is very robust, but worn. **I_3** - The third lower incisors exhibit a more or less intense labial protuberance. Some of them have either open or very long

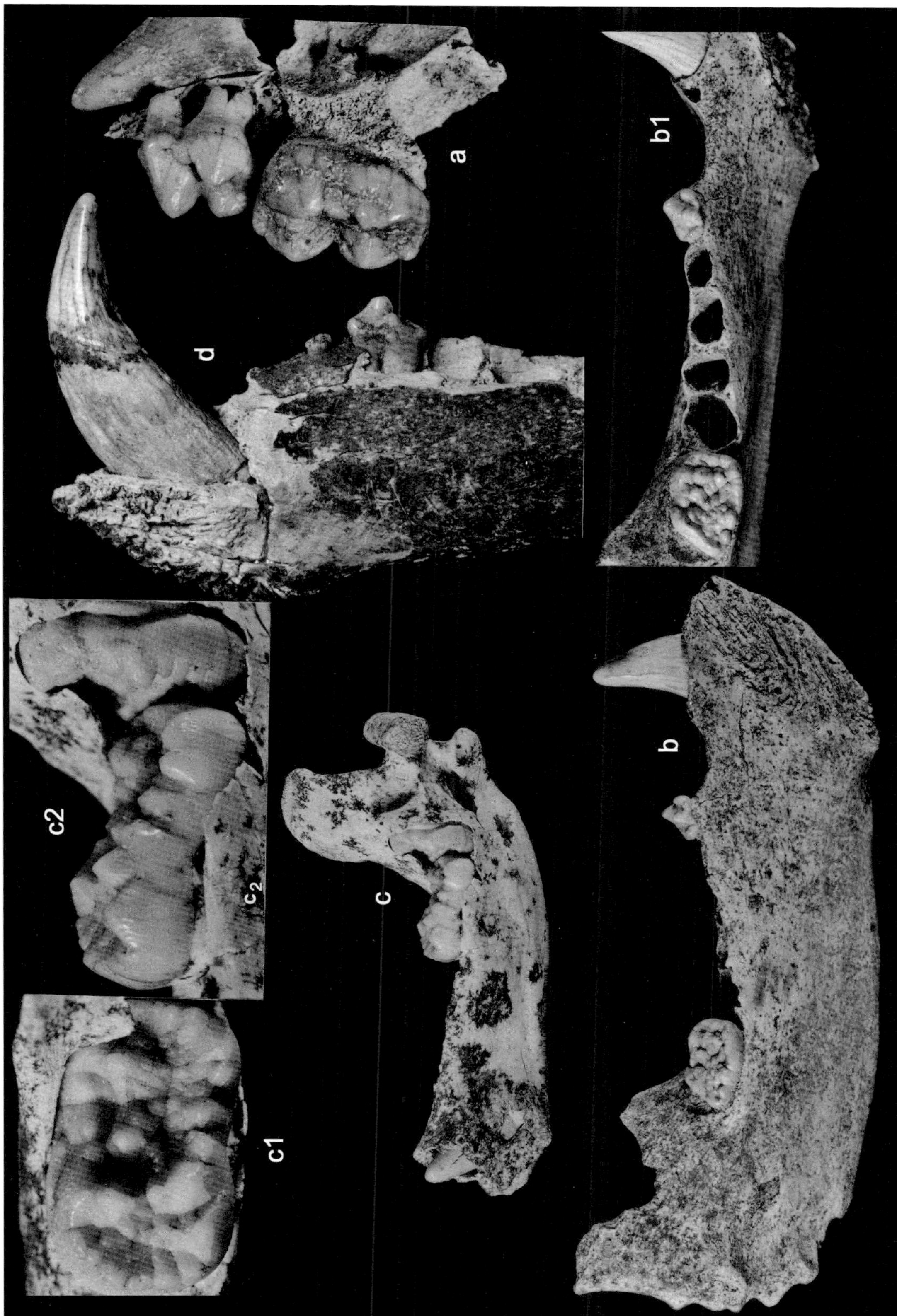

Plate 7.2. *U. deningeri* YAR

a) Maxilla fragm. R1059d with C, P^4, M^1 palatinal (nat.); b) Mandible Y3805s with P_4, M_3, lingual (1/2 nat.), b_1) occlusal. *U. spelaeus* YAR c) Mandible P306d with M_2, M_3 lingual (1/2 nat.); c_1) M_2 occlusal, c_2) M_2, M_3, (d1/2 nat.). *U.* cf. *arctos* YAR d) Mandible P131d with C, P_3, P_4, (nat.).

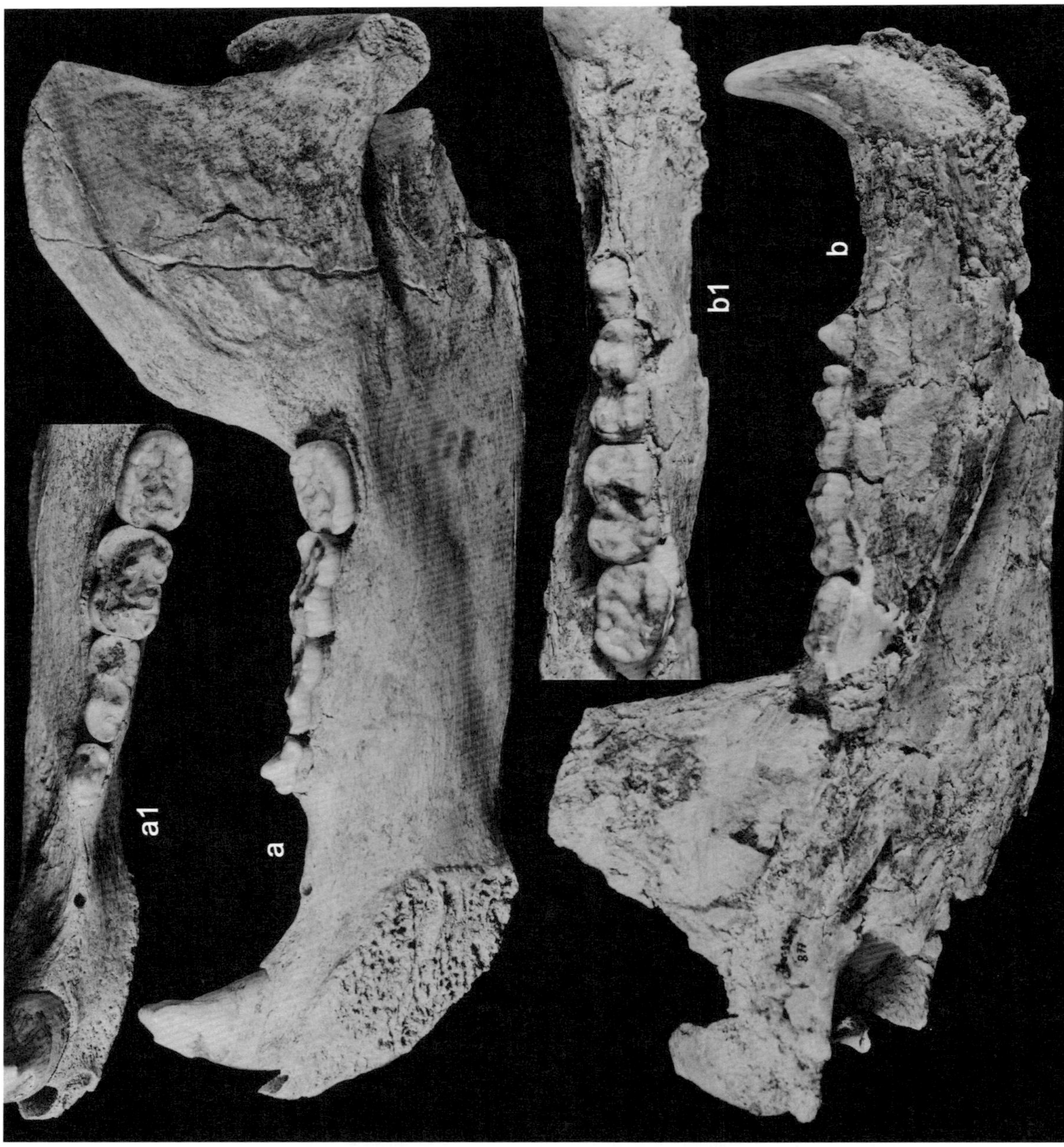

Plate 7.3. *U. deningeri* YAR a) Mandible Y258d with full dentition; b) Mandible T877s with full dentition; a,b) lingual, a_1, b_1) occlusal (1/2 nat.).

roots, others are germs. The speleoid forms are very robust (R82, R573), while the arctoid ones are short and slender.

Canines: $\mathbf{C}^{s}_{i}$ - The permanent canine teeth are mostly well preserved and unworn or lightly worn. They show a great variation in size, from very slender to very robust (Fig. 7.6). This variation is attributed mainly to sexual dimorphism, as well as to the presence of more than one species.

Anterior premolars: $\mathbf{P}_3$, $\mathbf{P}^3$ - Third premolars, which are an arctoid character, are very rare. The single-root teeth are small with low, slender crowns. The main cuspid has two crests, and there is a well distinct cingulum at the base of the crown (Fig. 7.5).

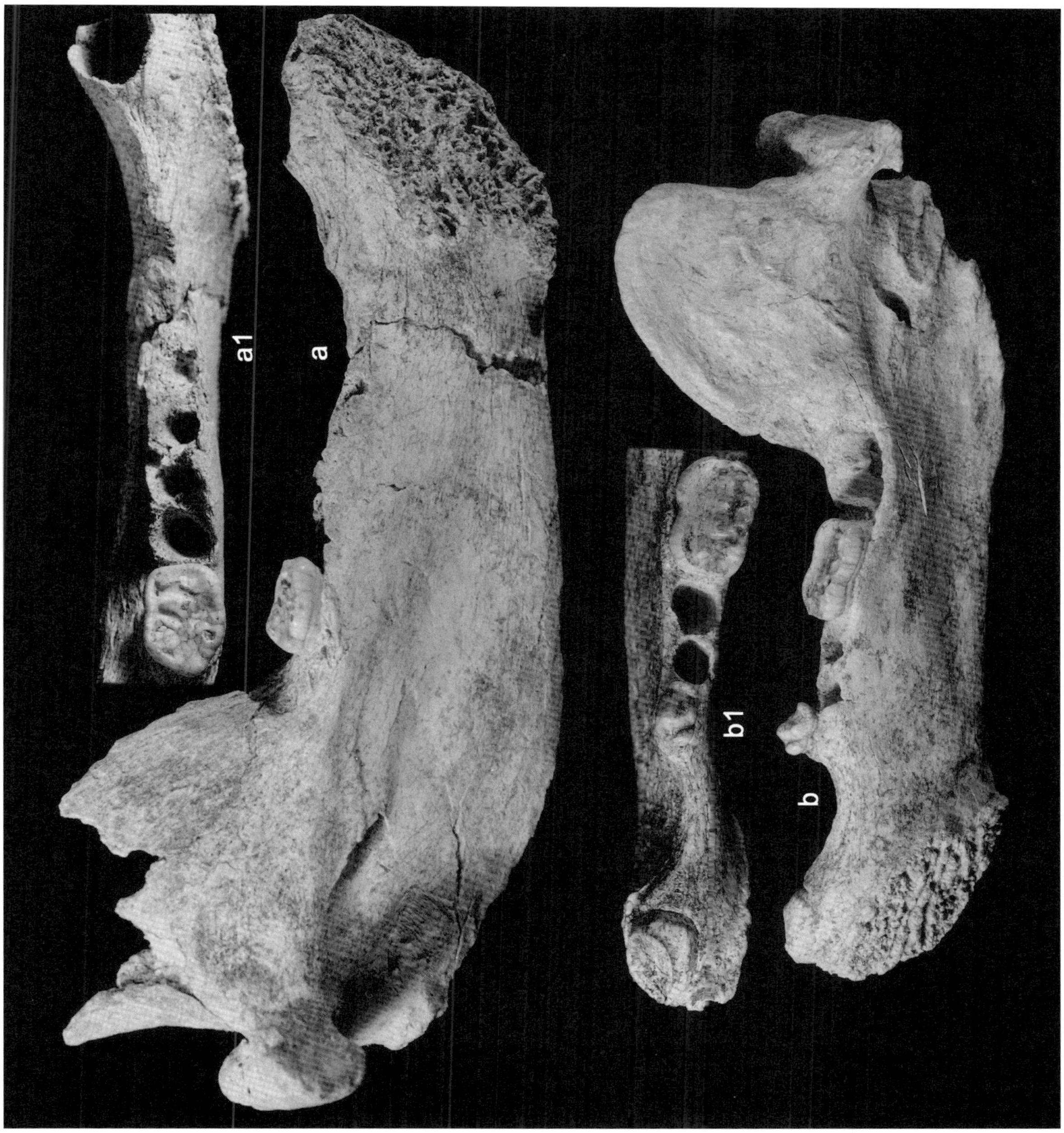

Plate 7.4. *U. deningeri* YAR a) Mandible T349s with M_3; b) Mandible P457d with C frag., P_4, M_2 ; a,b) lingual a_1, b_1) occlusal (1/2 nat.).

Upper premolars: **P**[4] - The fourth premolars (Fig. 7.7) have either a marked labial cingulum (Mx R1059, R240, R1202, Y548) or a slightly developed one. Some specimens have a well-developed cingulum between the protocone and paracone (R382, T241), while others show a tubercular one between the protocone and metacone (Mx R1059, R1042, R1202, T382). The paracone is generally very well developed, with a palatinal crest. The metacone varies from undeveloped to well developed (R1202, R1205). About half of the specimens exhibit a metastyle and a metaconulus. The former is either well developed (R173, R833) or hardly distinguishable; likewise, the latter is generally well developed (R1202, R1205) or hardly distinguishable (T361, Cr Y220 Fig. 7.9). The protocone is single (R41, R342), either with an anterior accessory cuspid (R418, T361, T612)—which

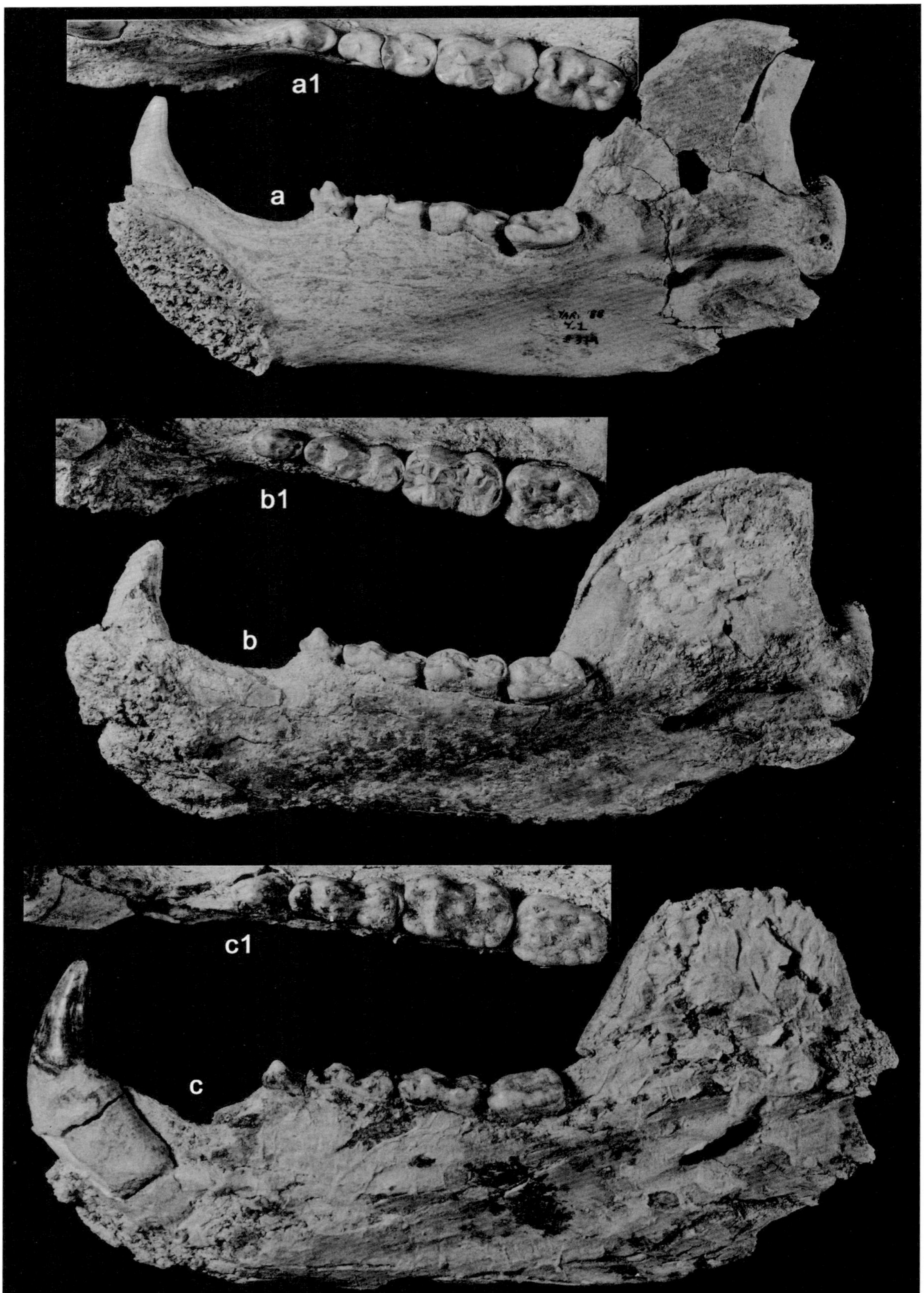

Plate 7.5. *U. deningeri* YAR a) Mandible Y379d with full dentition, lingual; b) Mandible T1137d with full dentition, lingual; c) Mandible T876 d with full dentition, o labial, a_1-c_1) occlusal (1/2 nat.).

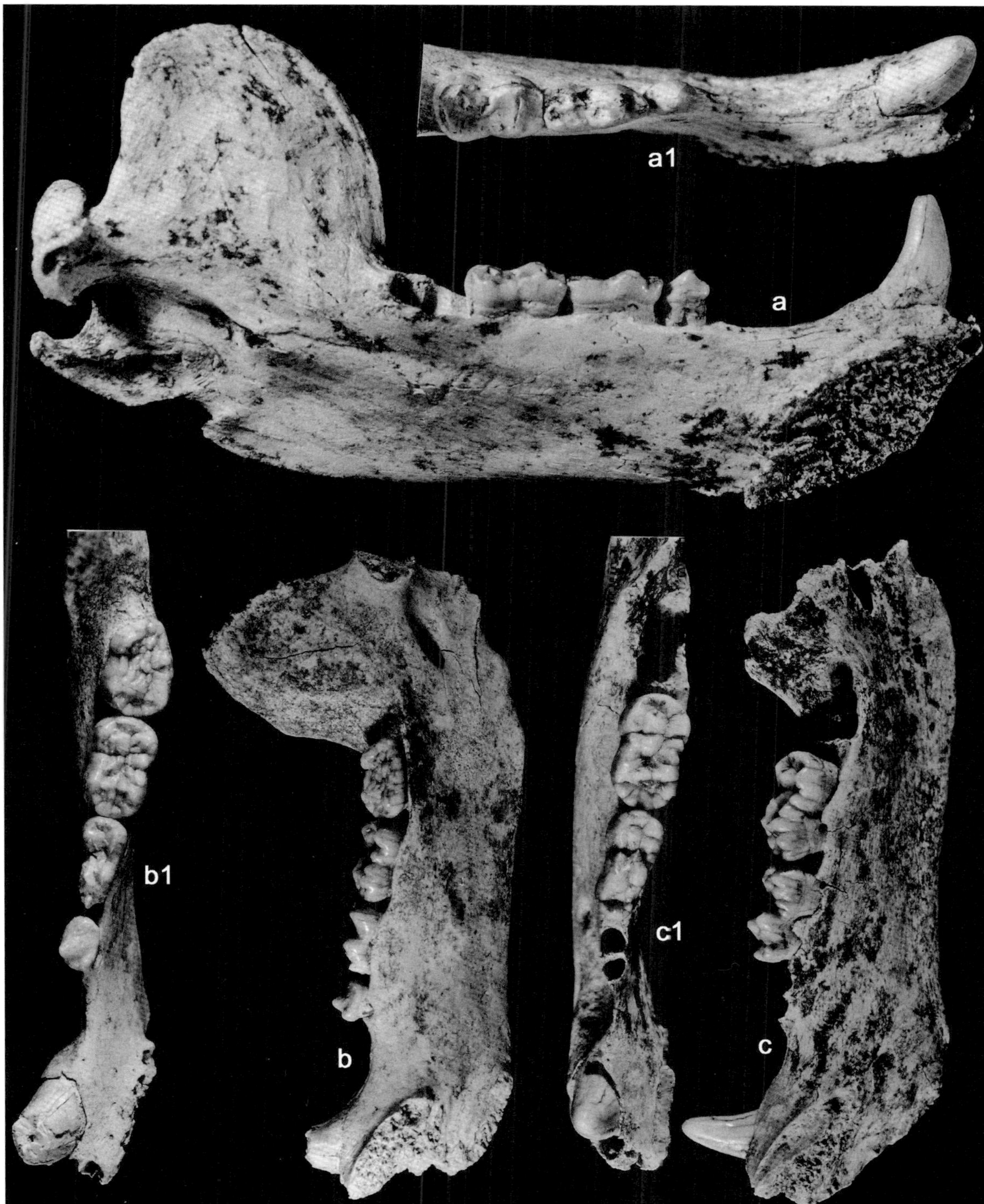

Plate 7.6. *U. deningeri* YAR a) Mandible P414s, with C, P_4, M_1, M_2,; b) Mandible P288d with full dentition; c) Mandible T786d, juv., with C, M_1, M_2,; a-c) lingual, a_1-c_1) occlusal (1/2 nat.)

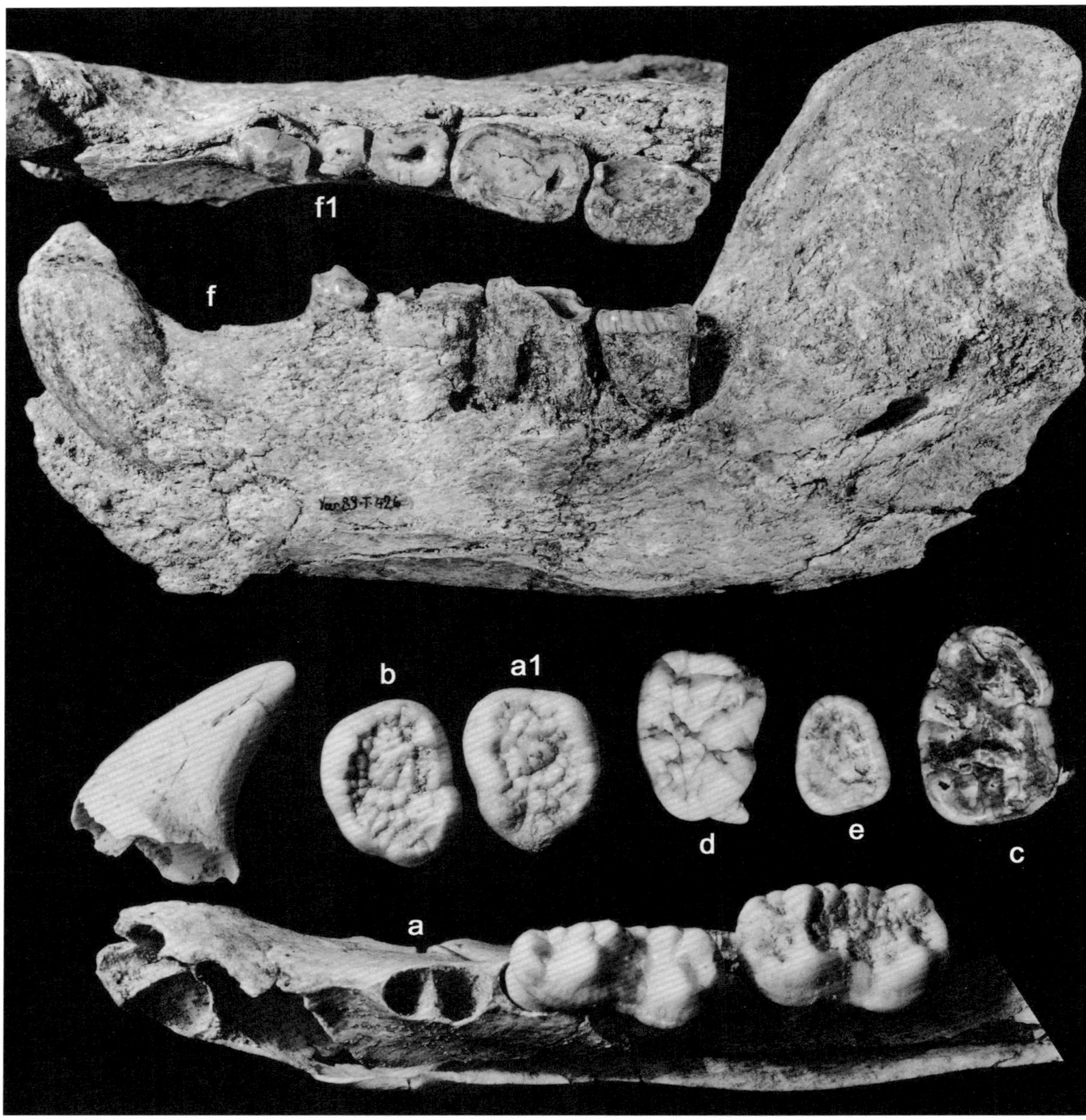

Plate 7.7. *U. deningeri* YAR a) Mandible frag. T878s, juv., with C, M_1, M_2, a_1) M_3; b) M_3 T292d, d) M_3 T440d; a-d) occlusal (nat.). *U. spelaeus* YAR c) M_3 T784s, occlusal; f) Mandible T426d with full dentition, lingual, f_1) occlusal (1/2 nat.). *U.* cf. *arctos* YAR e) M_3 T784s, occlusal (nat.).

may be pronounced (R833)—with two accessory cuspids on one (R1042) or both sides (R388), or with a posterior cuspid (T382). The robust speleoid premolars (T241, T361, T612, R1042) have a protocone with a well developed anterior accessory cuspid, a tubercular cingulum between the protocone-metacone, and a pronounced labial cingulum. Two of them have a distinct metastyle.

Upper molars: **M^1** - The first molars have a sub-quadrate shape (Fig. 7.8), with a developed talon, or they are wide (T30). They generally have wrinkled palatinal enamel (R507). The middle field is mostly lightly or finely (S122) granulated (Cr Y447, Mx R1059, T21, T711). In most cases the labial cingulum is slight and short; more rarely it is shaped like a cuspid (T30). The palatinal cingulum may be pronounced and tubercular, or in other cases else hardly distinguishable (R671, R1048). With the exception of one specimen (S122), the paracone is more developed than the metacone.

There are pronounced crests on the palatinal sides of both the paracone and metacone (Cr Y447) or just the metacone (Mx R1059). Between the paracone and metacone there is one (T504) or two (Mx R1059, T30) accessory cuspids. Two types of paracone-protocone profile (palatinal-labial section) were observed (Fig. 7.8). One type shows a marked development of the palatinal cingulum, causing the protocone and paracone to be comparatively close to each other (R35, R1048, R461). The other type shows a less developed cingulum, and a protocone that is more palatinal (R507, R671) (see Appendix 7.4, measurement n. 10). The middle field of the speleoid robust molars is coarsely granulated (R397, R556). **M**^{2} - The second molars are elongated (Fig. 7.9), with slight to pronounced (P141, R507) torsion of the talon, and wrinkled palatinal enamel (R507, R1060). The middle field varies from little or finely granulated (P147, R507) to coarsely granulated (T264, T676, T1059, T1138). The periphery of the crown is in most cases slightly tubercular or very tubercular (P141, R1227, T1059, T1138). The palatinal cingulum is generally either long, pronounced and tubercular, or hardly distinguishable (CR Y220, R1227). The paracone is well developed, whereas the parastyle is either barely distinguishable or very well developed and double (T1059). The metacone is less developed than the paracone, mostly with an accessory cuspid that may be anterior (Cr Y220), posterior (T310+315, R474, R507) or present on both sides (R83, T264). The protocone and hypocone are either double (R507) and tubercular (P141, R1227)—in some cases, the former has an accessory cuspid (R1060, T1059, T1138)—or they are undeveloped (R474, T310+315, T1004). The mesoconule (R474, T1004) and the hypoconule (R1060, T1059, T1138) are long and well developed. Two forms of paracone-protocone profile (palatinal-labial section) (Fig. 7.9) were noted, one with a pronounced profile (P141) and the other with a more elevated middle field (R507, Y149). One unworn, very robust and large speleoid molar (R 600) displays a slight torsion of the talon and a posterolabial fold of the palatinally-wrinkled enamel. The middle field is very coarsely granulated, and the periphery of the crown is very tubercular. The paracone is well developed with an anterior accessory cuspid, while the protocone and hypocone are single.

Lower premolars: **P**$_4$ - There is great variation in the morphology of the fourth premolar (Fig. 7.7). It ranges from a very primitive, simple morphotype to very complicated types, indicating a tendency for molarization in the latter group. The fourth premolars are mostly unworn with separate, distally separated (R435) or unseparated (R160, R423, R813, R894) roots. The protoconid is rarely developed (R38, R160), or it is short (R435) with a posterior crest (R894, T1141). The paraconid is generally developed, rarely double (R853, T968), or it is highly developed (Md P457, R48) with an accessory cuspid between it and the metaconid (R818). The latter feature can be either absent (R16, R253, R421, R435, R50, R853, R894, T1141, T835, Md S275, Md T573, Md P288, Md R414, Md Y379), highly developed (R38, R48, R160, R423, R818, Md P457, U42, Md Y380) or very lightly developed (R645, T968, Md Y258, Md T876). The hypoconid is either undeveloped or slightly developed, but the endoconid is developed (R698, R818, R853, R894). The speleoid premolar (P463) is extremely developed and very robust. Part of the talonid is has been destroyed post-morten. The protoconid and paraconid are well developed, but the metaconid is not so well developed. On one specimen (P337) the protoconid is highly developed, the paraconid well developed, and the metaconid undeveloped and displaying a tubercular crown and crests. On another specimen (Md T426) the slightly worn protoconid and metaconid are developed, while the paraconid is crested. The arctoid premolar (Md P31) is slender; its protoconid and metaconid are well developed, whereas the paraconid is less developed.

Lower molars: **M**$_1$ - Most of the first molars are worn or slightly worn. A few are unworn, or deeply worn (Fig. 7.10). Some of them are more robust (R994, U352) than the rest, so they may belong to

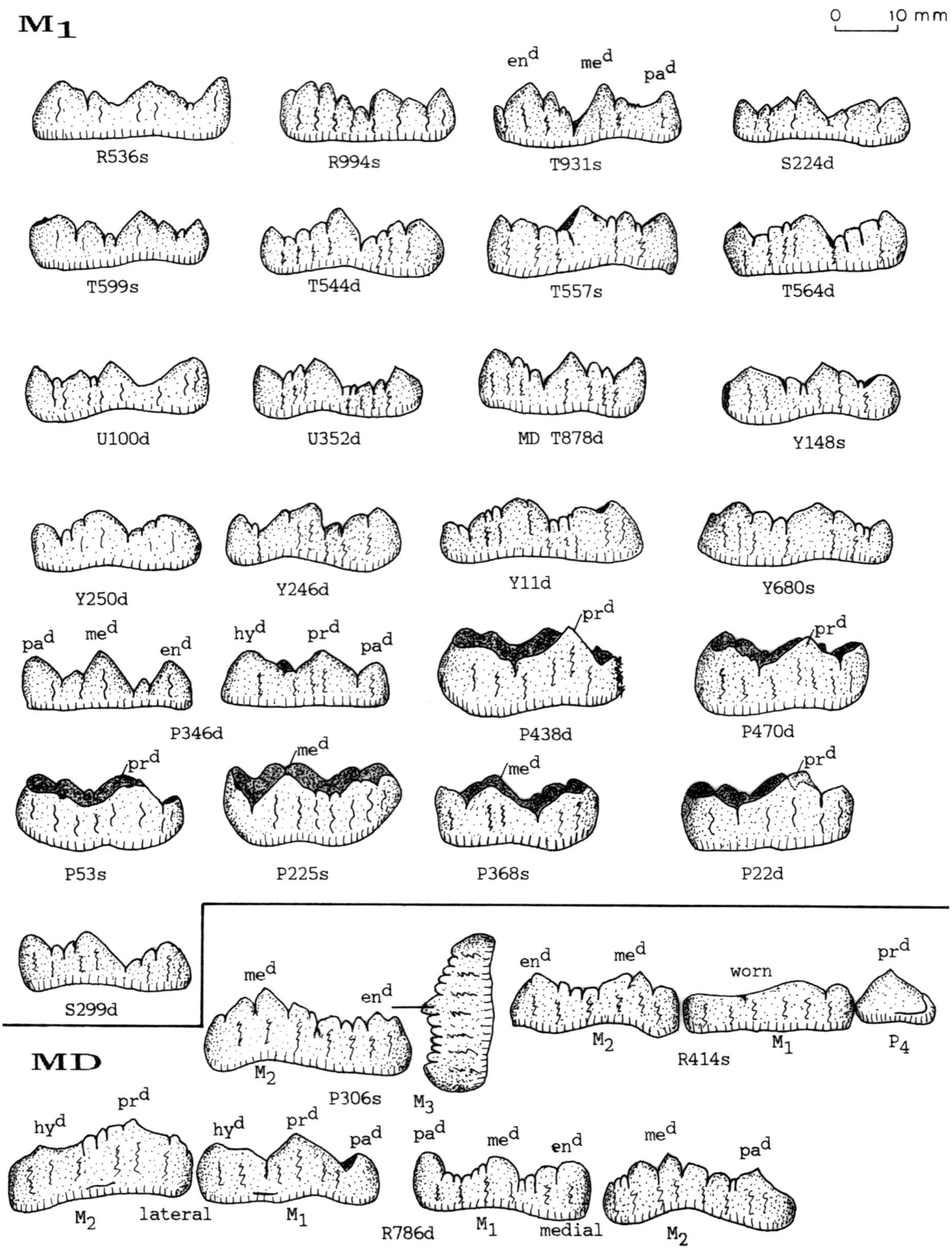

Figure 7.10. Yarımburgaz ursids. Upper: first lower molars, lower: tooth-rows of mandibles

males. Wrinkled enamel is present on the labial side (T564, Md T878, U352). The short cingulum at the base of the hypoconid, close to the constriction in most teeth, is very developed in a few specimens (S299, U352). The middle field of the talonid generally is not very granulated, but a few specimens display coarsely (T544, Y246) or finely (R994, Y11) granulated surfaces. The periphery of the crown is not very tubercular on most specimens, but in a few it is very much so (R994, T327, T544, Y680). The protoconid is generally simple or possesses an accessory cuspid (P53, T544, Y680). The hypoconid is also simple or has an accessory cuspid (P225). The paraconid is well developed, and the metaconid is either single or double (P225, R994, Y246) with an accessory cuspid. The entoconid is double or single with an accessory cuspid. The speleoid molars are very robust and broad (T557), with high crowns. Intensely wrinkled enamel is noted on the labial side (R536, Y11), and there is a pronounced labial cingulum (T557). The middle field of the talonid is very coarsely granulated, and the periphery of the crown is very tubercular. The protoconid and hypoconid possess an accessory cuspid (T557, Y11), the paraconid is very well developed, the metaconid complicated, and the entoconid is double with an accessory cuspid. The single arctoid molar (P1) is of a different morphology, and probably belongs to the same individual as the M_2 (P2). This specimen is very small and slender, with a less tubercular crown. The paraconid is less developed, while the metaconid and entoconid are double.

M_2 - Most of the second molars are unworn to slightly worn, while a few are deeply worn. They are mostly well preserved (Fig. 7.11), relatively long and rectangular in shape, and with a constriction between the trigonid and talonid. The constriction may be expressed both labially and lingually (P11, P209, T877) and be very pronounced (R1316, Y270), or it may be slight lingually but pronounced labially (P174, P385, P624, Md T878). Some specimens are robust (R912, S215, U33, U204), while others are gracile (Y270). Very wrinkled labial enamel is rare but occasionally present (T1120). The middle field is greatly variable. Normally it is coarsely granulated (R151, R161, R501, R1064, S149, U203), but sometimes scarcely granulated (R912, R1012, T932, T1141, Md P288) or finely granulated (R241, R838, T878). It also varies from very (T1120) or moderately granulated (R516, R725, Md T658, S215) to ungranulated (R419, R1316, U33, U204). The periphery of the crown is grossly tubercular, but occasionally finely tubercular (R241, T878). The protoconid is always the most developed cuspid (except for U33, on which it is undeveloped). It is usually long and simple, or with a small accessory cuspid. The hypoconid is developed (except for U33), and it may be simple (R241, R725) or possess an accessory cuspid (T1120), sometimes of a more posterior position (R501, R838, U203). The metaconid and entoconid are mostly double with accessory cuspids, or else triple (P174, Md T877). The speleoid molars are very robust and high-crowned (T115) and, like other teeth described above, they are broader posteriorly than anteriorly, a speleoid character. Specimen Y156 has a vertical fold of the enamel between protoconid and hypoconid. Only one example of a short triangular cingulum on the labial constriction is noted (Y156). The middle field is coarsely (T47) or finely (T1289) granulated. The periphery of the crown is very tubercular because the cuspids are more complex: for example, the metaconid is triple in specimens P306, Y156, Y367, with an accessory cuspid on P306 (Fig. 7.10). The arctoid molar (P2) is of a different morphology. It is slightly worn, very small and gracile, with no granulated middle field and a less tubercular periphery on the crown. The trigonid is slightly broader than the talonid, and the cuspids are weakly developed.

M_3 - Most of the third molars are unworn or slightly worn, and more or less robust (Fig. 7.12). Most are of an oval shape, but a few have angular-oval crowns (Md P288), which rarely show a slight torsion (T703). Also rare is a vertical fold of the enamel on the talonid (S324, T703). The middle field

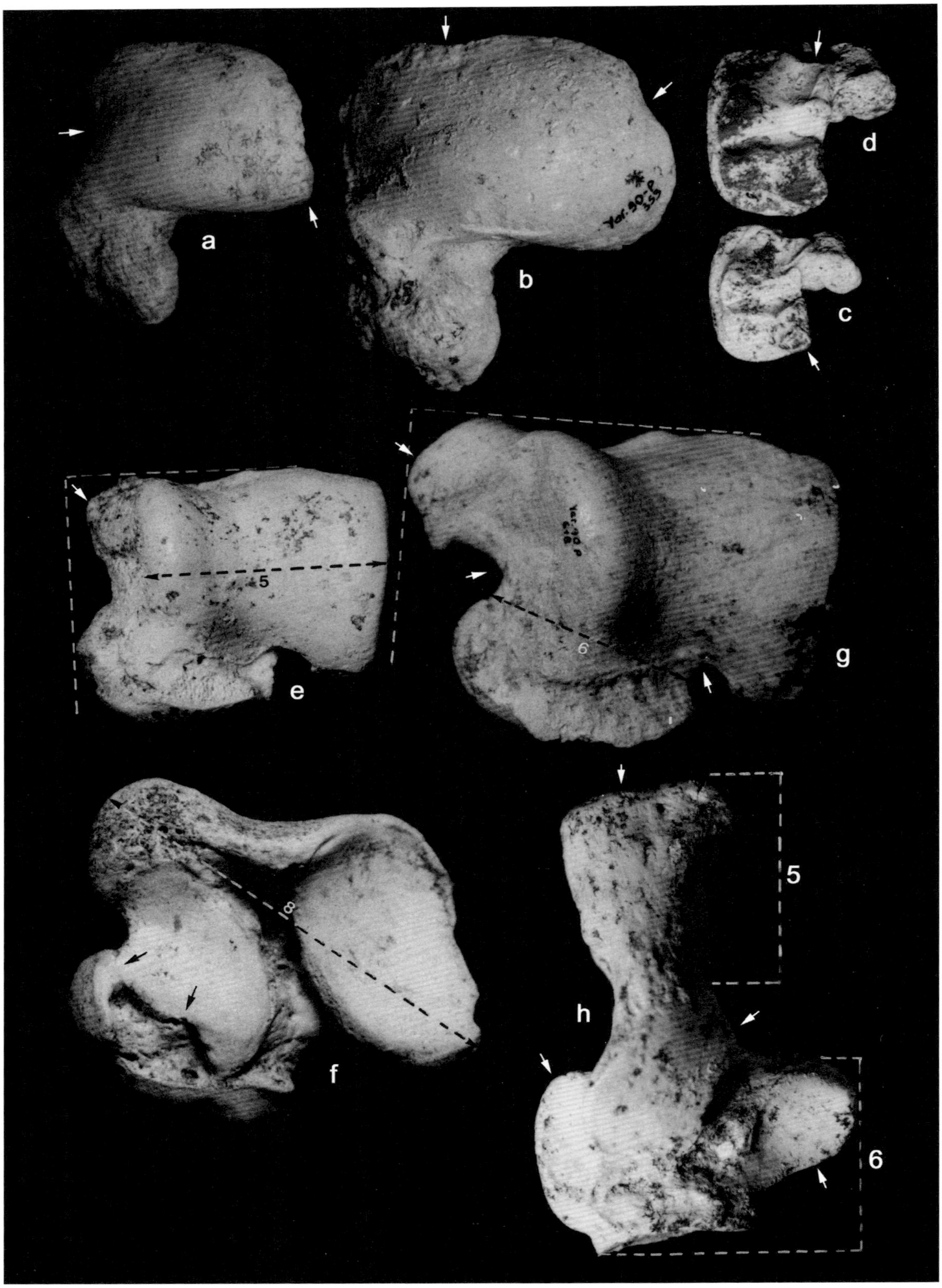

Plate 7.10. *U. deningeri* YAR a) Scapholunar P489d, proximal (nat.); c) Scapholunar T277d, distal (1/2 nat.); e) Astragalus P348s, anterior (nat.); f) Astragalus T388d, posterior (nat.); h) Calcaneus P433d, anterior (nat.). *U. spelaeus* YAR b) Scapholunar P353d, proximal (nat.); d) Scapholunar T157d, distal (1/2 nat.); g) Astragalus P618s, anterior (nat.).

in the zone of the insertion of the pisi-crochu or pisi-metacarpal ligament. Others are less robust or very gracile, with arctoid and speleoid characters (see arrows in the plates).

Pyramidal (ulnar, triquetrum): These are mostly whole, although seems to have been abraded or rolled. The very robust speleoid specimens (T424, T448, Plate 7.11e) have a perfectly formed, very well developed articular surface for the Mc5.

Scapholunatum: These elements are generally whole, and very small (P146). Gracile to very robust (P638) specimens are present in the assemblage. The morphological analysis shows some differences among them (see arrows in the plate): some specimens could be classed as arctoid (P489, T277, Plate 10a, c) or speleoid (P353, Plate 7.10b, T157), while a few (R1018) combine arctoid and speleoid characters.

Unciform (hamatum, crochu): Specimens are mostly whole. They vary from gracile to robust, with some morphological differences (see arrows in the plate) indicating either an arctoid (R751, T25 Plate 7.11b, f) or speleoid (R705, T378, Plate 7.11h-i) affiliation.

Magnum (capitatum). The magnums are mostly whole. They are either gracile (P543, T88, Plate 7.11j, T91) or very robust (P424, R682, T447, Plate 7.11k). Some morphological features indicate arctoid or speleoid forms (see arrows in the plate).

Trapezoid and Trapezium. The second and first carpals are rare in the collection but whole or nearly whole. Their morphology is varies little among the ursid species. Only one trapezoid exhibits speleoid characters, while the rest exhibit arctoid characters.

Astragalus. Astragali are mostly whole. A few of the specimens are from juvenile individuals. About half of the measured specimens are very robust (compare metrical analyses by Caillat 1986; Chagneau 1985; Pratt and Thibault 1976), exhibiting clearly speleoid characters that include an elongated shape, widening of the plantar face, very developed medial apophysis and incisure, and a distinctively shaped postero-medial articular surface to the calcaneus (see arrows in the Plate 7.10e-g). The arctoid specimen (P18) is very small and slender (L = 37.5, DAP = 29.0, D-oblique = 35.5mm). It is abraded and damaged, and of this reason its *collum* and *caput* are missing.

Calcaneus: Few calcaneii in the collection are whole, and most of these belong to juvenile individuals. Very few of the calcaneii are robust and display speleoid characters. Rather, most of the specimens are comparatively gracile, with arctoid characters such as a larger and shorter *sustentaculum tali* (see arrows in the Plate 7.10h), a distinctive shape of the articular surface, a more developed *manubrium*, tuber, and insertions of the ligaments (Plate 7.10h).

Scaphoid (navicular): Scaphoid bones are mostly well preserved. Two of them are very robust (T465, T511, Plate 7.11l), the proximal articular surface of the former is slightly more concave than that of the latter. The latero-palmar projection of the bone is distinct (see arrows in the Plate). The other specimens are less robust, with arctoid and speleoid characters.

Cuboid: These elements are mostly whole. A small number of them are more robust (T449, Plate 7.11n, T488), while most are gracile (T527, Plate 7.11m). The robust specimens exhibit some morphological differences: the proximal articular surface distinctly wider on the medial articular surface, which is separate, and the distal articular surface has a more convex border (see arrows in the plate).

Cuneiforms: The tarsals 1, 2 and 3 are mostly whole. Some are robust, others more gracile. Distinct morphological elements on T3 are the separate medial articular surface (T455, Plate 7.11p) and the constriction of the distal articular surface (see arrows in the plate).

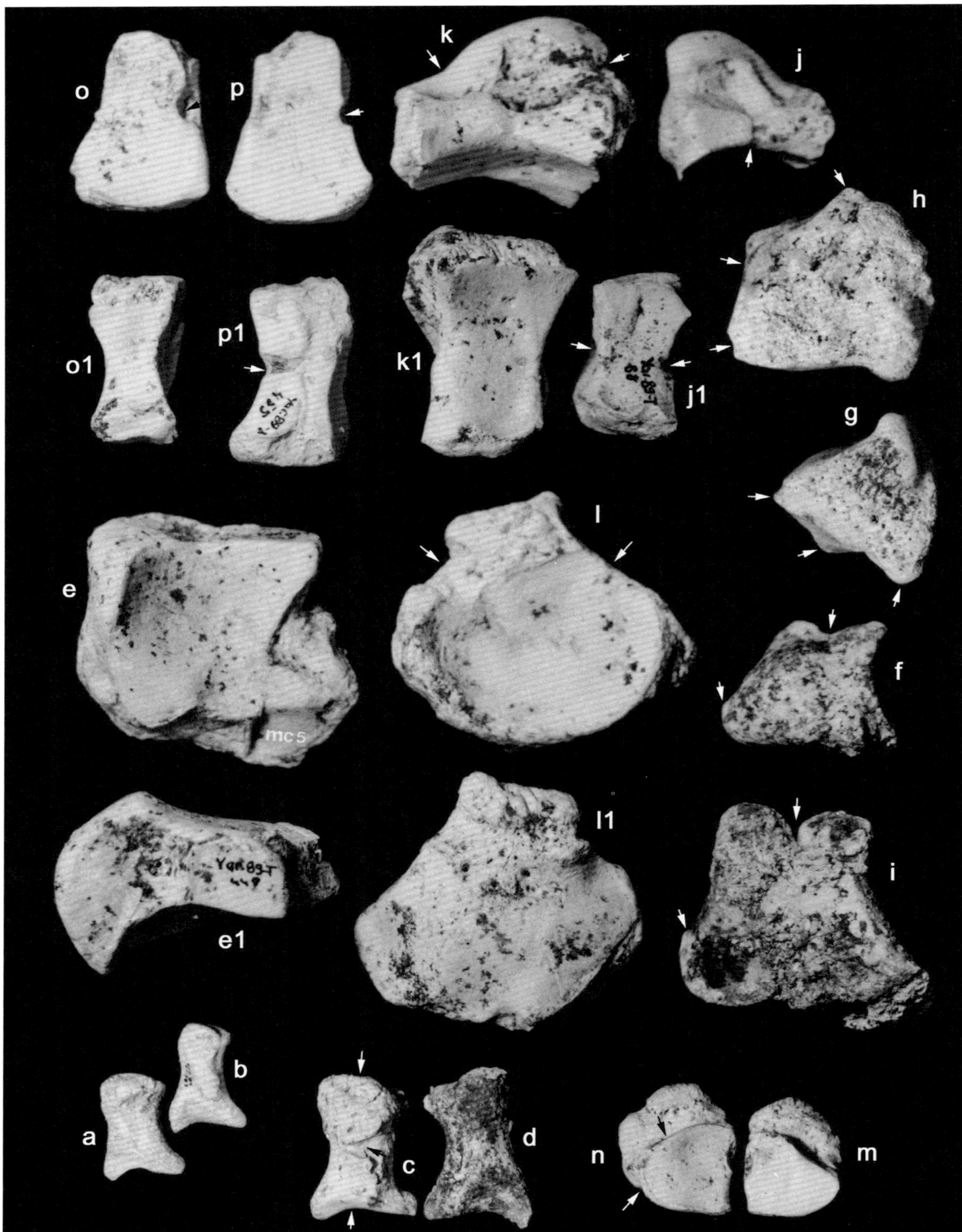

Plate 7.11. *U. deningeri* YAR a) Pisiform T78d, proximal (1/2 nat.); b) Pisiform T783s, distal (1/2 nat.); f) Unciform T25s, lateral (nat.); g) Unciform R751s, anterior (nat.); j) Magnum T88s, medial, j_1) distal (nat.); m) Cuboid T527d, distal (1/2 nat.); o) Cuneiform 3 T405d, proximal, o_1) lateral (nat.). *U. spelaeus* YAR; c) Pisiform T291s, distal (1/2 nat.); d) Pisiform T 220s, proximal (1/2 nat.); e) Pyramidal T448d, distal, e_1) medial (nat.); h) Unciform R705d, anterior (nat.); i) Unciform T378d, lateral (nat.); k) Magnum T447d, medial, k_1) distal (nat.); l) Scaphoid T511s, proximal, l_1) distal (nat.); n) Cuboid T449s, distal (1/2 nat.); p) Cuneiform 3 T455s, proximal, p_1) lateral (nat.).

Metapodials: Most of the metapodials are in good condition, though very few are complete. They show great variation in size, ranging from very slender (Mt5 R906, Plate 7.9e) to very robust and large (Mc5 R495, Plate 7.9d). Rarely, specimens were found in articulation. A deep depression for the insertion of the muscles is noted on the anteromesioproximal diaphysis of some Mc3 (R520, T214, T611, T694, U336, Y488). Likewise, there is a very deep depression on the posterodistal diaphysis of some Mc2 (T185, U336) and Mc5 (P452, P571). There is considerable variation in the forms of the metapodials. One (Mt1 P487) is remarkably short, and a few seem to have pathological deformations (Mc3 S270, Mc4 T954+998, T893, T1058, Mc5 T857). Some of the Mt2 have a proximal articulation that is more laterally inclined than the others.

Arctoid characters are expressed in the long and slender metapodials, especially the first ones, whereas speleoid characters expressed as short and more robust metapodials. This distinction can be captured in the ratio between the lengths of Mt1 and Mt5. This ratio is >70.0 in *U. arctos* (Altuna 1973), 71.3 in *U. arctos* from Jaurens (Ballesio 1983), 70.5 in *U. deningeri* from Hundsheim (Zapfe 1946), <67.0 in *U. spelaeus,* and 67.5 in *U. deningeri* from Nauterie (Prat and Thibault 1976). For the Yarımburgaz ursids this ratio is 67.6 placing them in the range of *U. spelaeus* or *U. deningeri* rather than *U. arctos.*

Phalanges: Phalanges are abundant and often whole. The elements display great size variation. The first and second phalanges range from very short and slender to very robust and elongated. The third phalanges range from long and relatively slender (P7, T410, U117, U168, i.e., arctoid) to very short and robust (P342, P352, P600, T190, speleoid), although some specimens are intermediate in these characters.

Ossa sesamoidea: Sesamoids are well preserved and display great variation in size. Some of them are very elongated and gracile, others are very robust.

DISCUSSION

Distinguishing species of the genus *Ursus* from Yarımburgaz Cave is very challenging for the *deningeri/spelaeus* group due to the polymorphy of *U. deningeri.* The dimensions *U. deningeri* skeletal elements generally range between those of robust *U. arctos* and gracile *U. spelaeus* (Reichenau 1908; Zapfe 1946). The earliest member of this speleoid lineage has been recently suggested to be represented by *U. dolinensis* in Spain (Garcia and Arsuaga 2001).

In the Yarımburgaz collection, dental characteristics (Rabeder 1983, 1999; Rode 1934) and the great size ranges of the teeth (Fig. 7.18) and bones clearly show the presence of the three species. It is difficult or impossible to determine fine temporal divisions in the stratigraphy, and there is good reason this believe that the sediments accumulation over a long period during the Middle Pleistocene. The premolars of the Yarımburgaz ursids range from very primitive to much evolved types. There is a tendency for molarization in the lower premolars of the *deningeri-spelaeus* group, and a trend from broadened to elongated occlusals in the upper premolars. The P_4 specimens range from the evolutionary stage of a single cuspid to the stage of four cuspids. The P^4 also varies from the simpler form to a more complex morphology. Concerning the morphotype, the Yarımburgaz premolars are distributed between Rabeder's (1983, 1999) primitive type A and the more complicated speleoid types C and D, with intermediate types present. By way of comparison, the prevailing morphotype among the Petralona (Greece) bears is A-A/C (Rabeder and Tsoukala 1990). Sexual dimorphism of P_4 has been described by Reffienna (1986), and dimorphism of the other cheek teeth is

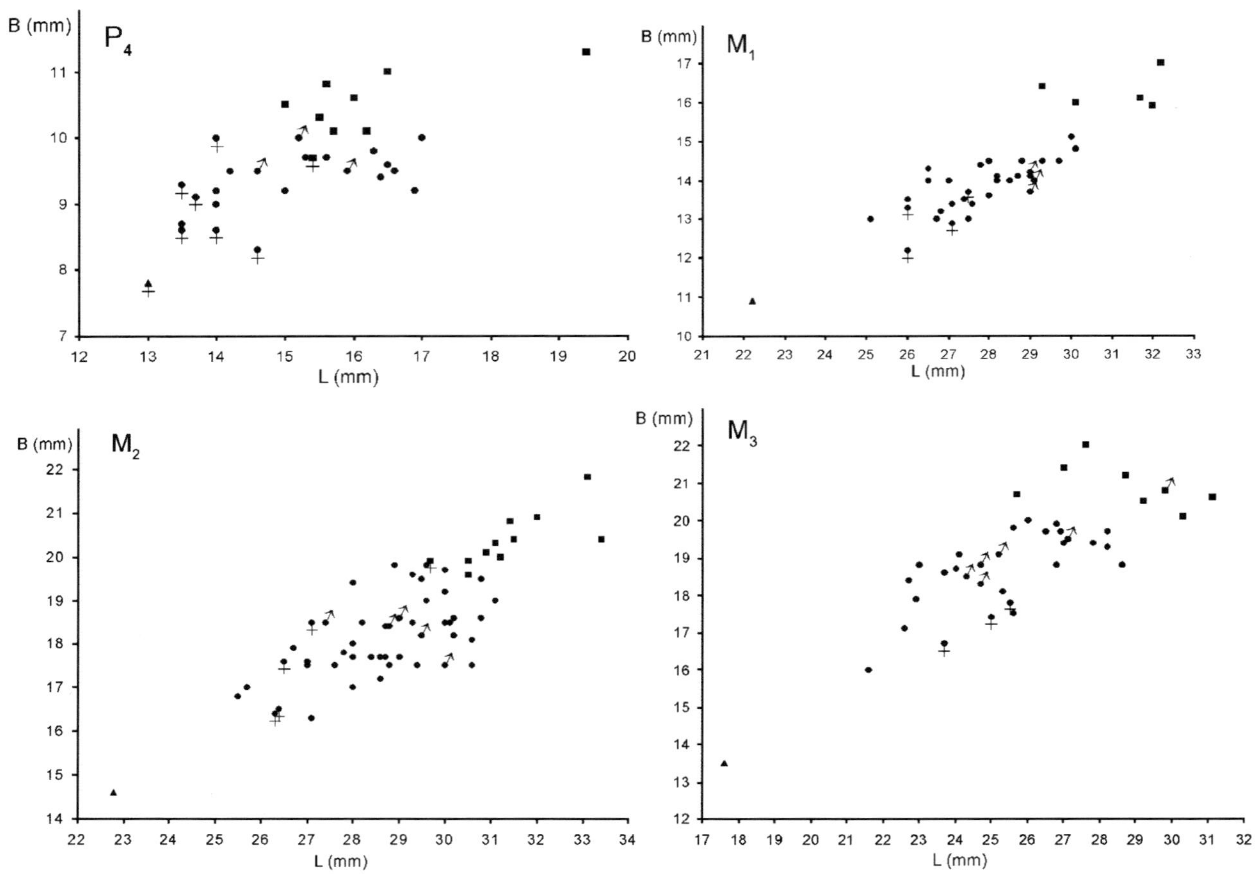

Figure 7.16. Yarımburgaz ursids: Scatter diagrams with the length (L) and breadth (B) of the lower cheek teeth with the indication of the sex of certain specimens belonging to mandibles with distinguished sexual dimorphism. Solid square: *Ursus spelaeus*, solid circle: *U. deningeri* and solid triangle *U. arctos.*

discussed by Baryshnikov and colleagues (Baryshnikov and David 2000; Baryshnikov et al. 2003). As for morphometric and morphogenetic consideration of the molars (Paunovic 1988), microevolution is expressed as increasing size with time, the presence of the cuspids that evolve from simple and primitive to more complex forms, or both of these. For the upper molars M^1 and M^2, two types of paracone-protocone profile (palatinal-labial section) are present, one with great development of the palatinal cingulum, resulting in the two cones being closer to each other, and another type with less development of the cingulum and where the protocone is closer to the palate. This morphological differentiation probably reflects differences in feeding adaptations. On the other hand, in the brown bear group, the presence of the fore single-rooted premolars, as well as the extremely small size and morphology of some teeth, in spite the long and gracile metapodials, testify to the presence of *U. arctos* in the Yarımburgaz sample.

Bears and other carnivores show sexual size dimorphism throughout their dentition and post-crania, although it is often difficult to distinguish between the sexes, not to mention between closely related species in time-averaged accumulations (Argenti and Mazza 2006). Sexual dimorphism in the skull of the cave bears is expressed at an early ontogenetic stage (Grandal d'Anglade and López-González 2005). Kurtén (1976) noted that, among permanent dental elements, the canines are the most different between males and females. The width of adult cave bear canines produced a markedly

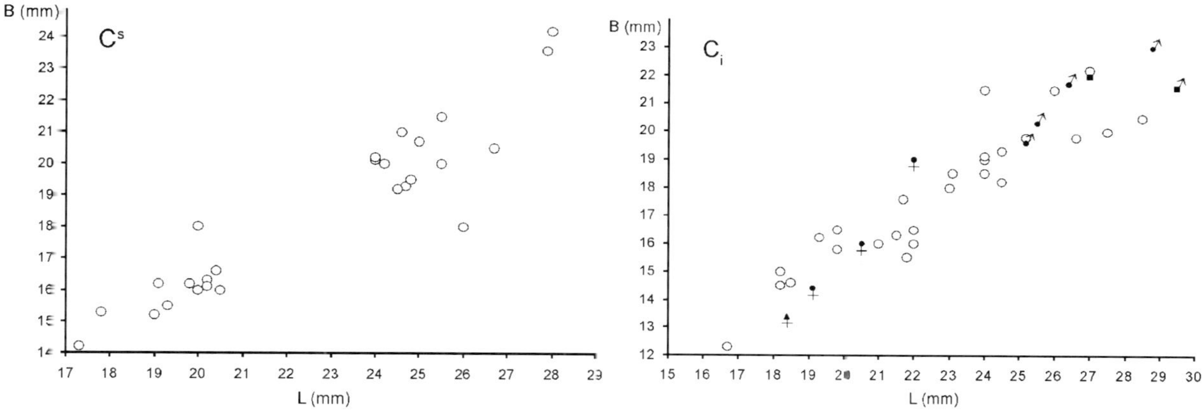

Figure 7.17. Yarımburgaz ursids: Scatter diagrams with the length (L) and breadth (B) of the canines (open circles). The distinction of the two groups (with special notes on the two extremes of a male *U. spelaeus)* is well shown in the left diagram. In the right diagram, especially for the lower canines, there is indication of the sex of certain specimens belonging to mandibles with distinguished sexual dimorphism. Solid square: *Ursus spelaeus*, solid circle: *U. deningeri* and solid triangle *U. arctos.*

bimodal distribution, suggesting male and female groups. This method generally presents a near-complete separation of the ranges of variation for each sex, and thus it is an accurate method for sexing intact skulls and jaws in a cave bear fossil population (Grandal d'Anglade 1993; Grandal d'Anglade and Lopez-Gonzalez 2005; Grandal d'Anglade and Romani 1997). Mandibles also exhibit great variation in morphological and metrical features due to sexual dimorphism. In addition to broader canines, the mandible of a male bears tends to possess a long diastema, a robust, hypsognathic and distally curved corpus, and a large condylus and symphysis. A sample of canines in mandibles in the Yarımburgaz collection is the basis for examining the sex distribution of cheek teeth specimens in Fig. 7.17. The single arctoid mandible fragment (P131) with a slender brachygnathic corpus probably represents a female individual.

In the post-cranial skeleton the great robustness of the humerus and radius, the long *collum* of the *caput femori*, the morphology of the olecranon, as indicated by the development of the proximal articulation and the intense torsion of the tibia (>45° for *U. spelaeus*, < 45° for *U. arctos* [Altuna 1973] and 49° for Yarımburgaz), demonstrate the presence of cave bears. However, in *U. deningeri hundsheimensis* the torsion of the tibia is 50° (Zapfe 1946), and for *U. spelaeus deningeroides* it is 48°-52° (Mottl 1964), so this feature cannot distinguish between the two taxa. The study of the carpal and tarsal morphology (based mainly on Caillat 1986; Chagneau 1985; Chagneau and Prat 1983; Prat and Thibault 1976; Quilès and Monchot 2004) reveals clear speleoid and arctoid characters in the Yarımburgaz assemblage, as well as many intermediate forms, showing thus great variability. The shortness and breadth of the metapodials and phalanges, and the relative shortness and heavier build of Ph3 elements are typical of speleoid bears (Kurtén 1968, 1976).

In comparison with the Pleistocene ursids of Western Europe (Bonifay and Bussiere 1989; Quilès 2003; Torres Perez-Hildago 1988), the *U. spelaeus* in Yarımburgaz is rather robust, though in the range of the typical *spelaeus* (the width of the lower carnassial is notable, Fig. 7.18, middle). The *U. deningeri* in Yarımburgaz is also robust (Fig. 7.18, upper), within the range of deningeroid or intermediate characters. The *U. arctos* from Yarımburgaz is clearly smaller than the *U. arctos* from Spain (Altuna 1972), showing great similarity in dimensions of the lower cheek teeth to *U. etruscus* from West Europe (Quilès 2003) (Fig. 7.18, lower). Also, there is a great

1. L **Ulna**	1	335.00	-	2	425.00	420.0-430.0	
2. DAP olec.	1	67.30	-	2	88.25	86.5-90.0	
3. DT olec.	1	33.00	-	1	48.00	-	
4. DT pr.ar.	2	53.65	50.0-57.3	3	67.90	62.8-72.6	
5. D int.inc.sem.	2	35.75	35.0-36.5	3	40.67	39.0-42.0	
6. D ext.inc.sem.	2	73.75	73.5-74.0	3	86.43	81.0-93.3	
7. DAP dia.min.	2	27.65	27.1-28.2	5	37.48	36.0-42.1	
8. DT dia.	2	21.30	20.5-22.1	4	28.80	27.4-30.0	
9. DAP dist.	3	44.33	42.2-47.3	5	55.20	53.5-57.0	
10. DT dist.	3	26.53	25.6-31.5	4	34.07	29.5-37.0	

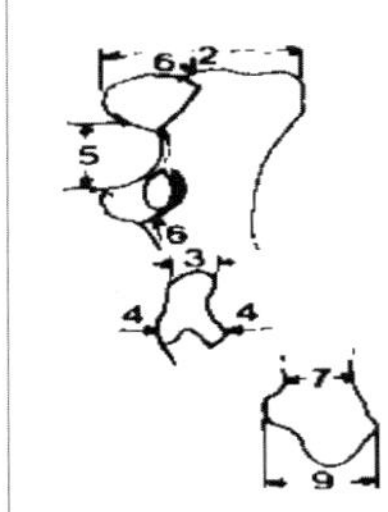

1. L **Tibia**	2	272.25	254.5-290.0	2	345.00	340.0-350.0
2. DT pr.	2	87.40	85.0-89.8	2	107.95	107.5-108.4
3. DAP pr.	2	65.40	64.8-66.0	1	77.50	-
4. DT dia.min.	3	31.30	26.2-37.5	1	35.50	-
5. DAP dia.	3	34.03	26.8-39.8	1	37.80	-
6. DT dist.	4	74.68	67.0-81.6	2	90.00	84.2-95.8
7. DAP dist.	4	42.40	38.5-46.0	1	51.70	-
8. DT d.art.	4	58.60	49.2-64.2	2	78.55	73.3-83.8
9. DAP d.art.	4	32.12	31.0-33.0	1	40.00	-
10. φ° torsion	2	38.50	37.0-40.0	1	49.00	-

1. L **Fibula**	2	234.75	231.5-238.0	1	(270.0)	
2. D prox.max.	2	27.75	26.5-29.0	-	-	
3. D dist. max.	2	31.70	31.2-32.2	1	41.0	

Appendix 7.7. Ursid carpal and tarsal bone measurements.

	U. deningeri					***U. spelaeus***				
Carpal bones	**n**	**x**	**min-max**	s_{n-1}	**v**	**n**	**x**	**min-max**	s_{n-1}	**v**
1. L **Pisiform**	8	41.65	38.5-44.0	1.82	4.38	4	52.25	46.5-56.5	4.19	8.02
2. DT art.	9	29.50	27.5-32.5	1.45	4.92	4	38.45	37.3-41.0	1.72	4.48
3. DAP art.	10	16.68	14.2-18.0	1.14	6.86	4	23.15	22.0-24.0	0.93	4.01
4. D tuber	9	24.95	23.0-29.5	1.88	7.54	5	32.76	29.5-38.0	3.35	10.24
5. L min.	9	37.78	34.3-42.0	2.44	6.46	5	47.60	43.5-52.5	3.21	6.74

1. L **Pyramidal**	5	19.50	16.3-22.4	2.52	12.95	2	28.15	28.0-28.3	* after Chagneau [16]	
2. DT	5	38.70	31.5-43.5	5.25	13.56	2	41.35	41.2-41.5		
3. DAP	5	35.94	28.0-45.5	6.36	17.71	2	46.00	46.0-46.0		
4. L max*	4	42.80	37.0-46.0	3.99	9.34	2	52.75	52.5-53.0		
5. B max*	4	40.55	35.0-44.0	4.03	9.94	2	47.55	47.1-48.0		

1. L **Scapholunatum**	8	29.74	23.5-33.3	4.04	13.59	6	37.68	36.2-43.0	2.62	6.96
2. DT	8	49.65	43.5-55.0	4.18	8.41	6	61.32	58.0-63.8	2.00	3.26
3. DAP	8	54.26	49.0-63.8	6.22	11.47	5	66.16	63.3-68.5	2.14	3.24
4. DT pr.art.	8	49.07	41.0-53.0	4.45	9.08	6	60.18	56.0-62.4	2.38	3.96
5. DAP pr.art.	8	33.08	28.8-39.0	3.77	11.39	6	41.08	39.2-42.5	1.51	3.67
6. DT d.art.	8	42.94	30.5-49.5	6.22	14.49	6	54.72	52.3-58.5	2.75	5.02
7. DAP d.art.	8	30.58	24.2-37.5	4.82	15.75	6	37.30	36.0-38.5	0.97	2.59
1. L **Unciform**	10	30.93	29.0-33.5	1.54	5.00	4	37.80	36.5-38.5	0.89	2.35
2. DT	10	29.06	26.5-34.5	2.31	7.96	4	39.92	33.2-43.5	4.53	11.96
3. DAP	10	27.35	25.0-29.9	1.62	5.93	4	33.12	28.0-37.0	3.84	11.59
4. Dt dorsal	8	26.01	24.0-28.6	1.67	6.44	3	31.37	30.1-32.5	1.20	3.84
1. L **Magnum**	7	30.04	24.4-33.8	3.47	11.57	3	32.70	32.2-33.2	0.50	1.52
2. DT	7	23.38	17.2-26.0	3.31	14.16	3	25.30	23.5-27.4	1.96	7.77
3. DAP	7	35.75	31.0-39.6	5.09	14.25	3	42.93	41.2-45.5	2.26	5.28
1. L **Trapezoid**	4	16.45	15.5-18.2	1.47	8.94	1	25.0			
2. DT	4	19.50	16.5-23.0	3.02	15.52	1	20.0			
3. DAP	4	27.82	25.6-30.0	2.24	8.07	1	34.0			
1. L **Trapezium**	2	17.95	17.7-18.2							
2. DT	2	16.55	16.1-17.0							
3. DAP	2	24.95	24.8-25.1							

Tarsal bones	**n**	**x**	**min-max**	s_{n-1}	**v**	**n**	**x**	**min-max**	s_{n-1}	**v**
1. L **Astragalus**	11	52.33	47.5-58.5	5.34	10.21	8	59.33	58.0-62.5	1.65	2.79
2. DT	11	57.30	52.0-66.5	5.83	10.18	8	67.62	57.5-76.5	5.80	8.58
3. DAP	12	32.85	28.5-38.5	3.43	10.43	8	39.05	36.2-41.2	1.91	4.90
4. DAP min	12	27.83	24.0-31.5	2.46	8.84	9	34.08	32.5-36.6	1.18	3.48
5. DT trochlea	11	43.56	38.0-49.8	4.55	10.45	10	50.29	45.0-58.2	3.76	7.47
6. DT collum*	12	30.47	22.5-35.5	3.94	12.94	8	36.87	32.5-40.4	2.74	7.44
7. DT caput	10	36.13	29.0-42.0	4.25	11.78	5	41.88	39.9-45.3	2.09	5.00
8. D oblique*	11	51.96	46.0-60.0	5.18	9.96	6	61.80	55.5-67.5	3.87	6.27
1. L **Calcaneus**	5	86.14	80.0-95.0	5.65	6.55	2	104.35	98.2-110.5		
2. DT	8	60.62	53.6-67.2	5.27	8.69	-	-	-		
3. DAP	7	45.80	43.5-52.5	3.11	6.80	1	48.50	-		
4. D tuber	5	36.02	34.0-40.8	2.77	7.69	2	46.90	46.8-47.0		
5. L manubrium*	5	42.46	35.2-47.1	5.65	13.30	2	52.00	48.5-55.5		
6. L sustent.tali*	7	32.35	29.3-36.0	2.51	7.76	2	42.30	41.5-43.1	* Plate 10f, h	
1. L **Scaphoid**	6	16.73	14.2-19.3	1.93	11.53	4	20.30	19.8-20.6	0.35	1.75
2. DT	6	35.73	30.0-38.5	3.37	9.43	4	41.87	40.5-43.0	1.10	2.64
3. DAP	6	38.38	33.0-42.0	3.73	9.73	4	44.65	43.0-47.5	2.11	4.73

	n	x	min-max	s_{n-1}	v	n	x	min-max	s_{n-1}	v
L **Cuboid**	15	26.38	22.3-28.3	1.99	7.56	3	31.00	30.0-32.5	1.32	4.26
2. DT	16	34.58	29.5-39.0	3.26	9.43	3	41.27	39.0-43.0	2.05	4.97
3. DAP	16	40.65	35.1-46.4	3.89	9.56	3	48.00	47.0-48.8	0.91	1.90
1. L **Cuneiform 1**	6	19.85	18.2-22.0	1.28	6.44	2	22.10	20.7-23.5		
2. DT	6	17.85	17.0-19.6	0.97	5.46	2	18.00	17.5-18.5		
3. DAP	6	32.73	31.5-33.8	0.94	2.89	2	35.00	35.0-35.0		
1. L **Cuneiform 2**	4	15.05	13.2-15.8	1.24	8.23	1	17.8			
2. DT	4	17.85	16.5-19.7	1.61	9.02	1	16.0			
3. DAP	4	28.50	26.0-31.5	2.48	8.71	1	31.0			
1. L **Cuneiform 3**	6	15.35	13.6-16.8	1.36	8.91	5	18.78	18.2-19.5	0.56	3.02
2. DT	6	20.25	16.2-23.8	2.99	14.77	5	23.60	21.0-24.6	1.49	6.33
3. DAP	6	28.83	25.0-32.5	3.04	10.54	5	34.58	32.5-37.0	1.68	4.87

Appendix 7.8. Ursid metacarpal (Mc) and metatarsal (Mt) measurements.

	U. deningeri					***U. spelaeus***					***U. cf. arctos***
	n	**x**	**min-max**	s_{n-1}	**v**	**n**	**x**	**min-max**	s_{n-1}	**v**	**n=1**
1. L **Mc1**	7	59.97	55.3-68.5	4.27	7.13	1	70.20	-			
2. DT prox.	10	23.49	20.8-26.5	1.95	8.33	2	28.95	27.4-30.5			
3. DAP prox.	10	20.88	18.5-25.5	2.51	12.06	2	25.25	24.5-26.0			
4. DT dia.min.	7	11.72	10.6-12.5	0.65	5.58	2	16.00	14.5-17.5			
5. DAP dia.	7	11.18	8.8-13.0	1.37	12.29	1	12.80	-			
6. DT dist.	7	18.00	16.4-19.0	0.96	5.35	1	20.60	-			
7. DAP dist.	7	15.41	13.7-16.0	0.86	5.62	1	18.30	-			
8. DT d.art.	7	16.61	15.5-17.6	0.78	4.74	1	19.50	-			
1. L **Mc2**	8	70.70	66.5-76.0	3.31	4.68	6	85.10	81.0-89.0	2.95	3.46	
2. DT prox.	8	18.17	17.5-19.5	0.73	4.02	7	22.43	19.5-24.7	1.94	8.65	
3. DAP prox.	8	26.43	23.0-28.5	1.91	7.23	7	32.68	29.4-36.0	2.12	6.51	
4. DT dia.min.	8	15.02	13.9-16.2	0.92	6.17	7	17.87	17.2-20.2	1.06	5.97	
5. DAP dia.	8	12.53	12.1-13.0	0.39	3.18	7	16.43	16.0-17.0	0.37	2.29	
6. DT dist.	8	22.38	20.8-23.5	0.78	3.52	6	27.27	25.5-29.0	1.22	4.47	
7. DAP dist.	8	18.28	17.5-19.2	0.58	3.20	6	22.65	21.0-24.0	1.16	5.13	
8. DT d.art.	8	18.76	17.5-20.3	0.97	5.20	6	23.88	22.5-26.8	1.53	6.40	
1. L **Mc3**	12	74.49	69.5-81.7	4.62	6.20	8	86.87	83.5-92.0	3.18	3.66	
2. DT prox.	18	18.31	15.9-20.7	1.71	9.35	9	21.64	19.8-26.5	2.10	9.73	
3. DAP prox.	17	28.34	26.0-33.0	3.06	10.81	9	34.50	32.6-35.8	1.28	3.73	
4. DT dia.min.	14	14.35	13.0-16.7	1.13	7.87	8	17.57	16.2-19.0	0.94	5.34	
5. DAP dia.	14	12.50	10.5-15.0	1.41	11.32	8	15.92	14.7-18.5	1.27	7.98	
6. DT dist.	12	22.38	19.8-25.5	1.83	8.17	8	26.75	24.5-29.5	1.45	5.45	
7. DAP dist.	11	19.88	18.0-23.5	1.88	9.47	8	23.45	21.8-25.0	1.19	5.11	
8. DT d.art.	12	19.68	18.0-22.7	1.62	8.23	8	22.78	21.6-24.0	0.86	3.79	

1. L Mc4	13	74.46	71.3-79.5	3.02	4.06	8	90.37	84.0-95.0	3.60	3.98	
2. DT prox.	15	19.86	17.0-22.0	1.38	6.96	10	25.41	23.0-28.0	1.56	6.15	
3. DAP prox.	15	27.24	24.0-29.2	1.44	5.28	8	34.41	33.0-35.9	1.05	3.05	
4. DT dia.min.	15	15.25	14.0-18.0	1.07	7.05	9	19.46	17.8-21.0	1.03	5.28	
5. DAP dia.	15	12.86	11.5-15.5	1.01	7.91	8	17.50	16.0-19.5	1.16	6.65	
6. DT dist.	12	23.21	21.0-24.7	1.27	5.49	8	29.16	26.5-31.0	1.66	5.69	
7. DAP dist.	13	19.63	18.6-20.8	0.71	3.63	10	24.61	23.2-26.7	0.97	3.94	
8. DT d.art.	12	19.81	18.0-21.0	1.05	5.34	10	24.54	21.5-26.0	1.31	5.34	
1. L Mc5	5	75.90	74.5-77.0	1.14	1.50	12	90.27	80.0-97.0	4.55	5.03	72.5
2. DT prox.	6	26.46	24.8-28.2	1.13	4.26	14	31.98	25.5-35.0	2.63	8.23	20.2
3. DAP prox.	6	28.50	24.3-32.0	2.59	9.10	13	33.73	28.0-40.0	3.27	9.70	21.5
4. DT dia.min.	6	15.31	14.5-16.1	0.53	3.44	13	19.54	16.0-22.0	1.70	8.72	11.8
5. DAP dia.	6	12.80	12.0-14.2	0.70	5.50	12	18.35	14.2-20.6	1.77	9.65	9.3
6. DT dist.	5	24.26	22.5-25.5	1.60	6.63	12	29.21	25.5-32.5	2.06	7.05	18.3
7. DAP dist.	5	19.70	17.9-23.8	2.34	11.90	9	22.51	21.0-24.0	1.09	4.84	14.5
8. DT d.art.	5	23.94	21.5-29.0	3.04	12.70	12	27.81	24.5-31.0	2.51	9.03	16.7

1. L Mt1	14	57.75	42.5-65.2	5.89	10.20	6	65.80	62.5-67.8	2.08	3.16
2. DT prox.	13	22.88	20.5-25.5	1.49	6.50	6	27.93	26.3-29.5	1.19	4.28
3. DAP prox.	14	21.05	18.0-24.2	1.89	8.98	6	23.13	21.5-24.5	1.22	5.29
4. DT dia.min.	14	11.64	9.8-13.0	1.17	10.09	7	13.80	13.0-14.7	0.73	5.30
5. DAP dia.	14	11.55	9.8-13.5	0.95	8.24	5	12.76	11.7-14.0	1.08	8.52
6. DT dist.	14	17.26	15.3-19.6	1.24	7.21	6	20.31	19.5-21.1	0.65	3.22
7. DAP dist.	12	14.85	12.1-17.3	1.52	10.27	6	17.66	15.0-19.6	1.64	9.32
8. DT d.art.	13	15.99	13.0-17.7	1.34	8.39	6	18.53	17.7-19.2	0.63	3.40
1. L Mt2	9	64.73	60.5-68.3	2.86	4.42	6	73.85	69.3-80.5	4.38	5.94
2. DT prox.	10	14.32	11.7-16.0	1.10	7.69	6	17.00	15.8-19.3	1.21	7.12
3. DAP prox.	9	23.55	19.0-25.6	2.06	8.74	6	28.23	24.0-32.5	2.79	9.88
4. DT dia.min.	9	12.23	10.9-13.0	0.75	6.21	6	14.70	13.5-16.0	0.83	5.70
5. DAP dia.	9	10.01	7.5-10.6	0.95	9.57	6	12.53	11.6-13.6	0.73	5.85
6. DT dist.	9	18.91	17.3-20.4	0.84	4.44	6	22.51	21.5-24.7	1.18	5.26
7. DAP dist.	9	15.31	14.0-16.0	0.62	4.07	6	18.11	17.3-19.0	0.70	3.87
8. DT d.art.	9	17.38	16.0-18.3	0.87	5.02	6	20.81	19.5-22.8	1.21	5.82
1. L Mt3	15	70.58	67.5-80.0	4.56	6.19	8	84.92	75.4-89.0	4.21	4.96
2. DT prox.	16	17.74	14.6-20.3	1.53	8.62	12	21.42	19.5-25.5	1.69	7.88
3. DAP prox.	16	27.86	24.8-31.3	2.34	8.41	10	33.44	29.0-37.5	2.42	7.25
4. DT dia.min.	17	14.41	12.0-17.0	1.40	9.78	10	17.02	15.5-19.0	0.95	5.62
5. DAP dia.	17	11.93	10.3-13.9	1.27	10.6	10	14.98	13.5-17.3	1.29	8.66
6. DT dist.	16	20.87	19.0-23.5	1.69	8.10	8	24.81	23.2-26.8	1.31	5.29
7. DAP dist.	16	17.08	15.8-18.8	1.19	6.98	8	19.88	18.3-21.0	0.96	4.84
8. DT d.art.	15	18.10	16.5-20.8	1.36	7.49	8	21.15	19.4-23.0	1.05	4.99

1. L **Mt4**	10	80.19	74.7-85.5	4.28	5.34	11	91.98	87.5-96.0	2.84	3.09
2. DT prox .	12	19.80	18.0-22.0	1.24	6.27	13	23.27	20.0-27.5	1.87	8.03
3. DAP prox.	10	26.46	23.5-28.0	1.36	5.15	13	31.06	29.4-34.0	1.60	5.17
4. DT dia.min.	12	14.90	13.5-16.5	0.93	6.28	13	17.67	16.5-18.8	0.74	4.19
5. DAP dia.	12	12.67	11.8-13.5	0.49	3.93	13	15.00	13.5-16.0	0.79	5.29
6. DT dist.	9	22.48	20.0-24.0	1.46	6.51	11	25.95	24.0-28.0	1.18	4.56
7. DAP dist.	8	17.38	12.2-18.7	2.17	12.49	10	20.30	18.5-22.5	1.23	6.07
8. DT d.art.	9	19.59	17.4-20.6	1.07	5.45	11	22.39	20.2-24.6	1.09	4.88
1. L **Mt5**	9	85.88	76.0-94.8	6.84	7.97	1	96.50	-		
2. DT prox.	10	26.48	22.5-29.5	2.79	10.53	2	30.50	30.0-31.0		
3. DAP prox.	11	26.52	22.5-31.0	3.03	11.43	2	28.00	25.5-30.5		
4. DT dia.min.	10	13.53	12.0-15.2	1.19	8.80	1	16.00	-		
5. DAP dia.	10	13.86	10.5-16.0	2.01	14.51	1	16.30	-		
6. DT dist.	8	21.93	19.5-23.5	1.56	7.14	1	25.50	-		
7. DAP dist.	8	16.90	15.0-18.5	1.42	8.40	1	19.50	-		
8. DT d.art.	8	21.02	19.0-23.4	1.72	8.20	1	24.50	-		

Appendix 7.9. Ursid phalanx measurements.

Appendix 7.9.		***U. deningeri***				***U. spelaeus***				
Phalanges	**n**	**x**	**min-max**	**s_{n-1}**	**v**	**n**	**x**	**min-max**	**s_{n-1}**	**v**
1. L **Ph1**	103	40.51	30.5-48.0	3.96	9.79	42	49.05	43.2-53.0	2.91	5.98
2. DT prox.	102	21.53	15.3-26.5	2.49	11.60	42	26.70	22.2-32.2	1.82	6.83
3. DAP prox.	93	16.01	12.0-20.0	1.67	10.46	41	20.10	17.4-23.5	1.43	7.12
4. DT dia.min.	95	14.24	10.3-18.2	1.79	12.57	46	18.00	15.0-21.0	1.54	8.57
5. DAP dia.	94	10.24	8.0-13.2	1.22	11.89	46	12.71	11.0-15.0	0.99	7.79
6. DT dist.	91	16.21	12.5-20.0	1.76	10.89	38	20.04	17.0-24.5	1.48	7.41
7. DAP dist.	91	11.06	8.3-14.0	1.19	10.78	38	13.83	11.0-16.0	1.10	7.98
1. L **Ph2**	51	27.19	20.5-33.5	3.40	12.50	38	32.18	24.8-38.2	3.13	9.72
2. DT prox.	43	18.12	13.5-21.5	1.74	9.61	38	21.76	19.2-24.8	1.25	5.76
3. DAP prox.	39	14.26	11.2-17.0	1.30	9.11	38	17.54	14.8-21.0	1.48	8.49
4. DT dia.min.	40	13.50	10.0-15.5	1.21	8.97	38	16.23	13.1-19.0	1.34	8.28
5. DAP dia.	40	8.21	6.5-10.5	0.89	10.93	38	10.27	8.0-12.5	0.99	9.70
6. DT dist.	41	15.12	12.5-16.8	1.12	7.46	37	17.97	15.2-21.7	1.32	7.36
7. DAP dist.	41	11.68	9.8-14.6	1.08	9.25	37	14.82	12.0-17.3	1.35	9.12
1. L **Ph3**	44	37.26	30.0-50.0	4.34	11.66	32	41.06	33.0-50.0	4.98	12.13
2. H	48	23.59	18.4-26.8	2.05	8.70	25	28.24	21.5-33.5	2.94	10.42
3. DT max	49	14.54	11.3-18.0	1.34	9.23	34	17.71	14.4-19.3	1.06	5.98
4. DAP art.	47	12.83	10.5-16.0	1.11	8.71	38	14.90	12.5-16.5	1.16	7.84
5. DT art.	48	13.55	10.9-15.5	1.18	8.72	32	16.04	14.3-17.8	0.93	5.80

Appendix 7.10. Ursid sesamoid measurements.

Appendix 7.10.	Ursidae				
Ossa sesamoidea	**n**	**x**	**min-max**	s_{n-1}	**v**
1. L	23	20.38	15.8-27.0	2.24	10.72
2. DT	23	9.70	7.9-12.8	1.41	14.50
3. L art.	23	16.33	13.0-21.0	1.35	11.37
4. Dt art	23	8 96	7.5-11.7	1.36	15.16

CHAPTER 8

THE LARGER MAMMAL FAUNA (EXCLUSIVE OF URSIDAE) FROM YARIMBURGAZ CAVE

F. CLARK HOWELL

PREFACE

In May-June 1987, the author visited the University of Istanbul in order to examine faunal and artifact collections obtained from test excavations at Yarımburgaz cave the previous year. The visit was at the invitation of Prof. Güven Arsebük and his colleague Prof. Mehmet Özdoğan (Prehistory Section, Faculty of Letters, Istanbul University), the latter having been responsible for those excavations. In the course of several weeks it was possible to inventory the faunal collections in a preliminary but satisfactory fashion and to assign all specimens to the genus level, and most to the species level. The insights obtained from this endeavor led to the excavations over the course of three successive field seasons (1988-1990). That fieldwork focused almost exclusively on the infill of the Lower Chamber (LC), and the extensive excavations yielded much larger artifact and faunal assemblages with controlled recovery, recording and mapping procedures. The success of these efforts was surely due to the exceptional supervisory efforts of Mihriban Özbaşaran (now Professor of Prehistory at the University of Istanbul) in association with the work of a skilled group of devoted students. The inventory of all residues, both faunal and artifactual, and their stratigraphic context and coordinates were computerized by Oğuz Tanındı, thus enabling both control and ready recovery of data. All of the materials were inventoried by me over the course of the excavations, coupled with subsequent comparative studies elsewhere. This contribution presents a paleontological overview of all of the larger-mammal remains except Ursidae that were recovered from Yarımburgaz. Simple identification by colleagues in France of a few specimens of rhinoceros and lower vertebrates (fish, reptiles, birds) are also provided: the rhinoceros, fish and reptile (chelonian) remains were identified by Jean Desse (C.R.A., Sophia Antipolis, Valbonne), and the three bird specimens were identified by C. Mourer-Chauviré (Centre des Sciences de la Terre, Université Claude Bernard-Lyon I).

DESCRIPTIONS

PISCES

S-89, 4, 210. pharyngeal. *Cyprinus* sp. indet. (large)

This single specimen of a carp is the only occurrence of fish in the Yarımburgaz infill.

REPTILIA

Testudinae (Chelonia)

R-90, 1, 1345. plastron fragment. *Testudo* sp.

R-90, 1, 1346. plastron fragment. *Testudo* sp.

These two fragments of plastron of a tortoise are the only reptilian remains recognized in the Yarımburgaz infill.

AVES

Galliformes

Tetraonidae

Z-88, 12, x-1. left distal tarsometatarsus. *Tetrao urogallus.*

U-89, a-c/1-3, 473. left distal carpometatarsus. *Tetrao urogallus.*

U-89, b-d/2-3, x-2. right proximal humerus. Tetrao *urogallus.*

These few pieces of postcrania, all representing capercaillie, derive from two squares in which there are stratigraphic problems, either due to some later pit infilling (Z-88) or a large fallen roof-block (U-89). At least the latter square demonstrably samples the mid-Pleistocene infill. Specimen U-89, no. x-2 is of a size of extant *T. urogallus*, rather than more robust *T. praeurogallus* (D. Jánossy) of the mid-Pleistocene of central Europe. The extant species is a forest (deciduous and coniferous) resident, preferring dense and shady circumstances, including upland regions. Its presence at this elevation in Marmara may be an indication of cooler paleoclimatic conditions.

CARNIVORA

Canidae

(Figures 8.1-8.2)

Vulpes sp.

A.71 (86), 5, 343. proximal mt-III.

A.71 ('86), sieving ('88), level indeterminate:

- a = ph-1
- b = right proximal mt-IV, with diaphysis (larger species)
- c = left proximal mt-IV, with diaphysis
- d = complete mc-II (larger species)

Y-88, 1, 683. complete right M^1.

Y-88, 1, 688. left proximal mt-IV (larger species).

S89, 1, 12. left/C, crown only, incomplete (*V.* aff. *vulpes*).

S-89, 4, 95. ph-2 and 3, cf. pes. (larger species, cf. No. 101).

S-89, 4, 101. right femur, lacking proximal articulation. (larger species).

T-89, 2, 444, right C/ crown, germ. (small, ? *Alopex lagopus*)

T-89, 4, 207. left /C (?), with root (? *Vulpes* sp.).

T-89, 4, 587. left /C with root. (*V.* sp. indet.).

U-89, 1, 9. left scapula fragment, with glenoid and spine (smaller species).

U-89, 1 19. small innominate fragment, with sacral auricular surface. (?*Vulpes*).

U-89, 1, 28. right mandible body with P1 and P2; alveoli of P3 and P4; and M1 and M2. (*Vulpes* ? cf., *praeglacialis*).

U-89,1,85. right distal humerus (smaller species)

U-89, 1, c-1. left P_1 (?*Vulpes*)

U-89, 1, c-2. left P_1, root open and incomplete. (?*Vulpes*).

U-89, 1, c-3. cf. P_1, root open. (?*Vulpes*).

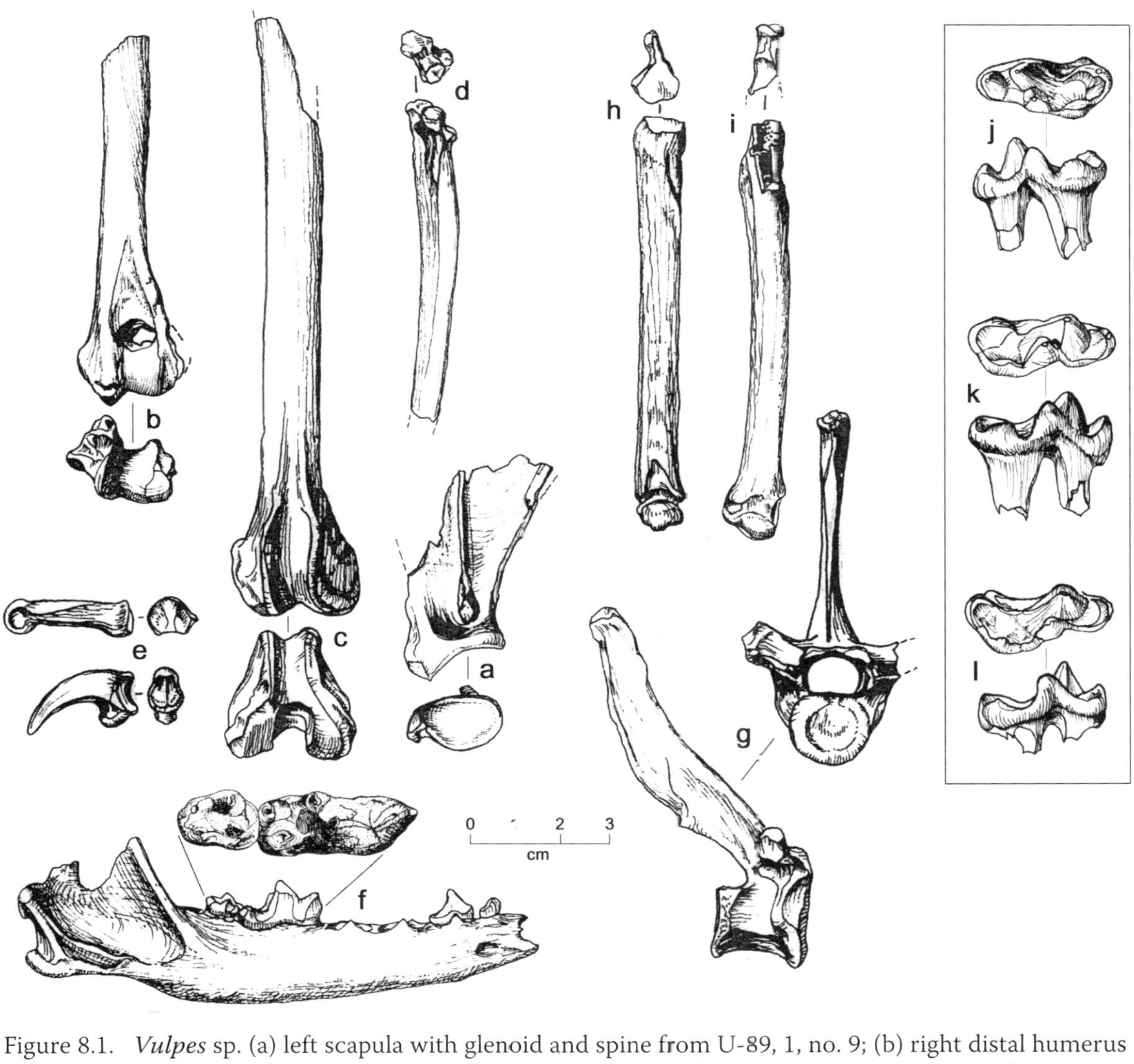

Figure 8.1. *Vulpes* sp. (a) left scapula with glenoid and spine from U-89, 1, no. 9; (b) right distal humerus from U-89, 1 no. 85; (c) right distal femur lacking proximal articulation from S-89, 4 no. 101; (d) proximal right IV metatarsal from A.71 (V-88); (e) phalanges 2 and 3, cf. pes from S-89, 4 no. 95; (f) right mandible with P_{1-2}, M_{1-2} (also enlarged) from U-89, 1 no. 28; *Canis lupus mosbachensis* (g) upper dorsal vertebra from T-89, 4, no. 643; (h) right III metatarsal from T-89, 4, no. 865; (i) left II metatarsal with damaged proximal epiphysis from cleaning of Y-88; *Canis aureus/lupaster* (j) right dP_4; (k) left dP_4; (l) left dP_4, all from U-89, 1, sieving.

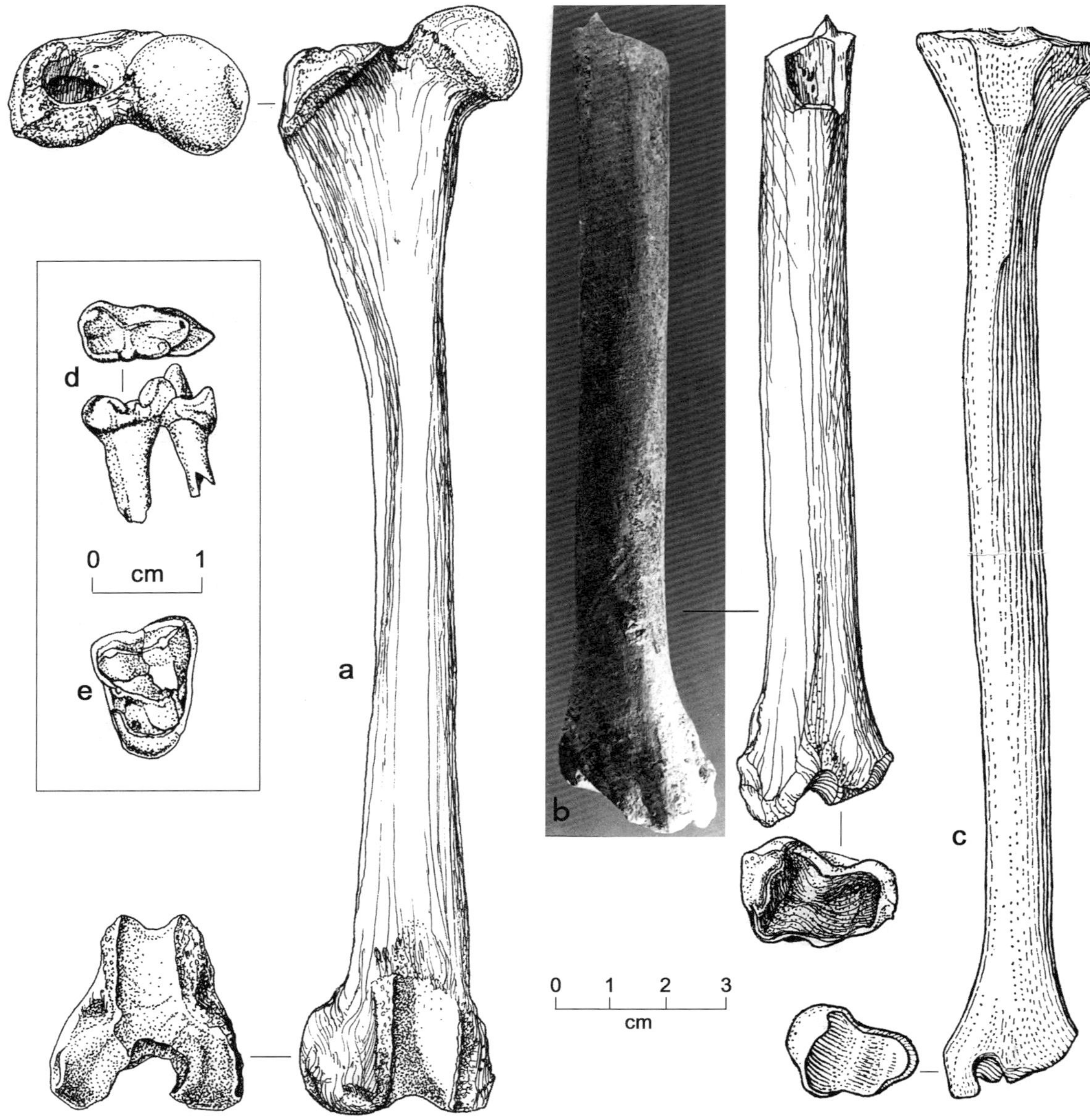

Figure 8.2. Some canid skeletal specimens from Yarımburgaz Cave: (a) right femur of *Canis lupus mosbachensis* from square A.71, ant-prox-dist views; (b) left distal tibia of *Cuon/Xenocyon* sp. from square A.40 level 14, ant-post-dist views; (c) left tibia of cf. *Cuon* sp. from Escale Cave (France), ant-dist views (after M.-F. Bonifay 1971); (d) detail of left dP_4 with roots of *Canis aureus/lupaster* from Y-88 level 1, crown-lat views; (e) detail of right M^1 with roots of *Vulpes vulpes* ssp. from Y-88 level 1, crown view.

A total of 20 specimens represent or are referred to smaller, vulpine-sized Canidae. There are both larger and smaller examples in this sample, suggestive of the presence of two fox species. These remains derive from five excavation squares, and from multiple levels in squares S-89 and T-89.

A complete right M^1 (Y-88, 1, 683) agrees broadly in size and overall occlusal morphology with *Vulpes vulpes* (Appendix 8.1). However, the crown is more slender and lacks the strong posterior incurvature. The mc^{1e} is only moderately developed, is not linked with the post-hypoconal crest (unlike *Alopex*), and is well separated on the post-protocrista from the pr (unlike *V. vulpes).* The pr^{le} is at best incipient, and distinct from the pr. The hy is substantial and bears a wear facet. The mesio-cingulum is strong and contiguous with the pre-hypoconal crest. These are all features reminiscent of *V. praeglacialis*, which is, however, of smaller size.

A right mandible (U-89, 1, 28) preserves the anterior premolars (P_1, P_2) and two molars (M_1 and M_2) and most of the ramus; there is a small alveolus for M_3. This is the size of *Vulpes vulpes*, but there are differences in the carnassial (M_1) and M_2. The principal cusps of M_1 are moderately worn. The anterior (pa^d) was originally high and well set off from the adjacent pr^d, the me^d was salient, robust, posteriorly directed, and well set off from the pr^d, the talonid basin is spacious and the hy^d distinct, enlarged and swollen labially, and the en^d also distinct, large and offset disto-lingually. M_2 is swollen mesio-labially (rather than lingually as in *Vulpes vulpes*), has a weak, but distinct and doubled pa^d, strong but separated pr^d and me^d, well-developed hy^d, a partial pro en^d adjacent to it, and enlarged disto-lingual margin with distinct en^d, the latter linked with a thick crest with hy^d. Some of these features are reminiscent of *Vulpes praeglacialis*, of which specimens from Eastern Europe may be of comparable size, including some from Petralona and Deutsch-Altenburg (Austria). This species was proposed by Kormos (1932) and attributed to a new species of arctic fox (*Alopex*), distinctive in some features of the mandibular dentition. The new taxon was based on remains from Villány-Kalkberg (4 mandible fragments, isolated teeth), Beremend (2 C's), Püspökfürdö (mandible fragment, isolated teeth), and Brassó (mandible fragment), all localities in Hungary or (subsequently) Slovakia. Other localities yielding the species include, from west to east, Venta Micena and Cullar-de Baza (Spain), Vallonet, L'Escale and Vergranne (France), Petralona (Greece), Hundsheim (probably) and Deutsch-Altenburg (Austria), Przelitice and Stránská skála (Czech Republic), and Gombaszög (Hungary). Although consistently smaller than red fox (*V. vulpes*), it differs in a number of respects from arctic fox (*A. lagopus*), and thus it is now generally considered as an extinct species of *Vulpes* (Bonifay 1971). Appendix 8.2 provides measurements of the lower dentition.

A femur, lacking the proximal end (S-89, 4, 101), is of substantial size, and thus more like *V. vulpes* than *V. praeglacialis*. It approximates in size *V. v. jansoni* from Escale. The distal humerus (No. 85) from U-89, level 1 is within the range of *V. praeglacialis* in its distal articulation dimensions, as well as some *V. vulpes* ssp., and its shaft robusticity is elevated by comparison with the Escale sample of the former subspecies. Appendix 8.3 affords measurements of the known postcranials.

Some postcranials (metapodials) and some isolated, canine teeth suggest a still smaller canid species as yet unidentified. It might be *V. praecorsac* or *V. corsac*, or even *Nyctereutes* sp.

Canis sp.

A-71 (= V-88), sieving, 1. left femur, complete.

Y-88, 5, 684. right dp_4, with roots.

Y-88, ?, 685. left mt-II, complete (damaged proximal epiphysis)

S-89, 4, 73. ph-1, damaged distally.

S-89, 4,103. left pisiform.

T-89, 4, 643. upper dorsal vertebra, largely complete (cf V^D3 or 4)

T-89, 4, 865. right mt-III, complete.

U-89, 1, 94. left proximal mc-III.

U-89, -, 485. distal fibula, damaged.

R-90, 1, 101. distal mc (cf. rt. mc-V).

R-90, 1, 441. proximal, right mt-I.

P-090, 1, 232, right mc-IV, complete.

P-90, 2,473. ph-1, distal only.

These dozen specimens, largely postcranial and particularly the metapodials, appear to represent a wolf-like form of *Canis*. However, the general absence of cranial or gnathic remains, particularly dentition, scarcely enables precise taxonomic assignment. Differences in size among the specimens do suggest that more than one taxon is represented in the collection, hereafter termed "species A", "B", and so on.

Appendixes 8.4 through 8.12 provide measurements of the most complete specimens, and measurements of other comparative samples. The largest, and most complete sample of associated cranial, dental and postcranial parts of the small species *C. lupus mosbachensis* is that recorded from L'Escale cave (St. Estève-Janson, Bouches-du-Rhone) (Bonifay 1971).

The single element of the dentition is a left dp_4 (Y-88, 5, 684) with roots (open). Morphologically it is unquestionably canid. The pr^d has a distinctive distal crest. The me^d is very strong and is set disto-lingually of pr^d. The pa^d is moderately high and well set off anterior of pr^d, with a distinct V-notch separating them. The talonid is elevated and bears distinct hy^d and en^d, disposed obliquely, the former more distal and the larger of the two. The distal border of the crown passes obliquely lingually. A tiny crestlet tends to join these talonid cusps, bifurcated by the talonid basin. There is a very low, but distinct and obliquely directed distal marginal ridge. Between the me^d and en^d is a tiny expression of me^{le} (pro-en^d of Thenius 1954). This specimen is distinctly lupine-like but small; its dimensions (l = 11.36 x talonid br = 4.78) are less than a *C. lupus* dp_4 from Taubach (l = 12.0 x talonid br. = 6.1) (Hemmer 1977) or an even larger example from Gerde (Clot 1980). In size it closely matches a dp_4 from Geula cave (Mt. Carmel) attributed by Petter and Heintz (1970) to *Canis* cf. *lupaster*, the wolf jackal, generally now considered a subspecies of *C. aureus*, the golden jackal. *C. aureus* is distributed now in the southern Levant (Sinai) and Egypt, but its remains are recognized in Levantine caves in Upper Pleistocene contexts by Kurtén (1965) and D.M.A. Bate (1937). The only notable distinction is the rather lower pa^d and its displacement from the pr^d in the Y-88, 5, 684 specimen. This specimen might presumably match *C. l. mosbachensis*, but about which there is lack of published information on the deciduous dentition. Thus, it must be considered Canidae indet cf. *Canis* aff. *lupus/aureus*.

A complete left femur (A.71 [=V-88] -1) with only slight damage to the greater trochanter, was recovered from screening (in 1988) of backdirt tips from the A.71 excavations from 1986. Thus, although of uncertain stratigraphic provenience, it must originally have derived from the deeper infill of that excavation square. The specimen, excellently preserved, is demonstrably canid, and closely resembles wolf, but is of small dimensions. Comparison with some values for *C. lupus mosbachensis* from three localities, two western (Escale; Gr. de l'Eglise) and one central European (Stránská skála), indicates a closer approximation in size to the latter (Stránská skála) than to the former on account of its small size. Thus, one is prompted to suggest attribution to *C. lupus* ex. gr. *mosbachensis.*

There are six metapodials, three complete and three comprising only articular ends. On comparison these fall into two size groups, which enables their exclusion from a single taxon, but does not necessarily enable their appropriate attribution.

Three are metacarpals.

- mc-III (proximal). This is smaller than *C.l. lupus*. It is close in size to *C. l. mosbachensis* from Arago, but smaller than that from Stránská skála, and larger than Escale and Hundsheim homologues. Species A (small).
- mc-IV (complete). This is longer and more robust than Escale and Stránská skála samples of *C. l. mosbachensis.* It is definitely smaller (diaphyseal diameter) than *Xenocyon* (Stránská skála). It most closely approximates in size *C. l. lupus* (for example, the Lazaret sample). Species B (larger).
- mcV (distal). This fragment is larger than samples (3) of *C. l. mosbachensis.* It approximates the size of *C. l. lupus* (Combe Grenal example). Species B. (larger).

Another three are metatarsals.

- mt-II (complete). This example is shorter and less robust than *C. l. lupus.* It is rather longer than samples (2) of *C. l. mosbachensis,* but approximates the largest record at Stránská skála, and the transverse diameters are comparable. Species A (small).
- mt-III (complete). This is smaller than *C. l. lupus,* and definitely smaller (shorter, transverse diameters) than a homologue of *Cuon priscus* (Hundsheim). It is shorter than a small sample of *C. l. mosbachensis* (Escale) and diameters are similar to, or just greater than, very limited data on two such samples. Species (?) A (small).
- mt-IV (proximal). This fragment is larger than several samples of *C. lupus* subspecies. It most closely approximates homologues of *Cuon priscus* or even *Xenocyon* (*e.g.*, Stránská skála). Species C (largest).

There is a single carpal, a pisiform (left). This example is distinctively canid in morphology. It is definitely larger (l = 20.2) than *C. aureus* (golden jackal, l = 12.8) and several examples of *C. l. mosbachensis* (l = 15.7-16.7, No. 3). However, it is smaller than an example of *Cuon priscus* (l = 24.5) from Escale. Its attribution is, of course, uncertain. Species (?) C (largest).

There are two first (Ph.1) phalanges attributable to *Canis* sp., one proximal (with diaphysis) and the other only the distal articulation. Neither of these is large and in fact they approximate in size *C. l. mosbachensis* rather than *C. l. lupus.* Although impossible to attribute to species, they demonstrably constitute a small species, whatever it may prove to be.

These few, largely postcranial and sometimes incomplete, specimens scarcely enable sound assessment of taxonomic affinities, beyond some exclusionary statements. This is unfortunate as the status of knowledge of differentiation and phylogeny among Canidae in west Asia and environs still leaves much to be desired. The situation as now known is briefly summarized below, according to conventionally distinguished species taxa, employing Kurtén's (1965) examination of Palestinian (Levant) carnivores as a point of departure.

The *Canis* taxa of concern are *Canis lupus* and *Canis aureus* (golden jackal). The former is generally considered to comprise the subspecies *C. l. pallipes,* a larger form and *C. l. arabs,* a smaller form. The former *(pallipes)* is broadly distributed from India westward into western Asia, including Anatolia, and southward into the Levant. There are size differences in the Levant, between more mesic and more xeric distributed populations, the former more coastal (≥ 400 mm. rainfall) and the latter more inland and southern (<400 mm), following Mendelssohn (1982). *C. l. arabs* is central/southern Arabian and Sinaian (Arava valley) in distribution. The extent to which these purported entities constitute a Bergmannian cline, disrupted by human and other agencies, comes to mind (Tchernov and Valla 1997) but has not yet been intensively investigated.

Canis aureus is often considered to constitute two principal subspecies within this broad geographic area. These are (a) the nominate subspecies, *C. a. aureus*, a normal small form; and, (b) a substantially larger form, *C. a. lupaster*, occurring in the lower Nile drainage and in Libya, apparently now extending into the Sinai peninsula. It has been claimed and formally proposed (Ferguson 1981) that the latter is indeed not a jackal, and hence not a subspecies of *C. aureus*, but a northeast African small subspecies of wolf, hence *C. lupus lupaster*, and which intergrades in Sinai with *C. l. arabs*, (first recognized by Pocock in 1935, as a distinct subspecies). There is a complex history to this problem, with various species attributions (to *aureus* or to *lupus*), and even acceptance of full species status as *C. lupaster* (large wolf-jackal) by Bate (1937) and subsequently by Kurtén (1965). However, in North Africa there is a definite west-to-east cline of size increase (skeletally, dentally), with purported (nominate) subspecies *C. a. moroccanus* (Morocco) and *C. a. algirensis* (Morocco, Tunisia, Algiers) grading into *lupaster* eastwards. The argument that such northeast African populations are lupine rather than jackal is stated strongly and seemingly convincingly by Ferguson (1981). However, my own experience is that there are definite, discrete morphological (and morphometrical) distinctions between wolf and golden jackal, and these still must be pursued in appropriate west Asian and northeast African populations if the issue of specific affinity is ever to be adequately resolved. In the literature known to me there is no fully convincing evidence of this resolution.

Measurements of upper (P^4) and lower (M_1) carnassial teeth in various *lupus* and *aureus* samples, extant and fossil, bear on this issue (Appendixes 8.13 and 8.14). The size of the P^4, the length of P^4 to $M^1 + M^2$ (combined) and the shape (and cusp dispositions) of M^1 have been sometimes employed to distinguish wolf (*lupus*) from jackal (*aureus*). These teeth also reflect, to an extent, body size and mass.

In respect to M_1 length: available measurements, unfortunately limited, apparently show *no* overlap between *aureus* and *lupaster* samples. The size of *C. l. arabs* overlaps, and exceeds that of *lupaster*; it is still greater in respect to *aureus*. In this regard, *lupaster* overlaps *pallipes* in size. Moreover Payne (1983) has shown a marked, discontinuous distinction between *arabs* and *lupaster* in the relative lengths of P_4 and M_2, the former having a significantly larger P_4 size.

In respect to P^4 length: In upper carnassal size, *arabs* overlaps *pallipes*, but does attain the upper size range of the latter; and, the mean values of *pallipes* are consistently slightly higher. Similarly, *lupaster* overlaps in size with *arabs*, but has a lower range, and hence lower mean value; the same is true in respect to *lupaster* relative to *pallipes*.

In respect to plots of lengths of P^4: M^{1+2} (as in Clutton-Brock 1962): maximum values are found, not surprisingly, in Palaearctic *C. l. lupus*, in which most (67%) have larger P^4, or rarely, equal P^4 size relative to combined upper molar length. The *pallipes* group is smaller overall, and the corresponding length relationships are even more dispersed, and with 54% having P^4 length in excess of molar length. The 'wolf-jackal' *lupaster* is smaller still, with predominance, contrariwise, of molar length (>80%) over P^4 length, and in contrast to the lupine condition.

In respect to M^1: The shape (with crown enlargement), cusp disposition (lingual elements), and cingulum (external distinct, and even enlarged) in *lupaster*, including fossil attributions, demonstrate its jackal rather than lupine morphology, and hence affinity.

These and other proportions and dentition characteristics, would indicate that the nomen *lupaster* designates a (large) subspecies of jackal, *C. a. lupaster*, rather than a subspecies of wolf.

The occurrence of a large jackal subspecies is sporadic and discontinuous over the later Pleistocene of the Levant. It is documented in Oumm-Qatafa (E2-E1, 1 specimen); in Zuttiyeh (basal level, 3 specimens); in Tabun (Ea, D/E, D and ?C, 7 specimens in all); and, in Mousterian contexts as well in Geula cave (B1, 1 specimen) and in Shukbah (D). A "normal" form of golden jackal (*C. a .aureus*), by comparison, is exceptionally uncommon and appears only to have been reported in terminal to post-Pleistocene levels at Abu Usba (with Natufian) and, similarly young, at Ksar 'Akil (level I, one tooth); it is also documented at Qafzeh (Level XI, 1 specimen). Middle Pleistocene antecedents of jackals are otherwise scarcely documented in Europe. However, more ancient, Lower Pleistocene antecedents have been sometimes proposed, perhaps wrongly, and to presumably include *C. arnensis* (Rook and Torre 1996) and the recently recognized (and proposed) *C. apolloniensis* of localities in Greece and, perhaps as well, Italy and even Spain, according to Koufos and Kostopoulos (1997).

By contrast the occurrence of *C. lupus* ssp. in Levantine localities is more common, although remains are few in number. The most important occurrences of later (including terminal) Pleistocene age are those of Kebara (levels B, C. E), El-Wad (B, ?C, only 6 teeth) and Tabun-B (1 specimen) on Mt. Carmel; Shukbah (B); Qafzeh (XVII, 1 specimen); and, repeatedly, from lower (XXVI) to higher (I) levels of the Middle and Upper Paleolithic occupation infills of Ksar 'Akil (Lebanon).

At Douara Cave (Palmyra basin, Syria), a small form of *C. lupus* is represented (by mandibles and some postcrania) in a Mousterian level . Payne (1983) has remarked on the close similarity to *C. l. arabs*, a similarity also found in the *Canis* from Palegawra (Iraq), in a Zarzian (terminal Paleolithic) context, and considered by Turnbull and Reed (1974) as a probably domesticated dog (not generally accepted, however). Moreover Payne has noted that some specimens from Mt. Carmel in Upper Paleolithic (Kebara-C) as well as Mousterian (Tabun-C) contexts are of the same size and morphology, distinct from the other larger *C. lupus* of the east Mediterranean littoral zone. The former may, as he suggests, constitute distinct populations from more xeric circumstances and times, and having a very substantial evolutionary history as well.

Perhaps the oldest such occurrence, and in age not unlike Yarımburgaz, is that of Bears' Cave in the upper Galilee from which an M_1 appropriate to *C. l.* cf. *mosbachensis* is known (Tchernov and Tsoukala 1997). Its more ancient precursor was *C. etruscus*, now known at Dmanisi (Georgia) in the Lower Pleistocene, and widespread and quite well known in many parts of Europe at this and later times.

Canidae indet. (large)

A.40 (1986), lev 14, no. 2, left distal fibula.

In my initial examination (1987) of the Yarımburgaz fauna, I noted the presence of an undetermined, large canid. Level 14, of square A.40 (UC) yielded from the 1986 excavations a well-preserved distal left tibia (no. 2) with its distal articulation. This is in the lowest level of the infill, and below the Eemian-aged fluvio-marine incursion. The specimen represents a large canid, having the morphology of the dhole (*Cuon*) or African hunting dog (*Lycaon*). Unfortunately, no additional remains have come to light to resolve the identification, but further comparisons have confirmed its distinctiveness.

The diaphysis is robust and more cylindrical than in wolf, the posterior supra-articular surface is flat, not thickened as in wolf, and the internal malleolus is stronger, broader and more extensive antero-posteriorly, and has a different relationship or 'set' relative to the adjacent articular surface (see also Musil 1971a). It is comparable in size to *C. lupus* and rather smaller than *Xenocyon* homologues from Koneprusy and Stránská skála. It does, however, closely match in size a distal tibia of *C. (X) falconeri* from Pirro Nord. A specimen (CD-625) from Escale cave, referred by Bonifay (1971) to cf. *Cuon stehlini* (= *C. priscus*), exhibits the very same overall morphology, but is smaller than the Yarımburgaz example, and considerably smaller than that from Stránská skála assigned to *Xenocyon* by Musil (1971a). *Cuon priscus*, rare in the early Middle Pleistocene (but present at Escale and Petralona), is known to co-occur with *Xenocyon* at Mosbach (main fauna) and at Hundsheim. *Canis* (*Xenocyon*) *falconeri* is the type species and occurs in Italy in the upper Valdarno, and at localities of Casa Fratta, Pirro Nord, Cueva dell 'Orba and Colle Curti. It also occurs in Spain at Puebla de Valverde, Venta Micena and Cueva Victoria. It also occurs in England (Westbury), France (Rosières), Germany (Untermassfeld and Würzburg-Schalksberg), in central/eastern Europe (Stránská skála, Prezelitçe, Gombasek, Betfia (?), and Naggharsany), and in Greece (Apollonia-1, Macedonia) as well as in Tadjikistan (Lakhuti locality). Elsewhere in western Asia it is worth noting the occurrence in the lower (Mousterian) infilling of Hayonim cave of a definitive specimen (hemi-mandible) of the African wild dog, *Lycaon pictus* (Stiner et al. 2001). This single and incomplete specimen from A.40, level 14 is insufficient for precise taxonomic attribution, but it clearly underscores the liklihood of additional canid diversity within the Yarımburgaz fauna.

FELIDAE

Panthera leo ssp.

(Figures 8.3-8.4)

A.71 = V-88), 7, 595. distal fibula

Y-88, 1b, 428a. right mesocuneiform

S-89, 4, 308. ph-2, cf. digit-II, pes. subadult

T-89, 1, 16. left ectocuneiform.

T-89, 1, 31. right distal fibula.

T-89, 4, 92. right proximal fibula.

T-89, 4, 753. astragalus, right, incomplete (medial trochleus).

T-89, 4, 787. ph-2, digit-V, manus, right.

T-89, 4, 799. right scapholunar.

T-89, 4, 801. left mesocuneiform

T-89, 4,804. sesamoid, ?

T-89, 4, 805. ?carpal, indet.

T-89, 4, 852. internal cuneiform.

T-89, 4, 853. ph-2, cf. digit-III, manus, left.

T-89, 4, 891. right mt-IV, complete. arthritic traces.

P-90, 1, 123. left radius, proximal half, articulation damaged (gnawed).

P-90, 1, 213. right mt-IV, complete.

P-90, 1, 583. right distal fibula, articulation complete.

P-90, 2, 523. right calcaneum, complete.

R-90, 1, 53. right proximal radius, articulation only.

R-90 1,326. distal fibula, articulation only, fragment.

R-90, 1, 358. right calcaneum, portion of body with subastragalar articulation (gnawed).

R-90, 1, 560. left P[4] carnassial, complete with roots.

Skeletal parts of large-bodied felids rare in the Yarımburgaz fauna. Reexamination in 1988 of the faunal remains from the original 1986 excavations in A.71 (=V-88), level 7, revealed a large fibula fragment of felid morphology. An astragalus from Y-88 (level 1b) also approximated felid structure. The 1989 excavations in Squares S-89 and particularly T-89 afforded postcranial parts of a large felid, which are unquestionably attributable to *Panthera leo* ssp. This sample was enlarged the next year with further specimens from P-90 and R-90 squares, at the back and toward the front of LC, respectively. Unfortunately, there are still no cranial or gnathic remains. Only a single upper carnassial derives from R-90 square. As these several occurrences derive from three successive infills, downwards beginning with Y (= T-89), X (= P-90, R-90) and W (T-89, P-90, S-89 and A.71-7), there are minimally three individuals, all adult.

The postcranial elements represented are proximal radii and portions of the manus (scapholunar, ph-2, digits, III and V, and carpal of the forelimb), several articular ends of fibula, and bones of the pes (astragalus, calcanei, cuneiforms, metatarsal-IV, and ph-2, digit 2 pes). Measurements of these specimens, and some comparative samples, are provided in Appendixes 8.16 through 8.24.

Two Eurasian subspecies of *P. leo* are generally recognized: *P. leo fossilis* from the earlier Middle Pleistocene, and *P. leo spelaea*, from the later Middle and Upper Pleistocene (Schütt 1969; Schütt and Hemmer 1978; also Hemmer 1967). Unavailable for comparison are a number of the postcranial elements of the

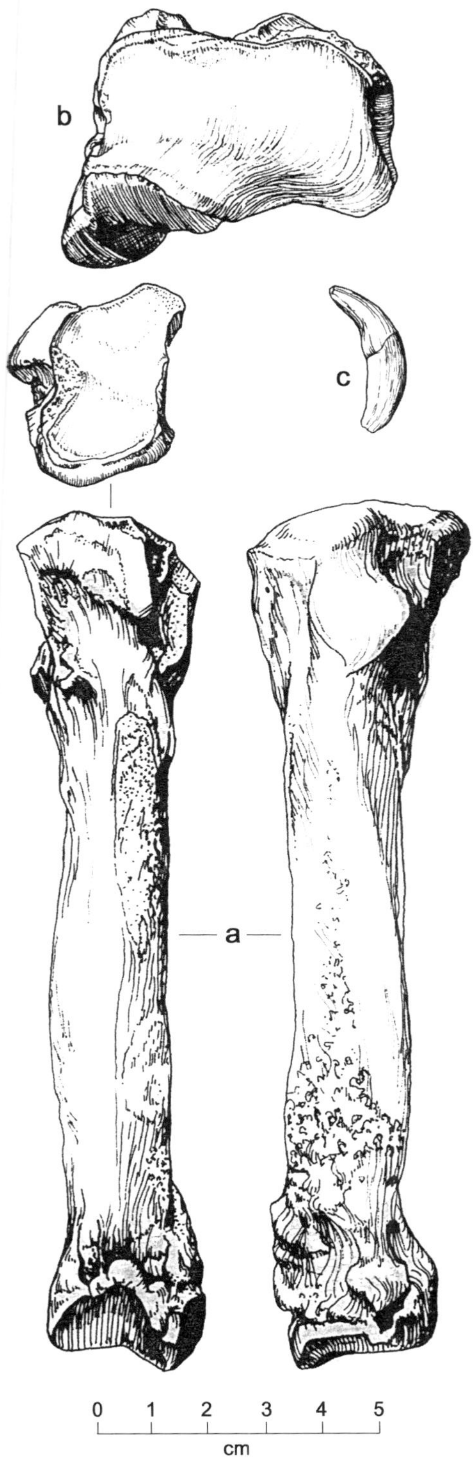

Figure 8.3.
Panthera leo ssp. (a) right IV metatarsal, arthritic distally, from T-89, 4, no. 891; (b) right scapholunar from T-89, 4, no. 799; *Felis silvestris* ssp. (c) left C_1 with root from T-89, 4, no. 587.

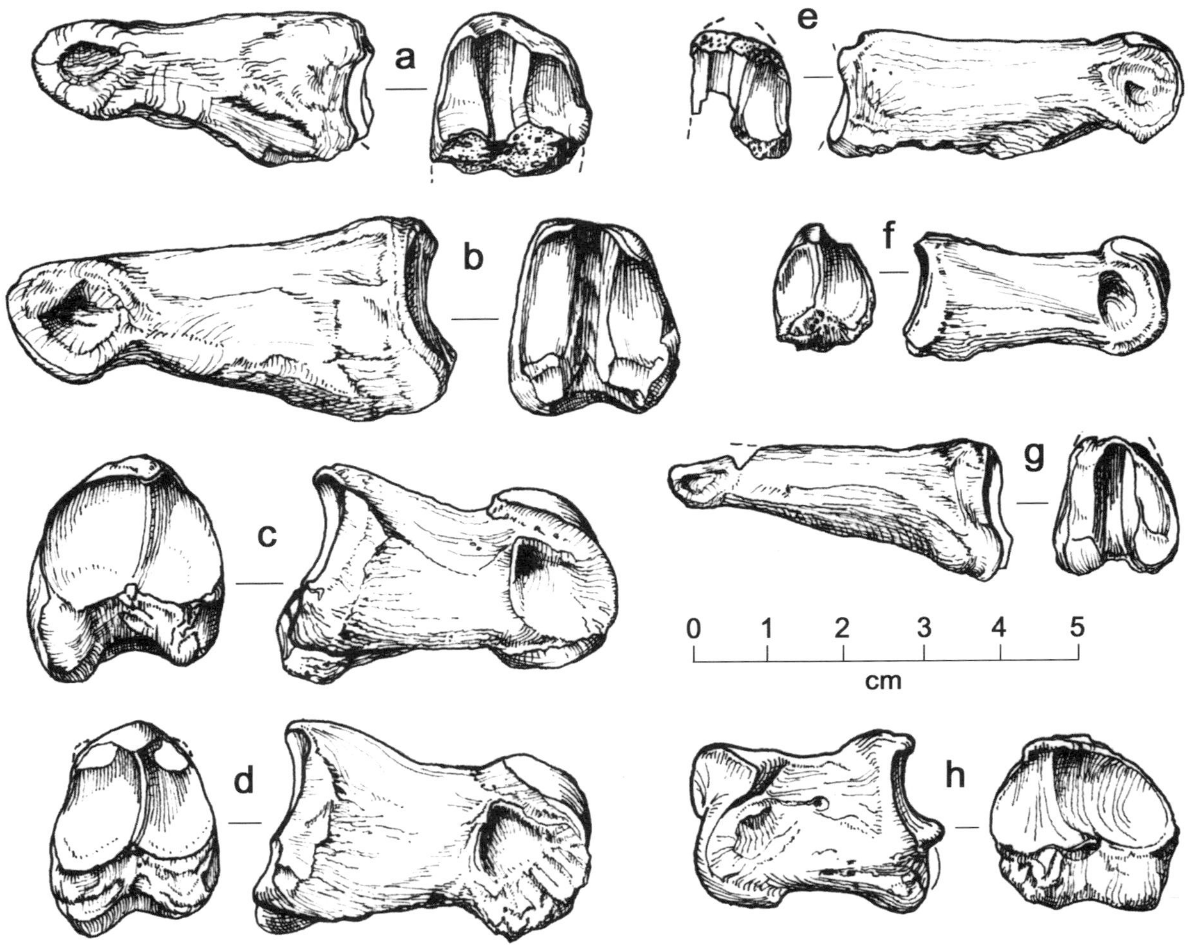

Figure 8.9. *Cervus elaphus* sspp. (a) left phalanx 1 from U-89, 1, no. 11; (b) left phalanx 1 from U-89, n, no. 48; (c) right phalanx 2 from U-89, 1, no. 389; (d) right phalanx 2 from T-89, 2, no. 129; (e) right phalanx 1, gnawed, from U-89, 1, no. 104; *Capreolus* aff. *süssenbornensis* (f) phalanx 2, cf. pes, from T-89, 4, no. 813; (g) phalanx 1, damaged distally, from T-89, 2, no. 575; *Sus scrofa* ssp. (h) left phalanx 2, manus, from U-89, 1, no. 35.

U-89, < 12 (prov. strat., -8.6 - 8.72), Bag E (a). right ph-1, manus.

R-90, 1, 1341. ph-1, proximal epiphysis.

The first few skeletal parts of wild boar (*Sus scrofa*) were encountered in 1989 in the lower fossiliferous level (4) of S-89 square. Both of these specimens are elements of the dentition. In addition, five forefoot elements were found in Square U-89. The dp^4, with roots formed but open, is 16.1 (l) x 13.2 (br); in a very worn specimen from Jaurens (Corrèze), the comparable values are 14.5 x 14.5. The dimensions of the manus phalanges are as follows:

ph-1: l = 45.8; proximal; br = 24.2; a-p = 22.1

ph-2: l = 32.8; proximal: br = 29.32; a-p = 24.26

The stratigraphic relationships of the U-89 square are uncertain due to the massive rockfall. However, the R-90.1 and S-89.4 occurrences derive from successive sedimentary events and hence represent two different individuals.

CERVIDAE
Dama sp. cf. *mesopotamica*
(Figures 8.10-8.12)

A.71, 5, 470. ph-1, manus.

T-89, 2, 97. Fragment of (cf.) mt-diaphysis; not large. cf. ?*Dama* sp.

U-89, 1, 91. Split segment, small mt-; with part of proximal articulation. *Dama*-size.

U-89 1, 104. Right ph-1; chewed (damaged proximally); and articulation incomplete. cf. *Dama.*

U-89 1, 299. Right proximal-most mt-, incomplete articular surface. *Dama* sp.

U-89, 1, 458. Right astragalus, complete; gnaw marks medial surface of body. *Dama*-size.

U-89, ?level (adj. 6, -8.02 - 8.06), 349. Right calcaneum, complete. cf. *Dama* sp.

Y-88, 5, 662. ph-1, pes.

P-90, 1, 21. maxilla fragment with dp^4 and germs of M^1 and M^2.

P-90, 1, 52. left mandible fragment(s) with M_1 (into occlusion) and M_2 (in crypt), and associated I_1, $-_3$, P_2, dp_3, dp_4.

P-90, 1, 70. ph-1, left, manus.

P-90, 1, 86. distal mt, complete articulation.

P-90, 1, 106. incomplete sacrum.

Fallow deer (*Dama)* are represented in five excavation squares (13 specimens) from two, successive sedimentary events (U, X). These are almost exclusively postcrania, with the exception of (subadult) upper/lower jaws (with dentition), the latter associated with three postcrania in P-90 (1evel-1). These remains are clearly that genus on the basis of

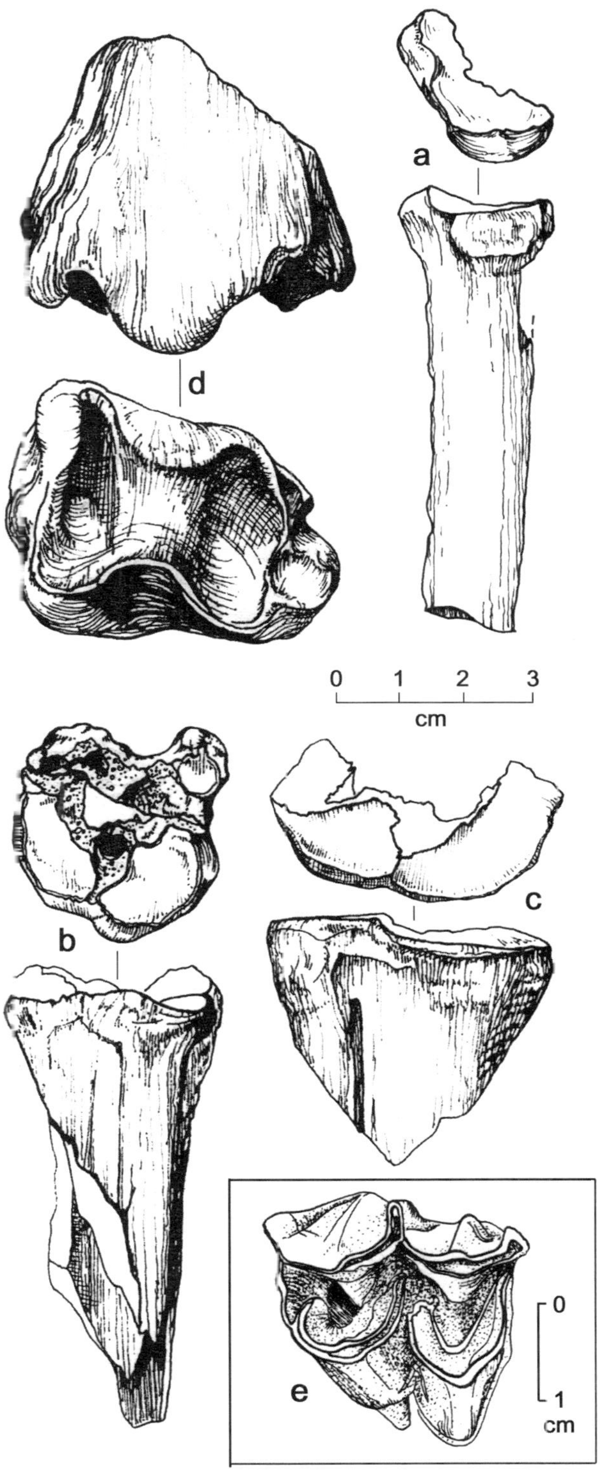

Figure 8.10.
Dama sp. (a) split metatarsal segment with incomplete proximal articulation from U-89, 1, no. 91; (b) right proximal-most metatarsal with incomplete articular surface from U-89, 1, no. 299; *Cervus elaphus* sspp. (c) left proximal metacarpal fragment from T-89, 4, no. 630; (d) left distal tibia with articulation from T-89, 3, no. 230; (e) enlarged view of M^2 from T-89, 4, no. 528.

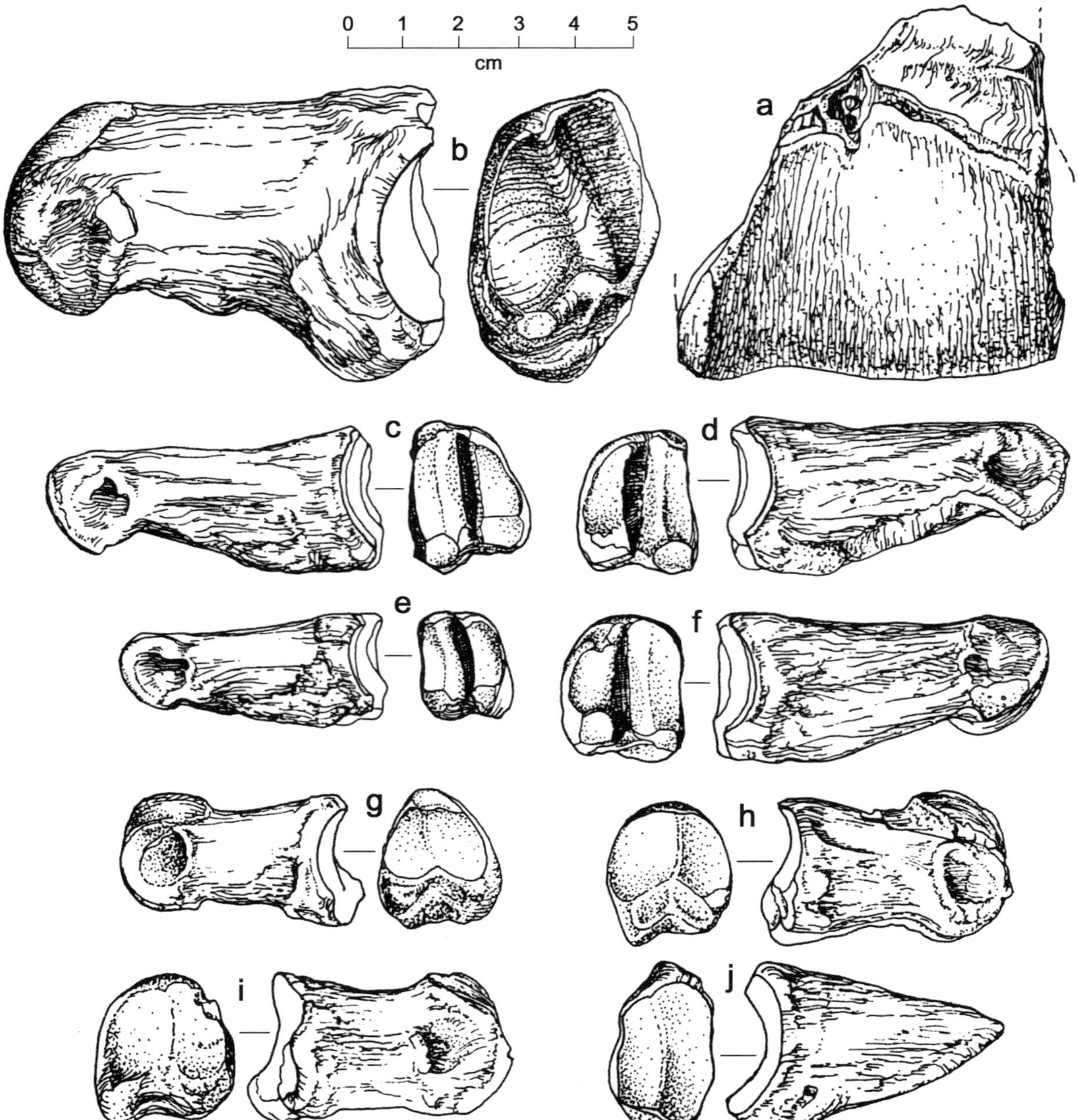

Figure 8.11. *Megaceros (P.)* aff. *verticornis* (a) partial medial trochlea of right distal humerus from Y-88, 1a, no. 130, ant view; (b) incomplete phalanx 1, pes, from Y-88, 1a, no. 131, lat-prox views; *Cervus elaphus* ssp. (c, f) phalanx 1, pes, from A.71, 4, no. 233; (d) phalanx 1, manus, from A.71 sieving in 1988, no. 33; (g) phalanx 2, pes, from A.71, 4, no. 234; (h) phalanx 2, pes, from A.71, 3 in 1986; (i) phalanx 2, pes, from Y-88, 1a, no. 157; (j) phalanx 3 from Y-88, 5, no. 664; *Dama* sp. (e) phalanx 1, pes, from Y-88, 5, no. 662.

both size and morphology (see Di Stefano 1996a, 1996). Measurements are set out in Appendixes 8.37A-B and 8.39A-B.

The few extremity remains of fallow deer are either within the modern *D. dama* range (Bosold 1968) or approach the larger size of *D. clactoniana*. The astragalus (U-89, no. 458) is nearly or quite comparable in size with *D. clactoniana* from Swanscombe (no values have been published

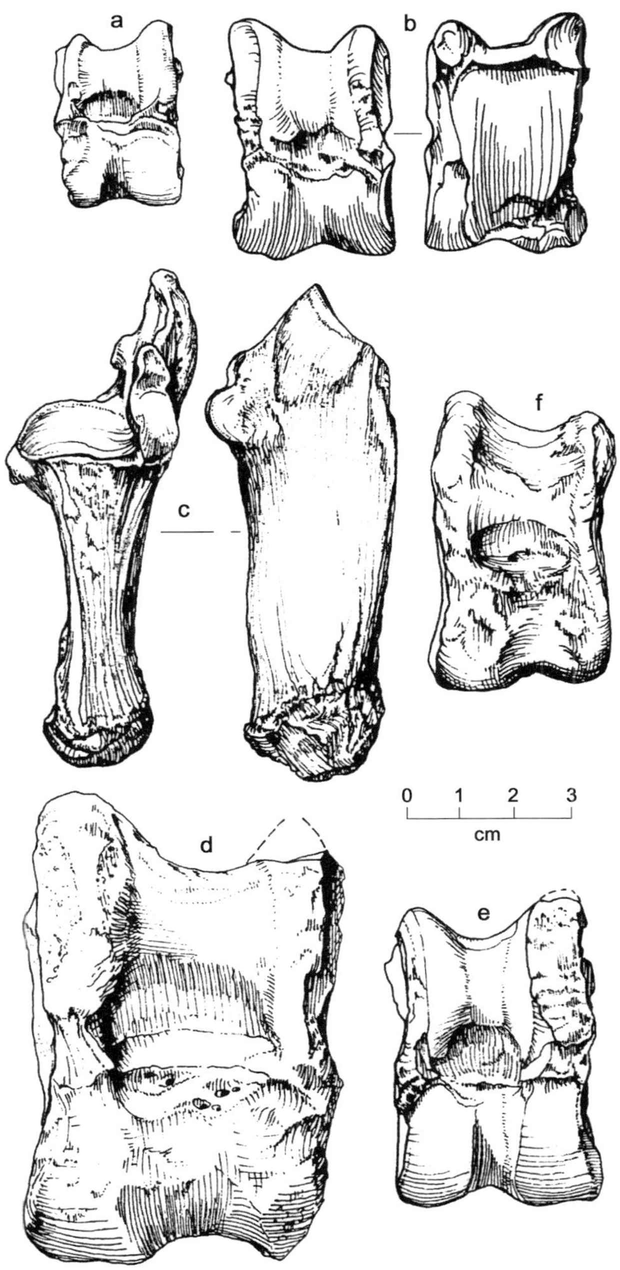

Figure 8.12. *Capreolus* aff. *süssenbornensis* (a) left astragalus from T-89, 4 no. 656; Dama sp. (b) right astragalus from U-89, 1 no. 458, and (c) right calcaneum from U-89, no. 349; *Megaceros (P.)* cf. *verticornis* (d) right astragalus from S-89, 4, no. 255; *Cervus elaphus* ssp. (e) left astragalus from S-89, 4, no. 253, and (f) right astragalus from T-89, 4, no. 1050.

for the subspecies from Riano in Latium, Italy). The calcaneum (U-89, no. 349) is rather shorter (92.1 vs. x 101., range 96 - 110) than the smallest of a few specimens of Swanscombe *D. clactoniana*, but within the range (/X 87, range 82 - 98) of Eemian-age *D. dama* from southern Britain. Unfortunately, no antler elements are available to enable a specific identification.

It has been occasionally noted among some geologically older Pleistocene samples of Levantine fallow deer that they are both larger and otherwise distinctive relative to those of later, demonstrably Upper Pleistocene age. In a recent, thorough quantitative study of the requisite assemblages, Di Stefano (1996) has demonstrated that such is in fact the case, and he has distinguished (and diagnosed) a new subspecies, *D. clactoniana mugharensis* for such occurrences. The type derives from Tabun-E (Mt. Carmel), and is distinctive in respect to antler form, cheek tooth morphology, and some postcranial dimensions. This is in contrast to the morphology characteristic of *D. mesopotamica* or to *D. clactoniana* aff. *mesopotamica*, which appear to exhibit a mosaic of primitive and more derived features (as at Oumm Qatafa). Probably at Evron and at Gesher Benot Ya'akov a form of *D.* cf. *clactoniana* is represented. Arguably, an early form of *Dama* may be represented at 'Ubeidiya, but it is still not very well known skeletally. An undetermined subspecies of *D. dama* is reported in the lower cultural horizons of Karain Cave, but is only numerous (20-26%) in the upper horizons (F, G) (Otte et al. 1998).

Cervus elaphus ssp.

(Figures 8.10-8.12)

A.71, 3, ?#. ph-2, pes.

A.71, 4, 231. Proximal tibial diaphysis, right; lacking articulation.

A.71, 4, 233a. Medial ph-1, pes.

A-71, 4, 233b. Lateral ph-1, pes.

A.71, 4, 234. ph-2, pes.

A.71, 6/8, 542. Proximal two-thirds of mt; strongly ravaged.

A.71, 12, 682. ph-1, cf. manus.

A.71, indet. (1988 sieving), 31. ph-1, manus.

S-89, 4, 71. Frontal frag. = cf. right pedicle and rosette. cf. *C. elaphus.*

S-89, 4, 140. Terminal antler with 2 tines; and base of another, below. = *C. e.* aff. *acoronatus.*

S-89, 4, 253. Left astragalus, complete; chewing damage inferior surface.

S-89, 4, 262. Single condyle (epiphysis), of distal metapodial.

T-809, 2, 129. Right ph-2.

T-89, 2, 208. Complete, terminal segment of antler; cf. trez tine; large.

T-89, 3, 230. Left distal tibia, articulation only; excellently preserved.

T-89, 4, 528. cf. right M^2.

T-89, 4, 630. Left proximal mc-, fragment. Smallish *C. elaphus.*

T-89, 4, 1012. cf. right carpal 3 = magnum/trapezoid. Cervidae, *C. elaphus* ssp.

T-89, 4, 1050. Right astragalus, complete.

T-89, 4, 1285. Right (?)distal tibia, subadult; lacking epiphysis. ?*C. elaphus.*

T-89, 5, 1062. Complete right humerus, slightly damaged proximally.

U-89, 1, 11. Left ph-2, small. cf. *Cervus elaphus.*

U-89, 1, 48. Left ph-1. cf. manus.

U-89, 1, 389. Right ph-2.

U-89, 1, 442. (sieving) P/. cf. *Cervus elaphus.*

Y-88, 1a, 157. ph-2, pes.

Y-88, 5, ?#. ph-2.

Y-88, 5, 664. ph-3

P-90, 1, 43. right P_2, partial roots, moderate wear.

P-90, 1, 364. left astragalus, complete.

P-90 1, 4. ph-2, pes, right.

P-90, 1, 93. ph-2, pes.

P-90, 1, 388. ph-2, pes.

R-90, 1, 558. left basal antler with rosette, and preserving brow and bez tines.

R-90. 1, 917. maxilla fragment(s) with M^2 and M^3, moderately worn.

R-90, 1, 200. left P_4 crown, broken roots; substantially worn.

R-90, 1, 223. cf. I_1, right; damaged crown, with root.

R-90, 1, 313/314. right I_2 and I_3, with roots; worn.

R-90, 1, 1330. incomplete /M, much worn.

R-90, 1, 1343. molar fragment, indet.

R-90, 1, 316. ph-2, manus.

R-90, 1, 399. ph-1, manus, cf. left.

R-90, 1, 850. ph-2, manus

R-90, 1, 268. left proximal tibia, with damaged lateral condyle, anterior articular surface and tibial crest.

R-90, 1, 360. left astragalus, complete.

R-90, 1, 228. posterior portion of calcaneum, right.

R-90, 1, 632. Distal (tuber) portion of calcaneum, left.

R-90, 1, 1340. mt., damaged and partially reconstructed.

The most abundant Cervidae is red deer, *Cervus elaphus*. In all there are 49 identifiable specimens. These occurred in all seven excavation squares and are represented through five sedimentary episodes, rarely in Squares T (N = 3), U (N = 1) and V (N = 1), and most numerously in Squares W (N = 5) and especially X (N = 28). Cranial parts are wholly lacking, but five parts of antlers were recovered from four squares (representing three sedimentary episodes). Three squares yielded isolated (3) cheek teeth. The R-90 (level 1) occurrence yielded a palate fragment (with $M^{2/3}$) and lower dentition (3 I's, P_4 and M) and some associated postcrania. The bulk of the collection comprises postcrania, rarely complete long bones (a humerus), and mostly autopodial elements.

The five fragments of antler all represent *C. elaphus*. They represent minimally three individuals on stratigraphic grounds.

a) R-90, 1, 558. This is the base of a (shed) left antler, without pedicle, and comprising the pearl-like rosette (burr) and the (larger) brow and (slighter) bez tines, each lacking its extremity. Rosette: a-p = 59.1, transverse = 66.2. Beam (between tines): a-p = 37.82, transverse 52.25. Brow tine (as preserved) = (87. - 88.). Bez tine (as preserved) = (92. - 93.)

b) T-89.2, 208. This is the upper segment of a large antler comprising only the trez tine. Base: a-p = 28.9, breadth = 27.14; 'mid-section': a-p = 26.4, br = 26.2.

c) S-89.4, 71. This is a fragment of right frontlet with antler pedicle and well-above it the rosette (or burr). Pedicle: min. l. = 41, max. l. = 74.75; dimensions below rosette: a-p = 33.2, br = 29.4.

d) T-89.4, 140. This is the terminal portion of a large antler. It preserves two upper tines; the right is largely complete, including tip, whereas the left preserves only its basal portion, broken a distance above the bifurcation. From the base of the latter another tine arose; its elongate oblate base being preserved and must have extended upwards and backwards at an acute or right angle to its source root. There is no apparent evidence of any other branching at the base of the complete tine, although it exhibits some breakage. The terminal part of the beam, from which these tines diverge, lacks any trace of palmation; it is incompletely preserved but is massive (transverse diam., below tine divergence = ~ 47.) and has a flattened (a-p) or rather ovate cross-section, its preserved length being ~ 142. I interpret this tine conformation to represent tine 3 or trez (complete) and tine 4 (base only), and tine 5 (origin only). This structure does not form a typical 'cup', as often occurs in modern red deer (Lister 1987), but may in fact be approximated in some extant examples (as in Heintz 1970, vol. 2, p. 23, fig. 6, 1876-1877-1878 examples) of some populations.

e) Y-88.5, 658. This is a segment of a large antler beam. It is straight below and curves upwards and backwards. It is split such that the 'upper' end has a flattish bevel against the outer surface, whereas the lower portion preserves more of the maximum beam cross-section. The maximum preserved length = 127.8, max. breadth lower end = 40.3, and max. breadth 'upper' end = 19.6. The 'upper' (or terminal) end bears usage traces terminally, and also along an adjacent salient margin.

From the evidence afforded by specimens a-d it is possible to provisionally construct an antler morphotype for the Yarımburgaz *C. elaphus.* It is characterized by the following features: a relatively long pedicle; small/moderate diameter rosette; stronger brow tine, more horizontally disposed, arising shortly above the rosette; rather upwardly angled (less than 90° re the beam axis), more slender and tapering bez tine; large, robust trez tine, only slightly curved except at the distal termination, and of unknown origin point from the beam; antler termination with less robust beam, without palmation, with divergent, nearly right-angled distal tines (3 and 4), the larger of which is the trez tine, and from the base of the lesser (4) of which arises a secondary tine.

Two important discussions of *Cervus elaphus* and of recognized Pleistocene subspecies are offered by Di Stefano and Petronio (1992, 1993). The character of *C. e. acoronatus* has been treated in some detail, and the validity of three (sequent) subspecies - *C. e. acoronatus, C. e. priscus,* and the newly proposed *C. e. eostephanoceros* - usefully examined. (Two other subspecies--*C. e. rianensis* [from Riano] and *C. e. aretinus* [localities in Tuscany]—have also been proposed). These lesser taxa have distinctive antler morphology, as well as some apparent differences in limb size and proportions. The antler structure in the Yarımburgaz cervid, although inadequately known, appears to bear a distinct resemblance to the last named subspecies, *eostephanoceros.* It is known from localities in central Italy, in the Latium, the type deriving from Cava Nera Molinaro, and the Basilicata, and in Austria (Hundsheim) and in southern France (Arago).

Principal measurements of the few dental elements of the Yarımburgaz red deer are provided in Appendixes 8.39A-C, and those of the postcrania in Appendix 8.38A. For comparative purposes, comparable dimensions of some *C. elaphus* subspecies are set out in Appendixes 8.38B (dentition), 8.39C and 8.42 (postcrania). These include available samples of *C. e. acoronatus* (Atapuerca, GD; Mosbach; Süssenborn; Tiraspol; and Voigtstedt), *C. e. priscus* (Mauer), *C. e. rianensis* (Riano), the newly proposed *C. e. eostephanoceros* (exemplified by Arago, Hundsheim), the recently proposed *C. e. simplicidens* (Combe Grenal, lower levels), and other unattributed samples from Petralona, Chatillon-St.-Jean and La Fage, as well as the large Burgäschisee (Switzerland) Neolithic sample of *C e. elaphus.* Unfortunately comparable quantitative data on Levantine cave samples of *C. elaphus* are almost non-existent.

The few elements of dentition from Yarımburgaz are not particularly revealing. The upper molars are within the size range of various mid-Pleistocene comparative samples, but are rather larger than in *C. e. acoronatus.* This is similarly the case with the only two lower premolars (P_2, P_4) in the Yarımburgaz sample.

Of the two major long bones only the humerus is complete as to length, although it is imperfectly preserved distally. It is on the short side, as in some other mid-Pleistocene examples. The proximal breadth is substantial, but its a-p diameter is relatively reduced, if compared to the greater dimension in a single Petralona specimen and in the Burgäschisee samples. Only the proximal (2) and distal (1) tibia is represented, and thus the length of this element is unknown. The proximal breadth is comparable to some mid-Pleistocene comparative samples cited here. The single distal example has both transverse and a-p diameters within the range of most such samples.

Metapodial elements (3) are surprisingly rare, and all specimens are ends only. The single proximal mc is within the size range of comparative samples. The single proximal mt is largely within the comparative sample range in breadth, but is again reduced in its a-p diameter, as noted already for the large fore- and hind-limb elements. Another distal metapodial preserves only a partial articulation.

Phalanges are substantially better represented by a total of 18, both manus and pes, ph-1 (6) and ph-2 (12) elements. The length and diameters of ph-1 are within the range of more recent (Holocene) and late mid-Pleistocene samples; however, they seem to be shorter and less robust than in *C. e. acoronatus* examples. This is broadly the case also for ph-2 in respect to length and proximal a-p dimensions, but the breadth is greater than in *C. e. simplicidens* (Combe Grenal) and many of the Holocene (Neolithic) samples. The distal dimensions are also comparable to other late mid-Pleistocene samples. The limited Atapuerca sample is of larger and more robust animals, and presumably in this respect approaches the *C. e. acoronatus* condition, although published comparative data is sparse.

There is a surprising lack of metric detail about the osteological features and dimensions of red deer represented in Pleistocene occurrences in western Asia. This is despite the substantial number of documented instances of the species, albeit in small numbers per occurrence and most often without antler presence or preservation. In Emirkaya-2 the species is probably represented, and older Levantine occurrences probably (cf. species) include those of Gesher Benot Ya'akov and Evron (single tooth), and the later cave fills at Oumm Qatafa, Zuttiyeh (lower), Qafzeh (discontinuously between 26 and 19, and in 6-5) and Tabun (C, B). Other Mousterian (and, Upper Paleolithic) occurrences include Kebara (12 to 6), Hayonim (E, D), Quneitra, Wad (G to C), Erq el Ahmar (H, F to D), and throughout Ksar 'Akil (except the basal levels 36 - 30). In southern Turkey, red deer of undetermined subspecies is uncommonly present in most lower cultural horizons of Karain Cave (Otte et al. 1998). The substantial distribution, both temporally and geographically, of this species affords a valuable but as yet little exploited resource for both biogeographic and paleobiological analyses.

Megalocerini indet.

(Figures 8.11-8.12)

A.71, 4, 232. Distal mt-, partial trochlea.

A.71, 12, 668. Proximal mc-, incomplete, with diaphysis.

Y-88, 1a, 130. Distal humerus, right; medial half of articulation only.

Y-88, 1a, 131. ph-1, pes; incomplete.

S-89, 4, 255. Right astragalus, complete; erosional + excavation damage.

Five specimens are here referred to Megalocerini (Appendix 8.41), and are presumed to be some form of *Megaloceros*. Only postcranial parts are represented. These derive from three excavation squares, and from three successive sedimentary episodes, U (A.71, 12), W (A.71; S-89, 4), and X

(Y.88, 1a). These are all demonstrably cervid, larger than red deer, and of morphology that suggests megalocerine affinity.

The astragalus from S-89 (level 4, 255) is within the range of *M. giganteus* from Chatillon-St. Jean (Drome) in all comparable measurements. It is rather larger than specimens of (*Praemegaceros*) from Süssenborn and Voigtstedt, and substantially larger than the specimens of 'Ubeidiya that are considered to be *M. (P.) verticornis* (Appendix 8.42). The proximal mc (A.71, 12) is rather smaller than specimens from Süssenborn and Voigtstedt attributed to *M. (P.) verticornis,* but close in size to specimens from 'Ubeidiya (Geraads in Tchernov 1986) that are referred to that taxon. The Y-88 level 1a specimens (ph-1, distal humerus) match *Megaloceros,* and even *M. (P.) verticornis,* in size. The ph-1 attributed to *Megaloceros (P.)* cf. *verticornis* is comparable in size to some other specimens referred to, or known to represent, this extinct form. However, ph-1's from 'Ubeidiya are substantially shorter (60, 64 and 67 mm) than specimens known from Europe. The material here is insufficient to ascertain species affinity. In any case it evidently is not *Megaloceros giganteus,* which remains unknown within western Asia, although it is recorded at a number of localities within the Caucasus province.

Capreolus sp. aff. *süssenbornensis*

(Figures 8.9 and 8.12)

T-89, 2, 575. ph-1, cf. pes, damaged distally. *Capreolus,* and large.

T-89, 4, 656. Left astragalus, complete. *Capreolus,* and large.

T-89, 4, 813. ph-2, cf. pes. *Capreolus,* and large.

The roe deer, *Capreolus,* is known from only 3 postcranial elements, all from Square T-89. These derive from upper (l.2) and lowermost (l.4) levels and hence successive sedimentary episodes. These distal limb specimens are demonstrably roe deer, but are large in comparison to extant or sub-recent *C. capreolus* (Appendix 8.40). This is evidenced in the astragalus dimensions (Appendix 8.42) in comparison with sub-recent and Upper Pleistocene samples of *C. capreolus,* and in contrast to similar sizes in the Voigtstedt and Süssenborn samples attributed to *C. süssenbornensis.* The known pes phalanges are also larger in the Yarımburgaz examples. Thus, they are reminiscent of the larger, Middle Pleistocene species, *C. süssenbornensis.* Roe deer, usually referred to *C. capreolus,* is not uncommonly recorded in west Asian Pleistocene localities, though never in any substantial numbers. A probable occurrence (cf. genus) in Anatolia is at Emirkaya-2 fissure fill. Older occurrences in the Levant are at Evron, and in infills of younger age at Tabun (Eb, D, C, B), Qafzeh (only levels 15, 5), Hayonim (E, D), Skhul (B), Shukbah (B), Kebara (levels 12 to 6), Wad (E to B), and throughout the Ksar 'Akil shelter succession. Roe deer occur in four of the lower cultural horizons of Karain Cave, and remains are quite numerous in the middle two horizons (Otte et al. 1998).

BOVIDAE

Skeletal parts attributed to Bovidae from the initial Yarımburgaz excavations (A.71) in 1986) were relatively few (only 6 specimens) and very fragmentary. This number increased in subsequent field seasons but not very substantially, and specimens are still generally incomplete. In all, five squares yielded bovid remains, only a single specimen deriving from U-89 (none from P-90 or R-90). The remains derive from four successive sedimentary units: T (Y-88, base): U (A.71.12); W (A.71.4, 5, 6; S-89.4; T-89.3); X (Y-88.1a; T-89.2; S-89.2; U-88.1); as well as possibly from Y (Y-88, topmost infill). The maximum number of specimens—at least 16—derive from the W and X units of the upper part of this infill.

BOVINI

Bos / Bison spp.

(Figure 8.13)

A.71, 5. 489. Very worn/damaged rt. M/, ? M^2. *Bos/Bison.*

A.71, 6, 551. Left (?) M_2, distal part; only part of distal lobe and median ext. column preserved. *Bos/Bison* indet.

S-89, 2, 116/117. V^C, upper (2), cf. = C3 + C4; associated and articulatable; largely complete, well preserved. cf. *Bos/Bison.*

S-89, 4, 137. Incomplete V^C, upper. cf. *Bos/Bison.*

T-89, 2, 547. Proximal glenoid frag. of right scapula. cf. *Bos/Bison.*

T-89, 3, 118. Segment of tooth wall, with enamel + dentine; M frag. *Bos/Bison* indet.

T-89, 3, 132. Basal section, with frontal contact of right horn core. *Bison (?priscus).*

Y-88, cleaning, 671. Left M^1 or M^2, complete (probably = M^2). *Bos* sp.

Y-88, cleaning, 671a. M, posterior lobe only. *Bos* sp.

Y-88, cleaning, 672. Left dp_4, complete. *Bos* sp.

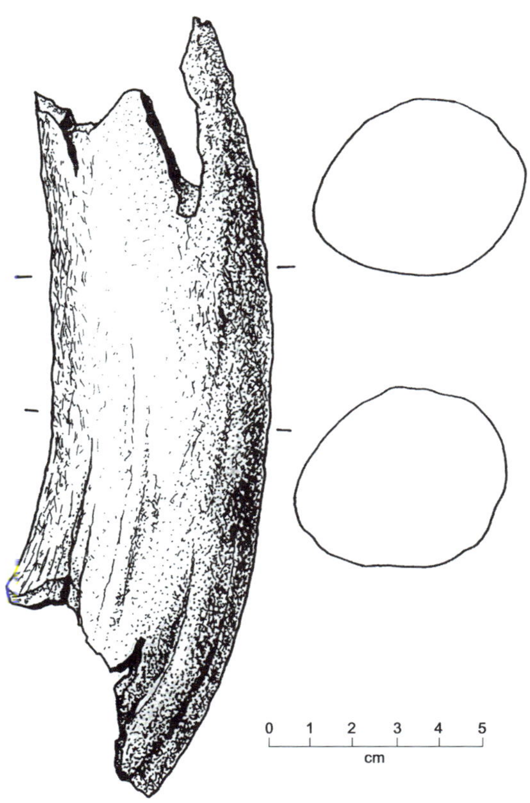

Figure 8.13. *Bison* cf. *priscus* partial right horn core from T-89, 3, no. 132.

Bos/Bison is represented in four squares (V, S, T and Y). Of the few teeth (from V=A.71 and Y-88), only two are complete (M^2 and dp_4) and most likely represent *Bos primigenius* (Appendix 8.43A). Two associated and articulatable cervical vertebrae from S-89 level 2, probably C3 and C4, represent another individual animal. The cranial remains constitute only a largely complete right horn core (no. 132) from T-89 level 3, which is *Bison*, probably *B. priscus*. Its dimensions are: basally, a-p = 49.8, br = 42.3; at mid-section, a-p = 51.1, br = 43.0. Unfortunately the fragmentary nature and rarity overall of the remains preclude any precision in identification of Bovini. *Bison* is most rarely identified in Pleistocene occurrences of west Asia, definitively only at Latamne and questionably at Gesher Benot Ya'akov. Essentially all such bovine occurrences are of aurochs, *Bos*, and specifically *B. primigenius*, ubiquitous at Paleolithic shelter and cave sites in Lebanon, Jordan, and particularly Israel, where early occurrences include Evron, Tabun (E upwards), and probably Our Qatafa (D) and (lower) Zuttiyeh. Aurochs occur at Karain Cave, in four levels of the lower cultural horizons, but mostly in low or moderate numbers in the upper three horizons (Otte et al. 1998).

CAPRINI

(Figures 8.14-8.17)

A.71, indet. (sieving 1988), 33. Left terminal horn core segment. *Capra aegagrus hircus.* (Considered "domesticated" *aegagrus*).

A.71, 4, 154. Frontlet, with both horn cores, incomplete distally. *Capra* aff. *aegagrus.*

S-89, 4, 274a. Tip of cf. right horn core. *Capra* cf. *ibex.*

T-89, 2, 282. Much of distal part of left horn core. *Capra* cf. *ibex.*

Y-88, 0 (cleaning), 7. Left horn core, missing tip. *Capra aegagrus hircus.* (Considered "domesticated").

Y-88, cleaning, 673. Left horn core, missing tip. *Capra aegagrus hircus.* (Considered "domesticated").

At least two species of Caprini are represented by horn cores (Appendix 8.43C). One (left) largely complete specimen was recovered from the uppermost level (O) of Y-88 (no. 7). Another (also left, no. 673) derived from cleaning that square, hence its provenience is uncertain. The latter is matched by a (left) terminal core fragment (no. 33) from the backdirt of Square A.71 obtained by re-sieving. None of these is heavily mineralized, and all are lightish in color. The no. 7 specimen has an antero-median keel, and a clockwise twist beginning some 5 cm above the base. The inferior outer surface is asymmetrically convex and the inner surface is flattened or (progressively) hollowed due to the keel and the twisted configuration. This morphology is characteristic of the domesticated goat, *Capra aegagrus hircus,* to which the specimen is reasonably referred.

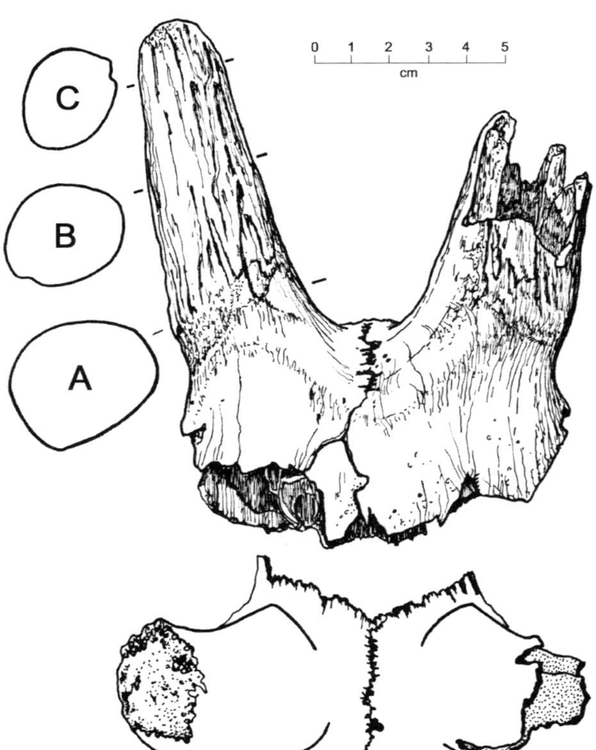

Figure 8.14. *Capra aegagrus* frontlet with incomplete horn cores from A.71 (1986), 4, no. 154.

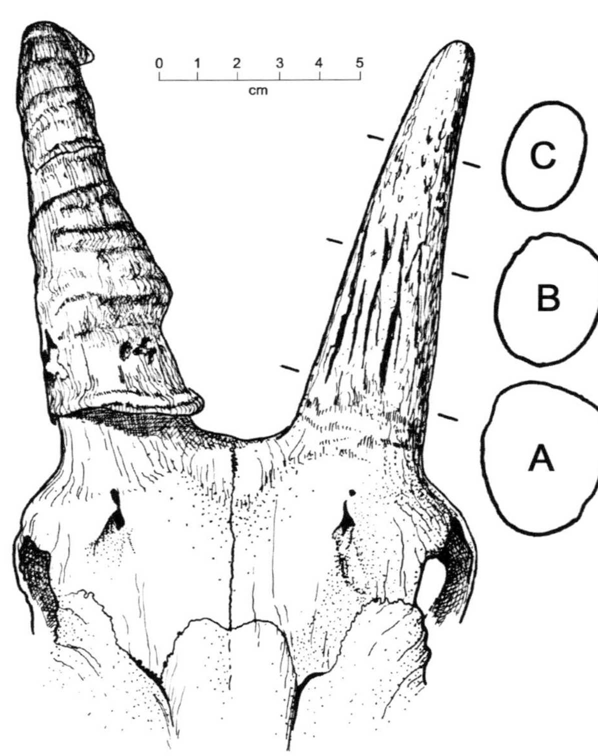

Figure 8.15. Modern *Capra ibex sibirica* cranium with horn cores, from Turkestan, USNM 62091.

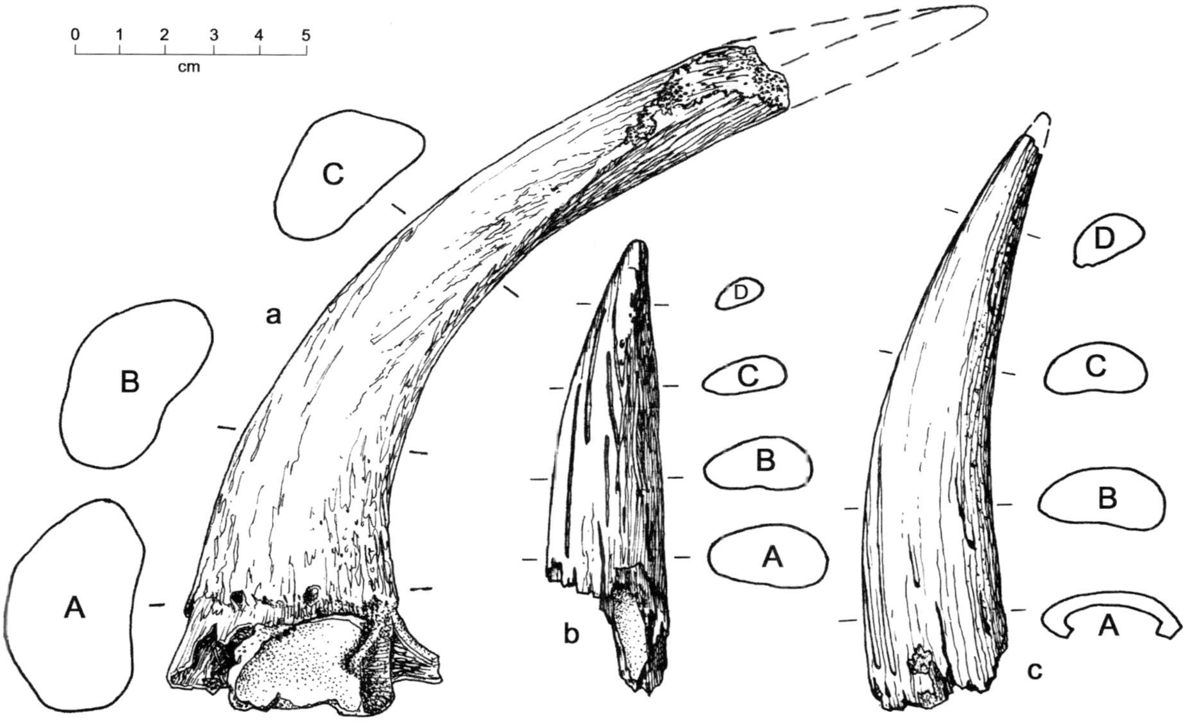

Figure 8.16. Horn cores of *Capra aegagrus hircus* from Yarımburgaz Cave: (a) left horn core missing distal tip from Y-88, 0, no. 7; (b) left terminal horn core fragment from A.71 sieving (1988), no. 33; (c) left distal horn core, broken near base, from Y-88 (level 0?, cleaning), no. 673.

The no. 673 specimen is substantially less robust than the above, is slender with gently convex outer surface and slightly depressed inner surface, and has an anterior keeled margin curving slightly, progressively upwards and inwards (clockwise); there is no posterior keel. The distal fragment (no. 33) matches no. 673 in that it shows a continuation (and termination) of slight clockwise torsion. Although smaller, with correspondingly rather different cross sections, these latter specimens presumably represent the same form as the aforementioned larger specimen (no. 7), and are certainly younger individuals and perhaps even female. None of these specimens constitute a part of the cave's Pleistocene fauna, but instead must derive from surficial remnants of a Holocene occupation (as is well-known in the UC).

The most complete specimen (no. 154) derives from Square A.71, level 4 (1986 excavations). It constitutes the posterior portion of a frontal with both horn cores present, each incomplete distally, but with intact bases. The specimen is broken at or above the superior orbital surface, with a portion of the supraorbital foramen preserved only on the left side. The coronal sutures join at bregma at an obtuse angle (sometimes said to be a distinction between *Capra* and *Ovis*, but in fact *C. ibex* has an intra-coronal angulation, whereas other species may not); the frontal suture is largely patent. The horn cores incurve upwards from the frontal, which is thus concave rather than flattened, and the cores diverge slightly (ca. 25% - 30%) outwardly/ upwardly, but only slightly backwardly. The basal horn core section (section a) is an irregular broad ovoid, almost quadrilateral, with a slightly flattened antero-internal surface and a rounded external surface. Superiorly, the horn cores are initially more circular (section b), and then more ovoid (section c). This specimen is both smaller and has a different horn core cross-section than *Capra libanotica* (Kretzoi) (ex-*C. primigenia* Fraas) from M. el-Zuttiyeh (Bate 1927) and elsewhere.

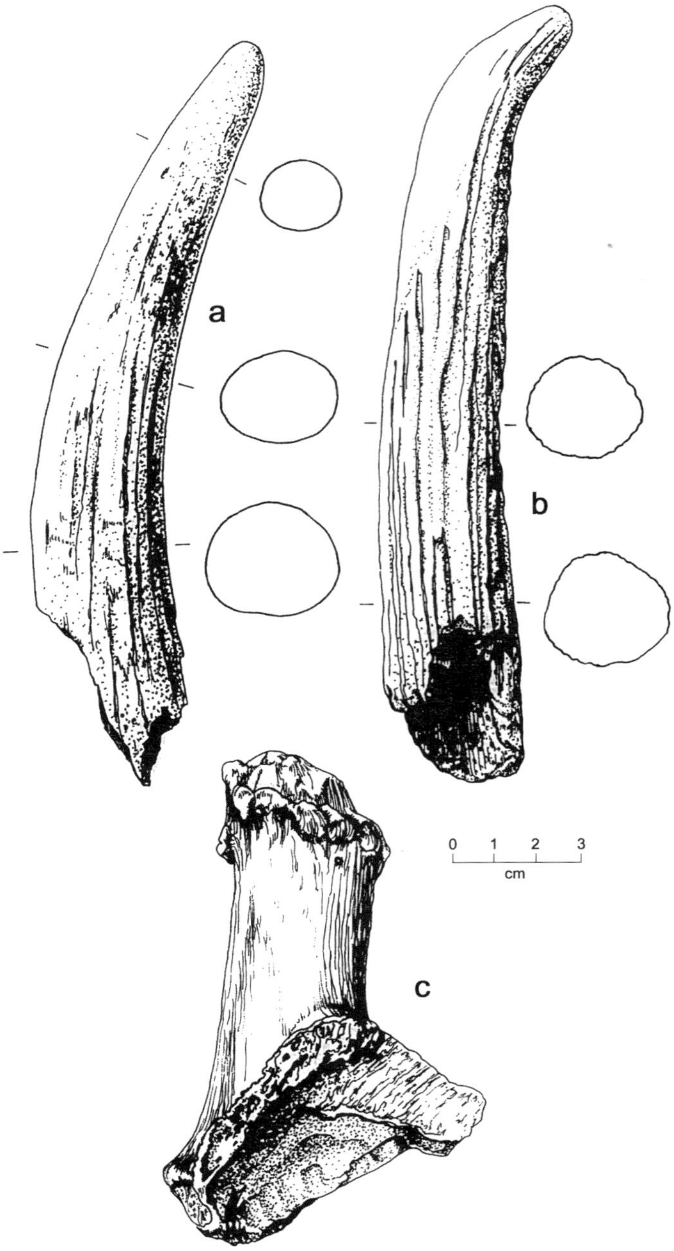

Figure 8.17. *Capra* cf. *ibex* (a) partial left horn core from T-89, 2, no. 282; *Cervus elaphus* ssp. (b) terminal antler segment including bez tine from T-89, 2, no. 208, and (c) frontlet fragment with right pedicle and rosette from S-89, 4, no. 71.

The latter is an extinct form of ibex (following Hooijer 1961a) and has an elliptical cross-section, lacking a sharp edge anteriorly, with the outer surface flattened. The Yarımburgaz specimen is generally larger in size than specimens from Ksar 'Akil, all of which were assigned by Hooijer (1961b) to *C. aegagrus,* and in which the horn core basal cross-section is oval, the interior surface flattened, the outer surface convex (throughout), and a weak anterior keel or edge is present somewhat above the base. Overall the Yarımburgaz specimen approximates this morphology and is attributed to *Capra* aff. *aegagrus* ssp. (bezoar goat). This is a very northward occurrence for this species. It occurs in southern and northeastern Turkey, and in Crete and Aegean islands, with a major distribution eastwards into Afghanistan and Baluchistan. It is considered ancestral to the domestic goat. It is known to have had a prior broader Levantine distribution and is identified in Mousterian and/or Upper Paleolithic localities in Lebanon (Ksar 'Akil, Antelias), perhaps Syria (Douara), and in Israel (Quneitra, Qafzeh, Kebara, Hayonim).

Two specimens of horn core represent an ibex (Appendix 8.43C). A tip of a right horn core (no. 274a) derives from S-89, level 4, and has the appropriate subcircular cross-section, and exhibits a slight medial incurvature. The other specimen (no. 282), from T-89 level 2, comprises most of the distal portion of a left horn core. It exhibits slight longitudinal furrows, has a more thickened, ovoid cross-section inferiorly, and shows progressive flattening of its inner surface terminally; there is slight, gentle internal-ward incurvature upwards to the tip. Ibex is quite commonly recorded in Levantine Mousterian (or antecedent) sites, including Zuttiyeh, Oumm Qatafa, Tabun (E, ?C, B), Wad (G to C), Erq el Ahmar (H to B), Abou Sif, Emireh, and in Lebanon at Antelias. Its distribution has not included Turkey in recent times. Unspecified Caprini (including *C. ibex, C. aegagrus, O. ammon)* are dominant ungulates in the lower occupation horizons of Karain Cave (Otte et al. 1998).

ANTILOPINI

(Figures 8.18-8.19)

A.71, 12 ,14. Right horn core, lacking distal portion. *Gazella* sp. cf. *subguttorosa*.

A.71, indet. level (sieving), 32. Left distal tibia. *Gazella* sp.

Y-88, < 5 (under stone fall), 668. Proximal right mt-frag., with ant.-lat. portion only of the proximal articulation. cf. *Gazella* sp.

U-89, 1, 29. Left mandible fragment, juvenile, preserving dp_3, dp_4 and M_1 and alveolus of M_2.*Gazella* sp. (small).

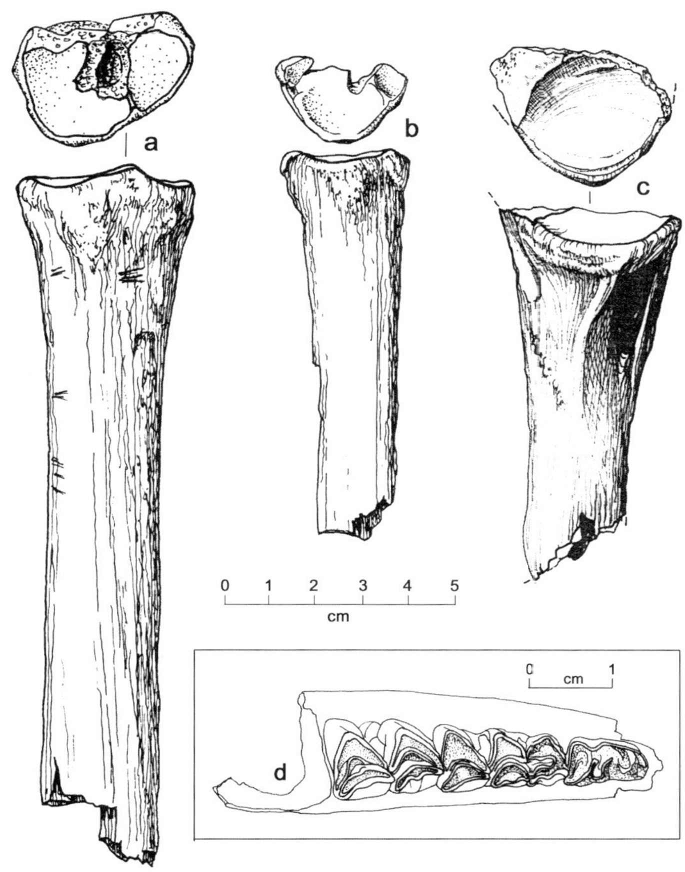

Figure 8.18. Large bovid cf. Alcelaphini (a) proximal left metacarpal with damaged articulation and cut marks on diaphysis from Y-88, 1a, no. 675; *Gazella* sp. indet. (b) proximal portion of right metatarsal with incomplete articulation from Y-88, 5, no. 668; Large bovid cf. *Bos/Bison* (c) proximal glenoid fragment of right scapula from T-89, 2, no. 547; (d) enlarged view of *Gazella* sp. (d) left mandible fragment, juvenile with dP_3, dP_4 and M_1, from U-89, 1, no. 29.

A single horn core (A.71, 12, 14) represents *Gazella*. It is a large species, rather more so than *Gazella arista* (Bate) from Mt. Carmel (which has been probably wrongly claimed to be merely a large *G. gazella*, and is well outside the range in horn core dimensions of that species) (Appendix 8.43C). The Yarımburgaz form is broadly comparable in size to *G. subgutturosa*, now present in southern and eastern Turkey and eastwards to southern central Asia. *Gazella* indet. is also represented by a distal tibia (no. 12) from this square. There is also a portion of proximal metatarsal (no. 668) from Y-88 (below level 5). A juvenile mandible fragment (no. 29), with dp_3 to M_1, from U-89 level 1 also is referable to *Gazella* sp. indet. (Appendix 8.43A). This is an unexpectedly northern distribution for *Gazella*. The characteristic form in the Levantine Pleistocene is *G. gazella*, its earliest occurrence probably at 'Ubeidiya and subsequently at Gesher Benot Ya'akov and Evron (all are cf. sp.) and definitively at Oumm Qatafa (E, D), Tabun (F upwards); it is a ubiquitous species at Mousterian sites in the southern Levant. In the northern Levant, it occurs at Ksar 'Akil, (once attributed to *G. dorcas)* and also at Douara (Syria) as well. Antilopini indet. has been reported from Latamne. No antilopini are recorded in the Karain Cave occupations.

Clearly felines and canines are the next most abundant carnivores after ursids. However, both groups are overall much less frequently represented than ursids. Lion occurs in three sedimentary phases, represented by one individual in each instance. Another probable pantherine may also occur. The bulk of this group, concentrated in one of two sedimentary phases, comprise lesser felines, most probably representing wild cat, swamp (or jungle) cat and caracal, and constituting very few individuals in each instance. Although the three lesser felines overlap in their natural distributions even to the extent of some sympatry in the southern Levant (see Dayan et al. 1990), such is not apparently the case now in Anatolia. Although both caracal and jungle cat are Anatolian in distribution, the latter farther northward than the former, neither definitively penetrates into Turkish Thrace. However, with reduction and transformation of water barriers with glacially-induced sea regressions, such distributions may well have been otherwise at times in the past (and as is documented in the case of some Rodentia).

The Canidae include both vulpines and canines. On the basis of both size and some morphological data, several taxa are represented within each group. The foxes appear to represent extant *V. vulpes*, and another smaller and perhaps extinct species. The canines appear to constitute wolf-like forms, both smaller and larger species; however, the presence of jackal is not to be excluded. Another quite large canid with cuonine affinities but otherwise unidentifiable is also represented in a basal part of UC infill.

Among the ungulates, artiodactyls dominate and perissodactyls are notably rare. A few fragments attest to rhino in the three upper sedimentary phases. A large caballine horse is the main perissodactyl in the assemblages, rare in V-phase and slightly less so in the succeeding W- (2 squares) and X- (4 squares) phases. There is no evidence for another Equine, particularly *E. hydruntinus/E. hemionus*.

Among artiodactyls the most uncommon are suids and bovids. *Sus* is rarely present in the W-, X- and Y-phases. *Bos/Bison* occurs in 2 phases (W and X), gazelle in 3 phases (only at the base and top of the sequence), and caprines represented by two species in W and/or X phases. Another bovid, perhaps (?) an alcelaphine, occurs in X-phase, but identification is quite tenuous.

Cervids are the most common artiodactyls. They include fallow deer (4-phases), red deer (5-phases), roe deer (1 phase, X only), and a form of Megalocerini (3-phases). In the oldest part (S-phase) of the infill, fallow and red deer and gazelle co-occur, along with Canidae indet. and traces of hyaena (coprolite). The next youngest (T) phase contains only a trace of gazelle. Most cervids occur in W- (3 spp.) and X- (4 spp.) phases, and rarely in the antecedent V-phase (1 sp.) and in the latest or Y-phase (2 spp.)

The composite faunal assemblage from Yarımburgaz LC is clearly indicative of a local fauna of mid-Pleistocene age. The salient biostratigraphic indicators are represented among each of the three orders of larger mammals. Among Carnivora these are *U. deningeri* ssp. (Ursidae), *P. leo* ssp., and probably *F. silvestris ssp. (Felidae), C. crocuta* cf. *praespelaea* and *P.* cf. *brevirostris* (Hyaenidae). Also suggestive are several canid species, only provisionally determinable, including small vulpine, small wolf, and cuonoid (only at the base of UC infill) forms. Among Perissodactyla, *E. c.* aff. *mosbachensis* (Equidae) is indicative. Among Artiodactyla (Cervidae) there are *Megaloceros* sp., *C. elaphus* ssp., and perhaps *Capreolus* aff. *süssenbornensis*. However, the biostratigraphic implications are neither straightforward (due to fragmentation and/or lack of critical diagnostic elements) nor fully confirmatory in respect to temporal assignment. Broadly considered a minimally mid-Middle Pleistocene age is congruent with the composite assemblage, whereas certain elements, among carnivores particularly, tend to suggest a still greater antiquity in some cases. These seeming incongruences cannot be readily resolved on the basis of the collections at hand. The application of isotopic geochronometry has also been insufficient or inconclusive to further such resolution.

The biogeographic affinities of the large mammal assemblage are wholly Palaearctic. There are no certain components representative of the Ethiopian realm, unlike Lower or even mid-Pleistocene occurrences in the Levantine area in which several such taxa may be variously represented. Unfortunately, the identity of a cuonoid-like canid and a presumptive medium-sized antelope are sufficiently indeterminate so as not to affect this conclusion. Several taxa afford links with the west Asian region in particular, rather than Europe proper.

The macrofauna lacks any taxa that are particularly indicative of past glacial paleoclimatic regime. This is not to suggest that such may not have been the case, only that the particular taxa in question are not confirmative of such. Rather the composition and diversity of the composite macrofauna is more a reflection of the local setting, including a very large cave interior and the presence of a diversity of provincial habitats and ecotones in the surrounding area. The (rodent) microfaunal remains do in fact reveal important evidence of altered range displacements, which can only have been a reflection of glacial paleoclimatic effects.

The admirable taphonomic studies by M. C. Stiner (this volume; Stiner et al. 1996) afford a full perspective on the overall composition, preservation condition, and differential make-up of skeletal element proportions within and between larger mammal taxa represented in the composite and individual (depositional phase) assemblages. These assemblages clearly constitute a palimpsest attendant upon the differential operation of various biotic processes in the course of the sediment infilling.

Micromammals constitute at least two such different processes (v. Koenigswald et al., this volume). Chiropterans (bats) were evidently autochthonous to the cavern setting, as some are still now. Rodentia (rodents) and (perhaps) Lagomorpha (hares, rabbits) were predominately the prey of large predatory birds, and perhaps particularly owls, that were periodic or frequent visitors to these karstic caverns. However, in the case of both and perhaps particularly lagomorphs, the predatory activities of canids and lesser felines may well have played significant roles in the accumulation of such residues (Stiner et al. 1996; and this volume).

Among the larger faunal elements other processes evidently prevailed. The Ursidae (bears) comprise at least two separate taxa, *U. deningeri* and *U. arctos*[1], the latter very rare (1-2 individuals) and almost exclusively from a single (X) depositional phase. *U. deningeri* is extraordinarily abundant, in excess of 90% of the total assemblage (if counted by element). It is represented by a full range of skeletal elements; however, there is some differential occurrence of body parts either due to preservation and/or recovery according to excavation extent and situation. Skeletal elements co-occur in abundance, but disarticulation of elements is very common. This species occurs from phase U through Y (5 units) in the LC, although most overwhelmingly in phases W and X. (The species also occurs in the lower infill of the UC, *e.g.*, square A.40.) The composite cave bear sample comprises over 40 individuals, perhaps as many as 45 in all, and is distinguished by a bi-modal (U-shaped) attritional mortality profile (Stiner this volume; Stiner 1998); there is a relative paucity of prime-age adults, and females slightly outnumber male individuals (Stiner et al. 1998). The composition of this aggregate population indicates normal, non-violent attritional process (including starvation, cannibalism, etc.) and matches that known from hibernation related mortality in other (including recent) circumstances.

The non-ursid Carnivora and the ungulate taxa are manifestly different in aspect and reflect distinct process(es) of accumulation. Appendix 8.46 provides a summary enumeration, at the genus or higher level, of the major categories of skeletal elements and their frequency. The contrast with skeletal representation of *Ursus* is pronounced and striking (see also Stiner, this volume). The major characteristics of the former two groups are:

- similar or same (12) number of taxa
- similar number of skeletal elements (99:109 NISP)
- dominance of autopodial elements in both samples (44:61)
- relative rarity of proximal and distal fore-/hind limb elements, with higher relative frequency in the (non-ursid) carnivores, and especially those of zygopodium.
- general rarity of axial elements and of girdle elements, the latter particularly in ungulate sample.
- head (skull s.l.) elements of broadly comparable frequency, and approximately one third of total NISP.
- high rates of fracture and hence generally low element incompleteness.
- moderately rate of damage from high predators (gnawing, tooth punctures/fractures) on both ungulates (23%) and (non-ursid) carnivores (18%).
- *very* rare signatures indicative of hominid-effected skeletal element damage.

During the infilling of the Yarımburgaz LC, the cavern experienced repetitive biotic events as evidenced by faunal and humanly-produced lithic assemblages. The diversity of ungulate taxa, represented by a limited array of skeletal elements, attests to substantial carnivore activity, particularly canid, in the cave and was largely a consequence of predation, scavenging and collecting activities in association with episodic denning. The human impact on the co-occurring faunal residues was correspondingly minimal, even negligible. However, the nearly 1700 lithic artifacts produced during successive intervals of human utilization of the chamber attest to both human presence and distinctive patterns of raw material exploitation, usage, knapping activities, and attendant tool and by-product production (see Kuhn, this volume).

Editors' note

1 At the time he wrote this chapter, F. C. Howell was not aware that some of the cave bear remains could be attributed to *U. spaelaeus* (see Tsoukala, this volume).

REFERENCES CITED

Alcalà, L., Morales, J., 1989. Los carnivoros del Pleistoceno medio de Cúllar de Baza-1 y Huéscar-1 (Cuenca de Guadix-Baza), in: Alberdi, M.T., Bonadonna, F.P., (Eds.), *Geologia y Paleontologia de la Cuenca de Guadix-Baza.* Trabajos Sobre el Neogeno-Cuaternario, Madrid: Museo Nacional de Ciencias Naturales, CSIC, Madrid, pp. 215-223.

Altuna, J., 1981. Fund eines Skeletts des Höhlenlöwen (*Panthera spelaea* Goldfuss) in Arrikrutz, Baskenland. *Bonner zoologische Beiträge* 32, 31-46.

Azanza, B., Sanchez, B., 1990. Les cervidés du Pléistocène moyen d'Atapuerca (Burgos, Espagne). *Quaternaire* (Paris) 1, 197-212.

Argant, A., 1991. Carnivores Quaternaires de Bourgogne. Documents des Laboratoires de Géologie, Lyon, No. 115. Departement des Sciences de la Terre, Université Claude-Bernard-Lyon 1.

Ballesio, R., 1975. Étude de *Panthera (Leo) spelaea* (Goldfuss) nov. subsp. (Mammalia, Carnivora, Felidae) du gisement Pléistocène moyen des Abimes de la Fage à Noailles (Corrèze). *Nouvelles du Archives Muséum d'Histoire Naturelle de Lyon* 13, 47-55.

Ballesio, R., 1986. Les carnivores du Pliéistocène d'Oubeidiye (Israel), in: Tchernov, E., (Ed.), *Les Mammiféres du Pléistocène inférieur de la vallée du Jourdain a Oubeidiyeh.* Mémoirs et Travaux du Centre de Récherche Français de Jerusalem, No. 5. Association Paléorient, Paris, pp. 63-92.

Bate, D.M.A., 1927. On the animal remains obtained from the Mugharet- el-Zuttiyeh in 1925-1926, in: Turville-Petre, F., (Ed.), *Researches in Prehistoric Galilee.* 1925-1926. British School of Archaeology in Jerusalem, London, pp. 27-52.

Bate, D.M.A., 1937. Palaeontology: The Fossil Fauna of the Wady el-Mughara Caves, in: Garrod, D.A.E., Bate, D.M.A., (Eds.), *The Stone Age of Mount Carmel,* Vol. I. Clarendon Press. Oxford, pp. 139-228.

Bishop, M.J., 1982. *The mammal fauna of the early Middle Pleistocene cavern infill site of Westbury-Sub-Mendip, Somerset.* The Palaeontological Association. London. Special Papers in Palaeontology, No. 28.

Bonifay, M.-F., 1971. Carnivores Quaternaires du Sud-Est de la France: Mémoirs du Muséum National d'Histoire Naturelle, Paris, n.s., Série C, *Sciences de la Terre,* 21 (fasc. 2), pp. 1-377.

Bonifay, M.-F., 1980. Le cheval du Pléistocène moyen des grottes de Lunel-Viel (Hérault). *Equus mosbachensis palustris* n. ssp. *Gallia Préhistoire* 23, 233-281.

Bouchud, J., 1972. Les grands herbivores rissiens des "Abimes de la Fage" en Corréze (Cervidés, Bovidés, Capridés, Rupicaprinés, Suidé et Equidés). *Nouvelles Archives du Museum d'Histoire Naturelle de Lyon* (fasc. 1), Lyon, pp. 33-59.

Bouchud, J., 1974. Etude préliminaire de la faune provenant de la grotte du Djebel Qafzeh près de Nazareth (Israel). *Paléorient* 2, 87-102.

Brinkmann, A., 1921. Canidenstudien. *Videnskabelige Meddelelser fra Dansk naturhistorisk Forening I Kobenhav* (I-III) 72, 1-43.

Cardoso, J.L., Eisenmann, V., 1989. *Equus caballus antunesi,* nouvelle sous-espèce Quaternaire du Portugal. *Palaeovertebrata* 19, 47-72.

Chauviré, C., 1962. Les gisements fossilières Quaternaires de Chatillon-Saint-Jean (Drôme). Université de Lyon, Faculté des Sciences (Géologie). Thèse 3éme Cycle.

Clot, A., 1980. La Grotte de la Carrière (Gerde. Htes Pyrénées). Stratigraphie et Paléontologie des Carnivores. Travaux du Laboratoire de Géologie de l'Université Paul Sabatier, Toulouse. Thèse 3éme Cycle. 2 vols.

Clutton-Brock, J., 1962. Near Eastern canids and the affinities of the Natufian dogs. *Zeitschrift für Tierzüchtung und Züchtungsbiologie* 76, 326-333.

Appendix 8.3. Measurements of femur and humerus in Pleistocene and Recent species of *Vulpes*.

Dimension	*Vulpes vulpes* Recent-Bigorre	*Vulpes vulpes* ssp. Gerde, H. Pyrenees	*V.v. jansoni* Escale	*V.v.* ssp. Lunel Viel	*Vulpes* *praeglacialis*	Yarımburgaz S-89,4,101
A. FEMUR	**(France) (1) (46/50)**	**(1) (N = 7-14)**	**(2) (N = 4)**	**(2)**	**(2)**	
Max. total length	124.3 ± 7.28 (112.6 - 18.2)		131., 124.5			
Mid-shaft breadth	9.04 ± 0.56 (7.7 - 10.5)	8.98 ± 0.54 (8.0 - 10.1)	8.75 (8.3 - 9.2)	8.8	7.6	9.34
Distal breadth	20.95 ± 1.30 (19.0 - 24.0)	21.35 ± 1.165 (19.7 - 23.7)	20.325 (19.6 - 21.0)		17.5	22.10
Distal a-p	22.21 ± 1.30 (20. - 25.)	22.31 ± 0.79 (21. - 23.2)			24.92	
	Vulpes vulpes **Recent-Bigorre**	***Vulpes vulpes* ssp.** **Gerde, H. Pyrenees**	***V.v. jansoni*** **Escale**	***V. praeglacialis*** **angustidens**	***V. praeglacialis*** **Escale**	**Yarımburgaz** **U-89, 1, 85**
B. HUMERUS	**(France) (1) (41-52)**	**(1) (N = 3 / 19)**	**(2) (N= 5)**		**(2) (N = 2 / 3)**	
Max. total length	120.6 ± 6.725 (108.8 - 134.3)	127.05 ± 6.61 (118. - 133.)	121.4 (119.3 - 122.8)			
Mid-shaft breadth	8.25 ± 0.55 (7.3-9.8)	8.4 ± 0.5 (7.5-9.2)	7.4 (7.2 - 7.5)		6.7 (6.4 - 6.9)	8.1
Distal breadth	20.7 ± 1.195 (18.5-23.0)	20.6 ± 1.0 (19.2-22.0)	19.7 (18.8 - 20.2)	Villány (3) 19.8 Hundsheim (4) 18.4	17.8 - 19.2	18.83
Distal a-p	15.1 ± 0.98 (13.1 - 17.0)	15.4 ± 1.175 (13.7 - 17.4)				15.16

(1) A. Clot, 1980; (2) M.-F. Bonifay, 1971; (3) T. Kormos, 1932; (4) E. Thenius, 1954.

Appendix 8.4. Femoral measurements in some *Canis lupus* subspecies and in *Canis aureus*.

		Canis lupus mosbachensis			*Canis lupus* ssp.				*C. aureus*
	Yarımburgaz	**Escale (1)**	**Stránská**	**Gr. L'Eglise (3)**	**Denmark (4)**	**Italy**	**India**	**Israel (6)**	**Israel (6)**
Measurement	**A.71,--, 1**	**(N = 3-7)**	**Skála (2)**	**(Grotte xiii)**	**(sub-fossil) (3)**	**(Recent) (N=2)**	**(pallipes) (5)**	**(Recent)**	**(Recent)**
Maximum length	199.3	171.7 (160.6-179.0)	194.7	179.5	216.-240.	214.;250.	203.0	192.9	164.2
Maximum breadth, proximal	42.7	36.2 (38.4-39.0)	42.0	37.0	51.-52.2	44.8; 55.6			
Mid-Diaphyseal breadth	13.5	12.13 (11.0-12.8)	11.8-12.4	13.0	15.3-16.7	15.8; 21.5	14.5		
Maximum breadth, distal	33.5	30.9 (22.3-33.0)	33.0 - 34.0	29.0	39.5 - 41.0	38.6; 48.9		34.57	28.0

(1) M.-F. Bonifay, 1971; (2) R. Musil, 1971; (3) C. Suire, 1969; (4) M. Degerbol, 1933; (5) A. Brinkmann, 1921; (6) Measurements per E. Tchernov and Y. Rak.

Appendix 8.5. Metacarpal III measurements in *Canis lupus mosbachensis* and some late mid-Pleistocene *C. lupus lupus* samples (France).

		Canis aureus	*Canis lupus mosbachensis*				*C. l. lupus*
	Yarımburgaz	Israel	Escale	Arago	Hundsheim	Stránská	Abri Suard
Measurement	**U-89, 1, 94**					**skála N=4/5**	**(2)**
Max. length		77.7	63.8 - 63.8		70.6	72.4 (71.7 -73.2)	
Prox. a - p	9.95	13.01	(8. - -)	10.8	6.0	11.3 - 11.8	14.0 - 15.0
Prox. breadth	7.46	8.28	(6. - -)	7.4	7.9	8.34 (8.0 - 8.6)	10.0 - 10.6
Distal a - p		10.05			9.0		
Distal breadth		10.9	7.7		8.2	8.475 (8.1 - 9.2)	

Appendix 8.6. Metacarpal IV measurements in some Pleistocene and extant canids.

		C. l. mosbachensis		*Xenocyon*	*Canis lupus* ssp.			*Canis aureus*	Cuon
	Yarımburgaz	Escale (1)	Stránská	Stránská	Lazaret (3)	Combe	Israel (5)	Israel	*alpinus*
Measurement	**P-90, 1, 232**		**skála (2)**	**skála (2)**		**Grenal (4)**	**(Recent)**	**(Recent)**	***europaeus***
Maximum length	7983	63.6; 65.8	72.5		78.8 (75.8-83.3)	89.0	87.7	76.35	70.0;77.9
Maximum breadth, prox.	7.92	(7.0)	7.1; 7.3	9.0; 11.1	9.1 (8.6 - 10.0)	10.5	7.48	8.52	
Ant.-post., prox.	14.59	(10.0)	11.5		12.8 (12.1-13.7)	13.75	13.75	12.77	
Maximum breadth, distal	10.45	7.7	9.0		10.35 (9.8 - 11.5)	11.5	9.05	10.64	10.3; 11.3
Ant. post.,dist.	12.86						10.19	11.33	

(1) M.-F. Bonifay, 1971; (2) R. Musil, 1971; (3) P. Vallensi, 1994; (4) C. Suire, 1969; (5) Measurements per E. Tchernov & Y. Rak.

Appendix 8.7. Metatarsal II measurements in some Pleistocene and Recent canids.

		C. l. mosbachensis		*Canis*	*lupus* ssp.			*Canis aureus*
	Yarımburgaz	Escale	Stránská	Lazaret (5)	Pech de	Abri Suard	Israel (4)	Israel (4)
Measurement	**Y-88, --, 685**	**(1)**	**skála (2)**	**(3-6)**	**L'Aze (3)**	**(3)**	**(Recent)**	**(Recent)**
Maximum length	78.5	(+64)	68.9; 74.2	77.22 (71.8 - 84.2)	87	84.5	69.75	59.9
Maximum breadth, proximal	(4.,8)	(+4.5)	5.1; 6.0	6.67 (5.9 - 8.2)	8.2 - 9.0	9.0; 10.5	4.94	4.94
Mid-diaphysial breadth			6.5; 6.3	7.18 6.6 - 8.3)	7.0; 8.5	7.6; 8.5		
Maximum breadth distal	9.6	(+8)	7.4; 8.2	10.38 (9.4 - 11.3)	11	11.7	9.09	7.03

(1) M.-F. Bonifay, 1971 (2) R. Musil, 1971; (3) C. Suire, 1969; (4) Measurements per E. Tchernov & Y. Rak; (5) P. Valensi, 1994.

Appendix 8.8. Metatarsal III measurements in some Pleistocene and Recent canids.

		C. l. mosbachensis			*Canis lupus* ssp.		*Cuon priscus*	*Canis aureus*
	Yarımburgaz	Escale	Stránská	Lazaret (6)	Abri Suard	Israel (4)	Hundsheim	Israel (4)
Measurement	**T-89, 4, 865**	**(1)**	**skála (2)**	**(3)**	**(3)**	**(Recent)**	**(5)**	**(Recent)**
Maximum length	76.88	71.4 - 72.5		85.3 - 85.4	93.7	78.28	94.8	65.59
Maximum breadth, prox.	9.86	(8)	8.57 (8.1-8.9)	7.0 - 7.5		9.21	14.4	7.17
Ant.-post.; prox.	13.82	(11)		8.2 - 13.8		12.85	17.9	11.21
Maximum breadth, dist.	7.98	(7.5)	8.8	9.4 - 9.6		8.39	10.4	6.43
Ant.-post.; distal	9.53					11.20	12.8	7.30

(1) M.-F. Bonifay, 1971; (2) R. Musil, 1971 (3) C. Suire, 1969; (4) Measurements per E. Tchernov & Y. Rak; (5) E. Thenius, 1954; (6) P. Valensi, 1994.

Appendix 8.9. Metatarsal IV measurements in some Pleistocene and Recent canids.

Measurement	Yarımburgaz R-90, 1, 441	*C. l. mosbachensis* Escale (1)	Stránská skála (2)	Lazaret (3) (2 - 3)	*Canis lupus* ssp. Pech de L'Aze (4)	Israel (5) (Recent)	*Canis aureus* Israel (5) (Recent)	*Xenocyon* Stránská skála (2)	*Cuon priscus (6)* Hundsheim
Maximum length		(72)		85.35 (85.3 - 85.4)	98.0	83.23	67.08		98.7
Maximum breadth, proximal	10.84	(9.5)	6.1 - 7.7	7.0 - 7.5		7.52	5,6	11.0 - 12.3	9.2
Ant.-post., proximal	16.72	(11)		11.8 (8.2 - 13.8)	11.5	13.67	10.33		16.3
Maximum breadth, distal		(8)		9.4 - 9.6	10.5	8.52	6.09		10.3
Ant.-post., distal					12.0	11.12	7.13		12.6

(1) M.-F. Bonifay, 1971; (2) R. Musil, 1971; (3) P. Valensi, 1994; (4) C. Suire, 1969; (5) Measurements per E. Tchernov & Y. Rak; (6) E. Thenius, 1954.

Appendix 8. 10. Measurements of dP_4 in some *Canis* species.

Measurement	Yarımburgaz Y-88, 1, 684	*Canis lupus lupus* Taubach (1)	Gerde (2)	*C. a. lupaster* Geula Cave level B1- (3)	*C. aureus* Israel (4) (Recent)
Maximum length, m-d	11.36	12.0	14.3	12.3	10.19
Maximum breadth, b-l	4.78	6.1	5.6		5.12

(1) H. Hemmer, 1977; (2) A. Clot, 1980; (3) G. Petter and E. Heintz, 1970; (4) Measurements per E. Tchernov and Y. Rak.

Appendix 8.11. Measurements of distal tibia in some canid species.

	Yarımburgaz	*Canis lupus mosbachensis* Mosbach;	Chatillon-	*Canis l. lupus* Lazaret (3)	Santenay (4)	Denmark (5)	*Xenocyon* Stránská	Koneprusy	Pirro	*Cuon priscus*
Measurement	A.40,14,2	Jagsthausen (1)	St. Jean (2)	(9; 1)	(6)	(Subfossil) (3 - 4)	skála (6)	(7) (C-718)	Nord (9)	Escale (8)
Mid-shaft, br	15.9	(12) ; 13.8	15.0 - 15.3			15.6 - 17.7	18.7		15.7	11.7
Mid-shaft, a-p	16.6			14.3 (13.5 -15.8)			12.7; 12.9		16.1	
Dist. articulation:										
max. br.	27.5	(21.); 24.0 Brassó = 21.0 P'furdo = 21.3	27.0 - 28.0	34.0	39.88 (37.6-42.6)	29.0 - 29.5	32.2 (23.3; 23.5)	32.29	26.6; 28.5	22.1
min., a-p	15.3						18.5 (17.5; 17.8)	22.4		12.5
max., a-p	20.0				36.65 (34.5-38.6)		24.3	24.85	19.6; 19.8	17.0

(1) E. Thenius, 1954; K. Weitzel, 1936; (2) C. Chauviré, 1962; (3) P. Valensi, 1994; (4) A. Argant, 1991; (5) M. Degerbol, 1933; (6) R. Musil, 1971; (7) personal observations/ measurements; (8) M.-F. Bonifay, 1971; (9) L. Rook, personal communication.

Appendix 8.12. Measurements of ph-1 in some *Canis lupus* ssp.

Measurement	Yarımburgaz S-89, 4, 73 (prox.)	*Canis lupus* P. de l'Azé (4) Gr. Suard (6)	Chateau	*Canis lupus mosbachensis* Escale *manus*, III, ph.1	Hundsheim	Stránská Skala
Max. length		27.0 - 33.5	23.8; 28.6	(24)	26.8 - 28.0	25.2 - 31.0
		22.0 - 36.0				
Proximal a-p	9.19	9.5	7.9; 8.8	(8.5)		
		8.5 - 11.0				
Proximal br	9.70	10.0 - 11.0	9.3; 9.0		(8.2)	
		10.5 - 11.5				
Mid-shaft br	5.84	6.0 - 7.0	5.4; 6.0	(4.5)	(5.0)	
		6.0 - 7.5				
	P-90, 2, 473 (dist)					
Distal, a-p	5.11	6.5 - 7.0	6.2; 6.5			
		7.5 - 8.5				
Distal, br	7.83	8.0 - 8.5	7.2; 7.5	(7)	(7.7)	
		8.1 - 10.0				

Appendix 8.13. P^4 length in *Canis* species/subspecies.

Taxa / author	No.	Dimensions
C. lupus lupus - N. Eurasia		
K. Stockhaus (1965)	51	24. 63. ± 1.318 (21.6 - 27.8)
H. Okarna & T. Buchalczyr (1993)		
Polish Carpathians (male)	49	26.8 ± 1.1 (24.4 - 29.2)
(female)	29	25.2 ± 1.3 (21.2 - 27.9)
C. lupus pallipes - West. Asia		
K. Stockhaus (1965)	10	19.94 ± 2.057 (17.4 - 22.8)
E. Hauck, (1947)	7	20.89 (19.0 - 22.5)
A. Brinkmann (1921)	8	21.19 (20.0 - 23.)
J. Clutton-Brock (1962)	22	--- (20.25 - 23.7)
T. Dayan (1994)	34	22.42 ± 1.10 (Sd)
Mediterranean sample	10	22.62 ± 1.31 (Sd)
Desert sample	24	22.34 ± 1.03 (Sd)
E. Tchernov & F.F. Valla (1997)	13	22.533 ± 1.027 (Sd)
C. lupus arabs		
J. Clutton-Brock (1962)	7	--- (19.55 - 21.0)
Bj. Kurtén, (1965)	5	20.7 ± 0.24 (19.9 - 21.3) (Sd = 0.53)
C. lupus - Palestine, U. Pleist.		
Postglacial (Bj. Kurtén, 1965)	4	21.3 ± 0.5 (20.3 - 22.5)
Last Glacial (Bj. Kurtén, 1965)	3	24.63 ± 0.24 (24.3 - 25.1)
C. aureus lupaster		
J. Clutton-Brock (1962)	5	--- (17.95 - 19.4)
C. lupaster (DMA-B, 1937)		
(after J. Clutton-Brock, 1962)	3	--- (18.05 - 21.9)

Appendix 8.14. M_1 length in *Canis* species/subspecies.

Taxa/Author	No.	Dimensions
C. lupus lupus - N. Eurasia		
K. Stockhaus, 1965	46	27.64 ± 1.530 (24.6 - 32.0)
H.I Koarma & T. Buchalcyzk, 1933		
Polish Carpathians	48 (male)	29.3 ± 1.2 (27.1 - 32.2)
	28 (fem.)	27.7 ± ??(24.6 - 30.5)
Turkey (S. Payne, 1983)	11	27.8 ± 1.08 (26.1 - 29.5)
C. lupus pallipes - W. Asia		
K. Stockhaus (1965)	10	22.52 ± 2.114 (20.2 - 23.8)
T. Hauck (1947)	5	24.2 (23.5 - 25.0)
T. Dayan (1994)	36	24.41 ± 1.13
Mediterranean sample	10	24.66 ± 1.15
Desert sample	26	24.32 ± 1.13
E. Tchernov & F.F. Valla (1997)	13	25.22 ± 1.388
C. lupus arabs		
B. Kurtén (1965)	5	23.1 ± 0.5 (21.7 - 24.2) (Sd = 1.1)
S. Payne, (1983)	7	23.5 ± 0.81 (22.0 - 24.2)
C. lupus - Palestine, U. Pleistocene		
Postglacial (B. Kurtén, 1965)	2	24.05 (23.8 - 24.3)
Last Glacial (B. Kurtén (1965)	6	28.2 ± 1.0 (25.1 - 31.6) (Sd = 2.5)
Douara cave (S. Payne, 1983)	2	22.7; 22.5
Palegawra (P. Turnbull & C. Reed, (1974)	1	21.9
Kebara - C; Tabun - C	2	23.9; 23.4
C. aureus (Israel - India)		
B. Kurtén (1963)	11	17.77 ± 0.23 (16.1-18.9) (Sd=0.93)
Turkey + Levant (S. Payne, 1983)	15	18.4 ± 0.98 (16.3-19.9)
C. lupaster (extant)		
B. Kurtén, (1965)	7	20.86 ± 0.37 (19.5-22.5) (Sd=0.99)
Egypt (S. Payne, 1983)	15	21.5 ±1.15 (20.3- 24.3)
C. lupaster - Palestine, U. Pleistocene		
B. Kurtén (1965)	6	21.75 ± 0.61 (20.4-23.7) (Sd=1.37)

Appendix 8.15. Measurements of upper carnassial (P^4) in some samples of *Panthera leo* ssp.

Locality	No.	maximum length	maximum breadth (at pr)	pa length	me length	blade length	me l x 100 / max l
Panthera leo fossilis							
Chateau (1)	1	40.7	- -	15.8	14.9	31.0	36.6
Isernia (2)	1	39.4	20.8	15.0	14.5	28.9	36.8
Mauer/Mosbach (3)	4-6	39.9 (36.4 - 45.1)	19.6 (18.3 - 21.7)	15.4 (14.0 - 17.5)	15.2 (13.8 - 16.8)	29.8 (27.3 - 32.7)	37,2.± 0.61 (35.9 - 39.3)
Petralona (4)	1	39.6	23.,3	15.8	14.4	30.2	36.4
Vértesszölös (5)	2	39.8; 36.7	22.0; 18.6				
Yarımburgaz (R-90, 1, 560)	1	41.37	(20.24)	15.64	15.62	31.01	37.76
Panthera leo spelaea							
Gerde (6)	3	38.5 (38.0-38.9)	19.0 (18.4 - 19.4)	15.1 (14.8 - 15.4)	15.7 (15.6 - 15.8)	28.5 (28.0 -28.7)	40.8 (40.1 - 41. 6)
Lazaret (7)	1	38.2	18.3	14.0	14.5	(29.3)	37.96
Lunel-Viel (8)	2	38.8; 36.4					
Azé (1)	2	41.3	22.8 - 23.0	15.5 - 15.7	15 5 - 15.7	31.3 - 31.6	37.5 - 38.1
La Fage (9)	2	36.9; 37.3	20.3; 20 6		14.3 ; 14.4		38.75; 38.6
L'Herm (6)	7/8	41.3 (40.3 - 42.0)	21.9 (20.8 - 22.8)	15.5 (14.8 - 16.4)	15.7 (15.0 - 16.7)	30.2 (29.0 - 31.3)	38.2 (36.8 - 39.8)
Germany occurrences (3)	6/7; 9	38.4 (33.8 - 42.4)	19.1 (18.0 - 20.7)	14.5 (13.0 - 15.4)	15.3 (14.0 - 17.3)	28.7 (25.6 - 31.0)	39.6 ± 0.43 (37.5 - 41.7)
Italy occurrances (2)	6/7	38.4 (33.7 - 41.7)	20.5 (17.3 - 22.1)	14.6 (13.0 - 15.8)	15.4 (13.8 - 16.9)	- - -	40.13 (38.3 - 41.1)
Repolust (10)	3	41.0 (40.3 - 41.4)	21.1 (20.7 - 21.4)	16.4 ((16) - 16.8)	16.4 (15.0 - 16.7)	31.5 (30.9 - 32.0)	40.0 (39.5 - 40.5)
Steiermark (10)	4/5	39.2 (38) - 40.3)		14.6 (14.3 - 15.0)	15.7 (14.5 - 16.8)	29.3 (28.4 - 30.4)	40.0 (38.0 - 41.5)

(1) A. Argant, 1991; (2) B. Sala, 1990; (3) G. Schütt, 1969; (4) E. Tsoukala, 1989; (5) D. Jánossy, 1990; (6) A. Clot, 1980; (7) P. Valensi, 1994; (8) M. -F. Bonifay, 1971; (9) R. Ballesio, 1975; (10) G. Schütt and H. Hemmer, 1978.

Appendix 8.16. Measurements of scapholunar in *Panthera leo fossilis* and *spelaea* subspecies samples.

Dimensions	Yarımburgaz T-89, 4, 799	Chateau (1) (N=3)	L'Herm (2) (N=11)	Caverna de qui (3) (N=2)	British Isles (4) (N=3)
Max. transv. br.	56.68	57.6 (55.3-59.5)	58.4 (53.0-61.0)	58.3; 59.5	53.1 (47.5-58.4)
Max. thickness, sup - inf	32.37	35.9 (35-37.60	35.(33.0-37.0)	31.6; 35.0	35. (28.7-40.6)
Max. thickness, a - p	45.97	52	34.9 (31.-39.8)	33.0; 34.0	35.6 (28.4-44.5)
Transv. diam. radial articular surface	51.94	(52.7)	50.6 (47.0-55.0)	48.8; 52.0	None

(1) A. Argant, 1980; (2) E/H Filhol, 1870; (3) D. Del Campana, 1919; (4) W. B. Dawkins & A. W. Sanford, 1867-1869.

Appendix 8.17. Proximal radius measurements in some samples of *Panthera leo* ssp.

Locality	N	Maximum breadth	Anterior-posterior breadth
Panthera leo fossilis			
Chateau (1)	2	51.0; 48.7	35.2; 34.8
Petralona (2)	1	42.2	30.5
Yarımburgaz:			
P-90, 1, 123	1	(54+)	(40 +)
R-90, 1, 53	1	58.8	38.7
Panthera leo spelaea			
Azé (1)	1	51.9	37.7
Jaurens (3)	5	43.2 (40.5-49.0)	
L'Herm (1)	2	51.0; 45.0	48.0; 38.0
Equi (4)	1	53	37.5
Tilde (5)	2	51.4; 48.5	35.9; 36.9
Wierzchowska (6)	6	45.3 (38.0-49.0)	30.3 (24.0-33.0)

(1) A. Argant, 1991; (2) E. Tsoukala, 1989; (3) R. Ballesio, 1980; (4) D. Del Campana, 1947; (5) A. Riedel, 1982; (6) K. Wojtusiak.

Appendix 8.18. Measurements of ph-2 (manus) digits of *Panthera leo spelaea* samples.

	Yarımburgaz	Chateau	L'Herm	British Isles
DIGIT III, MANUS				
Dimensions	T89, 4, 853	(1) (N=3)	(2) (N = 5)	(3)
Max. length	47.36	49. -(48.3 - 50.)	44.4 (43. - 46.)	50.8
Prox. breadth	23.02	22.3 (21.8 - 23.0)	20.0 (16. - 21.)	20.8
Prox. a - p	22.96	21.3 (20.3 - 22.5)	12.6 (12. - 13.)	17.8
Distal breadth	20.72	19.83 (19.5 - 20.00	23.2 (22. - 24.)	20.3
Distal a - p	15.01	14.9 (14.5 - 15.3)	16.2 (16. - 17.)	17.8
DIGIT V, MANUS				
Dimensions	T89, 4, 787	(1) (N=2)	(2) (N = 5)	**(3)**
Max. length	37.31	39.1; 39.9	38.6 (36. - 42.)	43.2
Max. breadth	18.71	19.0, 18.3	17.5 (17. -19.)	17.8
Prox. a - p	18.35	18, 17.2	11.8 (11. - 14.)	14.0
Distal breadth	16.95	17.7, 16.8	18.6 (17. - 20.)	16.5
Distal a - p	(14.03)	14.5, 13.2	14.8 (13. - 17.)	14.0

(1) A. Argant, 1980 (2) E./H. Filhol, 1870 (3) W. B.Dawkins & A. W. Sanford, 1867-1869.

Appendix 8.19. Measurements of proximal and distal fibula in *Panthera leo spelaea* samples.

Dimensions	Yarımburgaz T89, 4, 912	Yarımburgaz T89, 1, 31	Yarımburgaz P90, 2, 583	Chateau (1)	Lunel Viel (2)	British Isles (3)
Max. prox. br.	(34.8)			32.8, 34.1		
Max. prox. a - p	(17.0)			20.4, 20.0		16.5
Max. dist. br.		37.99	36.9	38.1, 36.1	31.0	31.2
Max. dist. a - p		18.06	18.0	21.3, 21.0	15.0	

(1) A. Argant, 1980 (2) M.-F. Bonifay, 1971 (3) W. B. Dawkins & A. W. Sanford, 1867-1869.

Appendix 8.20. Calcaneum measurements in some samples of *Panthera leo* ssp.

Locality	N	Maximum length	Maximum breadth	Tuber calcanei br.	Anterior articular br.
Panthera leo fossilis					
Chateau (1)	4	132.6 - 144.7	53.0 - 56.2	36.0 - 40.7	
Petralona (2)	1	(105.1)	46.8		
Yarımburgaz	1	140.34	60.85	42.1	38.28
(P90, 2, 523)					
Panthera leo spelaea					
Arrikrutz (3)	1	130.5	61.5	56.5	
Equi (4)	3	125.0; 132.0; 134.5	48.5; 50.0; 49.5	36.; 38.; 38.	
Germany localities (5)	3	112.0; 136.0, 136.0	45.0; 56.0; 64.0	29.; (32.); 39	
Tilde cave (6)	3	129.3; 130.5; 131.0	57.0; 56.0; 55.5		

(1) A. Argant, 1991; (2) E. Tsoukala, 1989 (3) J. Altuna, 1981; (4) D. Del Campana, 1947; (5) W. O. Dietrich, 1966; (6) A. Riedel, 1982.

Appendix 8.21. Measurements of pes cuneiforms in *Panthera leo spelaea* samples.

CUNEIFORMS (PES)					
Dimensions	**Yarımburgaz**	**Chateau**			
ENTOCUNEIFORM	T-89, 4, 852	(1)			
Max. transv. diam.	---	13.3			
Max. length, prox. dist.	28.57	28.8			
Height, dorso-plantar	---	20.2			
	Yarımburgaz	**Chateau**	**British Isles**		
MESOCUNEIFORM	T-89, 4, 801	(1)	(2)		
Max. transv. diam.	25.94	15.5	25.9		
Max. length, prox. dist.	34.33	19.1	13.97		
Height, dorso-plantar	34.17	32.7	----		
	Yarımburgaz	**Chateau**	**British Isles**	***P. gombaszoegensis***	
ECTOCUNEIFORM	T-89, 1, 16	(1)	(2) (N=3)	Chateau	Escale
Max. transv. diam.	31.76	30.3	33.9		
			(30.2 - 36.8)	22.4	21.0
Max. length, prox. dist.	26.74	28.4		19.2	18.0
			(20.1 - 24.4)		
Height, dorso-plantar	50.9	(39.6)	---	37.4	---

(1) A. Argant, 1980; (2) W. B. Dawkins and A. W. Sanford, 1867-1869.

Appendix 8.22. Measurements of metatarsal IV in *Panthera leo spelaea* and *Panthera leo fossilis* samples.

(P.l.f.)	***Panthera leo spelaea***					
Chateau (1)	**Lunel Viel (2)**	**L'Herm (3) (N = 7)**	**Nieder-lehme (4)**	**Thiede (4)**	**Wallertheim (4)**	**British Isles (5)**
--	133.0	153.9	(132.)	131.0	--	142.24
		(144.0 - 161.0)				
37.3, 37.7	29.0	38.6	--	32.0	38.0	21.34
		(38. - 40.)				
31.9, 31.6	26.8	27.6	(26.0)	28.0	23.0	40.6
		(25. - 30.)				
--	--	--	--	--	--	50.3
--	--	--	22.0	22.5	--	50.8

Bonifay 1971 (3); E/H Filhol 1870; W. Sanford 1867-69.

Appendix 8.23. Measurements of ph-2 pes in *Panthera leo spelaea* samples and in *Panthera gombaszoegensis.*

		Panthera leo spelaea				*P. gombaszoegensis*
	Yarımburgaz	**Chateau**	**L'Herm**	**Gerde**	**British Isles**	**Chateau - digit indet.**
Dimensions	**S-89, 4, 308**	**(1) (N = 3)**	**(2) (N = 5)**	**(3)**	**(4)**	**(1)**
Max. length	49.17	40.6 (40.2 - 41.)	54.3 (53. - 56.)	45.7	44.7	28.7, 36.I
Prox. breadth	21.58	20.6 (20.4 - 20.8)	21.0 (21. - 21.)	19.7	15.2	14.3, - -
Prox. a - p	20.68	21.13 (20.6 - 21.8)	12.67 (12. - 13.)	21.5	19.8	13.2, 16.2
Distal breadth	19.14	20.6 (20.0 - 20.8)	22.33 (22. - 23.)	20.3	23.1	12.6, 14.2
Distal a - p	15.38	14.7 (14.3 - 15.4)	16.0 (16. - 16.)	14.5	19.8	10.4, 10.4

(1) A. Argant, 1980; (2) E/H Filhol, 1870; (3) A. Clot, 1980; (4) W. B. Dawkins and A. W. Sanford, 1867-1869.

Appendix 8.24. Measurements of astragalus in some *Panthera leo spelaea* and in *Panthera gombaszoegensis* samples, and in *Homotherium crenatidens.*

		Panthera gombaszoegensis			*Homotherium crenatidens*	*Panthera leo spelaea*					
Dimensions	**Yarim-burgaz Y-88, 1b, 428**	**L'Escale (1)**	**Stránská skála (2)**	**Chateau (3)**	**Seneze (4)**	**Chateau (3) (N=4/6)**	**L'Herm (5) (N=13/14)**	**Caverna di Equi (6)**	**Gailen-reuth (7)**	**Westre-geln (7)**	**Brit. Isles (8) (N=4)**
Max. length	45.33	48.1	(50.0) (47.5)	50.0 - 51.2	49.0	78.6 (71.9- 81.4)	69.6 (66.0- 76.0	58.; 65.	68.0	63.0	79.825 (76.2 - 81.3)
Max. breadth	- - -	48.5	(47.5) (41.5)	(52.0+)	42.0	67.5 (63.0 - 71.9)	58.8 (52.0- 66.0)	42.5	56.0	54.0	- - -
Trochlear br.	41.0	(36.0)	(28.0) (27.5)	- - -	(33.)	44.8 (43.0-46.4)	- - -	30.0; 36.0	- - -	- - -	85.7 (80.0 - 93.5)
a - p head	- - -	25.0	- - -	27.6	- - -	31.7 (30.0 - 33.3)	- - -	- - -	- - -	- - -	39.65 (35.6 - 43.2)
br. head	30.0	19.0	(26.5) (24.0)	(21.0)	(30.)	41.3 (39.2 - 43.3)	40.15 (37.0-44.0)	25.;5; +I49 31.4	40.0	35.0	37.1 (32.0 - 50.8)

(1) M.-F. Bonifay, 1971; (2) R. Musil, 1971; (3) A. Agant, 1980; (4) R. Ballesio, 1963; (5) E/H Filhol, 1870; (6) D. Del Campana, 1919; (7) W. O. Dietrich, 1966; (8) W. B. Dawkins and A. W. Sanford, 1867-1869.

Appendix 8.25. Mandibular and lower dentition measurements in some (extant) smaller felines and in Yarımburgaz specimen.

	F. silvestris	*Felis chaus*		*Felis (C.) caracal*			
Measurements	(France) MNHN, Paris (N = 9)	(N. Afr.,W.Asia) SNG-F/M; ZSM (9-10)	(Egypt,Levant) SNG-F/M (N = 5)	(S.Asia,N.Afr., SNG-F/M (N = 5)	(E/W Afr.) MNHN, Paris (N = 12)	(E.Afr) (1) (N = 4)	Yarımburgaz A.71 - 5.
Body, depth	11.48	13.5	13.44	15.3	14.725	13.1	(+12)
behind M/1	(9.0 - 14.2)	(12.0 - 14.8)	(11.5 - 14.9)	(14.5 - 17.0)	(12.7 - 16.2)	(12.5 - 13.4)	
Body, depth	10.24	11.0	11.2	13.38	13.1	- -	10.8
at diastema	(9.2 - 11.7)	(10.4 - 12.5)	(8.9 - 12.6)	(11.9 - 15.7)	(11.4 - 14.5)		
Length of	6.68	7.1	7.24	7.86	5.54	6.7	6.7
diastema	(5.2 - 8.5)	(5.3 - 8.3	(6.1 - 8.9)	(57. - 9.8)	(4.0 - 6.8)	(6.0 - 7.2)	
P/3, length	5.73 (5.4 - 6.1)	7.9 (7.0 - 8.5)	7.70 (6.5 - 8.25)	7.26 (6.2 - 8.1)	7.32 (5.6 - 8.45)	7.5 (6.6 - 8.1)	6.4
P/3, breadth	- -	3.8 (3.5 - 4.0)	- -	- -	- -	3.9 (3.6 - 4.2)	3.0
P/4, length	7.14 (6.4 - 8.0)	10.1 (9.0 - 11.0)	10.23 (9.2 - 11.0)	10.2 (9.4 - 10.6)	10.14 (9.3 - 11.3)	10.2 (9.8 - 10.8)	8.4
P/4, breadth	- -	4.4 (4.0 - 4.6)	- -	- -	- -	4.7 (4.5 - 4.9)	3.6
M/1, length	7.93 (6.6 - 8.35)	10.0 (9.8 - 11.3)	10.42 (9.2 - 11.4)	11.56 (11.- 11.9)	11.25 (9.7 - 13.4)	11.8 (11.4 - 12.0)	8.7
M/1, breadth	- -	4.7 (4.4 - 5.0)	- -	- -	- -	5.3 (5.0 - 5.4)	3.75

MNHN = Muséum National d'Histoire Naturelle, Paris; SNG - F/M = Senckenberg Naturforschende Gessellschaft, Frankfurt/Main;
1 = Q. B. Hendey, 1974.

Appendix 8.26. Mandibular and lower dentition dimensions in *Felis silvestris* samples from Europe (Recent) and the Levant (Upper Pleistocene to Recent), and from Yarımburgaz cave.

Measurements	***Felis* Scotland *grampia***	***silvestris* Europe *silvestris***	**Recent (1) Spain *tartessia***	**Europe (Degerbol.'33) (40)**	**Recent (8)**	**LEVANT (2) (3) Postglacial (17 - 21)**	**Main Würm (12 - 13)**	**Ksar 'Akil (11/15/13)**	**Yarımburgaz A.71.5**
behind M/1									(+ 12.0)
body, depth at									10.8
diastema									
l. of diastema									6.7
l., P/3 thru M/1									23.25
/C l; br.									(6.3) - (5.2)
P/3 l	5.6 ± .05 (5.0 - 6.3)	5.73 ± .06 (5.3 - 6.2)	6.37 ± 0.22 (6.1 - 6.8)	6.2 (5.4-6.8)	5.39 ± 0.14	5.44 ± 0.15	6.41 ± 0.14	6.4 (5.5-7.1)	6.4
P/3 l br								3.07	3.0
P/3 l ht								(2.9-3.5)	4.5
P/4 l	6.78 ± 0.05 (5.7 - 7.5)	7.1 ± 0.00 (6.2 - 8.2)	7.57 ± 0.08 (2.3 - 7.9)	7.5 (6.8-8.5)	7.26 ± 0.15	7.49 ± 0.15	7.96 ± 0.15	8.0 (7.1-9.2)	8.4
P/4 br								3.69 (3.2-4.2)	3.6
P/4 pr d/ l									4.4
P/4 pr d/ ht									5.5
M/1 l	7.92 ± 0.06 (6.7 - 8.9)	8.26 ± 0.0 (7.4 - 9.0)	9.05 (8.4-9.7)	8.9 (7.5-10.0)	8.69 ± 0.22	8.74 ± .17	9.48 ± 0.15	9.5 (8.8-10.2)	8.7
M/1 br								4.18 (3.8-4.6)	3.75

(1) B. Kürten, 1965b (2) B. Kürten, 1965a; (3) D. A. Hooijer, 1961.

Appendix 8.27. Mandibule and lower dentition dimensions in *Felis silvestris* samples from (late) Lower and Middle Pleistocene localities in Europe and from Yarımburgaz cave.

Measurements	Olivola (1) (N=1)	W. Runton (1) (N=1)	PETRALONA (2) layer 11 type 137	PETRALONA (2) layer 11 TB.1115	Talus cone No. 2	Heppen-loch (1) (N=1)	Grotte de l'Eglise (3) (Dordogne) (N=1)	Lunel Viel (4) (N=2-3)	Arago (5) (N=1)	Yarımburgaz A.71.5
Length, /C - condyle			78.0	71.0						
Height, asc. ramus				33.1						
Height, condyle-angle				14.3						
Body, depth behind M/1	11.0		15.7	13.8				(13.03)	11.0	(+12.0)
Body, depth at diastema	9.5		13.7	10.7	12.0	12.2		(10.07)		10.8
Length of diastema	4.1		9.0		7.5			(7.5) (6.8-8.2)		6.7
Length, P/3-M/1, incl.	20.2		22.5	24.3	24.3			(21.8)	22.7	23.25
/C length	4.5		6.2	5.9				(4.57)		(6.3)**
								(4.3 - 4.8)		
/C breadth	3.3		5.0	4.0				(3.60) (3.4 - 3.7)		(5.2)**
P/3 length	5.7		7.0	7.1	6.3	7.4	7.0 - 7.5	6.45(6.4-6.5)	6.4	6.4
breadth	2.9		3.6	3.2	3.3	3.2		3.0 (2.8-3.2)		3.0
height	3.8			5.0	5.2	4.8				4.5
P/4 length	6.9	7.9	9.1	8.4	8.7	8.8	9.5	7.5(7.3-7.7)	8.1	8.4
breadth	3.2	3.6	3.8	3.8	3.9	3.8	4.5	3.5(3.4-3.6)		3.6
pr d/1	3.4	3.9	4.0	4.6	4.1	4.4		(3.97)		4.4
height	5.5	5.6		6.1	6.1	5.7	6.5	(4.83)		5.5
M/1 length	8.2		10.2	10.1	10.1			8.37(8.0-8.6)	8.3	8.7
breadth	3.5		4.7	4.3	4.6			3.8(3.8-3.8)		3.75
C/ length								5.9		6.8 (l. 4)
breadth								4.6		5.3 (l. 4)

B. Kurtén and Poalianos, 1977, 1981; (3) H. Laville, F. Prat and C. Thibault, 1972; (4) M.-F. Bonifay, 1971; (5) E. Cregut, 1979. **=alveolus.

Appendix 8.28. Some measurements of the mandible and lower dentition in *Felis silvestris* ssp. from Yarımburgaz , and from Weimar and Burgtonna, Germany (after Hemmer 1978, 1984).

	Yarımburgaz A.71 - 5	WEIMAR 1QW. Wei 4098	1QW. Wei 8082	Burgtonna MKB. Ma. 628
A. Measurements				
1. Mand. height, behind M/1	(12.0+)	12.3/ - -	- - / 13.3	13.5
2. Mand. height, at diastema	10.8	9.9/ - -	(10.4)/ (10.2)	- - -
3. Diastema length	6.7	7.5 / (6.5+)	- - / 7.5	- - -
4. Length P/3 - M/1 series	23.25	21.7 / 21.9	22.9 / 22.8	24.0
8. /C length	(6.3)	(5.0) / (5.1)	- - / (5.0)	- - -
9. P/3 length	6.4	5.9 / 6.0	- - / (6.5)	6.5
10. P/3 max. breadth	3.0	2.9 / 3.0	- - -	3.4
12 P/3 height	4.5	4.4 / 4.5	- - /4.5	4.6+
14. P/4 length	8.4	7.1 / 7.2	- - /(7.4)	7.8
15. P/4 max. breadth	3.6	3.3 / 3.4	- - -	3.8
17. Pr d/ length	4.4	3.7 / 3.8	(3.6) / (3.6)	4.0
19. Pr d/ height	5.5	5.0 / 5.0	- - / 4.9	4.8+
22 . M/1 length	8.7	9.0 / 9.0	- - /8.4	9.2
23. M/1 max. breadth	3.75	3.8 /3.8	- - -	4.1
25. Pa d/ height	- -	4.4 / 4.3	(4.6) / (4+)	3.5+
27. Pr d/ height	- -	5.7 / 5.3+	- - -	5.4
B. Indices				
13. P/3 cusp height in % crown length (9)	70.3	74.5 /75.0	- - / (69.)	71+
20. P/4 cusp height in % crown length (14)	65.5	70.5 / 69.5	- - / (66.)	61.5+
21. P/4 l. (14) in % P/3 l. (9)	131.25	120.5 / 120	- - / 114	120
29. M/1 l. (22) in % P/4 l. (9)	103.6	127 / 123	- - / 113.5	118
30. M/1 l. (22) in % P/3 l. (9)	135.9	152.5 / 150	- - /(129)	141.5

Appendix 8.29. Some dimensions of postcrania from Yarımburgaz attributed to *F. chaus* and to *Felis (C.) caracal*, and some related postcrania from Untertürkheim (Württemberg) referred to *Felis chaus* (after H. Helbing 1934).

Measurement	HUMERUS *F.c.* A.71.4 rt. distal	*F.(C).c.* A.71, indet. l. distal	ULNA *F.c.* A.71.6/9 l. compl.	*F. c.* A.71.4 rt.prox.	*F.(C.)c.* A.71. indet. l. prox.	RADIUS *F.(C.)c.* A.71.6 l. prox.	INNOM. *F. c.* A.71.4 right	MT-III *F.(C.) c.* Y-88, L.1	UNTER-TÜRKHEIM Nr. 16788 humerus	*F. cf. chaus* (1) (Stuttgart) ulna	innom.	MT-III
Maximum length			133.0		(61.7 inc.)					173.8		
Prox. artic., br.			21.7			10.5		9.42				9.9
Prox. artic., a-p			19.1			7.65		12.23				12.2
Mid-shaft, br.		10.5	9.8	5.5		4.85						
Mid-shaft, a-p		14.4	11.2	10.4		8.3						
Distal art., max. br.	22.5	27.5	23.1						27.2			
Distal art., artic. br.	17.2	19.5	16.6						19.0			
Distal art., a-p		15.7	12.5									
Olecranon, length					17.5					(17.5)		
Olecranon, height					15.6					(14.5)		
Semilunar notch, internal length					9.35					(10.0)		
Diaphysial ht., radial art.				15.9	14.01					(15.0)		
Diameter a-p, radial notch				9.9	10.2					(10.5)		
Height, ischium-iliac crest							100.2					
Acetabular br.							14.5				15.0	
Illiac br., above sac. aur. surface							21.7					
Ischium length							32.8					

(1) H. Helbing, 1934.

Appendix 8.29A. Some dimensions of humeri of *Felis silvestris* from Yarımburgaz cave, from Gerde, and from other recent samples.

Measurement	Yarımburgaz A.71.4 rt. distal	*FELIS SILVESTRIS* A.71 indet. l. distal	A.71.6/9 l. complete	Gerde (1)	Europe (2) (N = 19/20)	Bulgaria (3) (N = 23,15)	*F. chaus* W/S Asia (4) (N = 4)
			133.0		119.08 ± 3.142	116.11 ± 5.99	131.9
Maximum length					(109.4-130.3)	(101.0-126.3) Male	(120.1-142.3)
						111.44 ± 4.04	
						(107.0 - 123.6) Fem.	
			21.7		24.66 ± 3.043		
Prox. artic. breadth			19.1		(21.8 - 27.5)		
Prox. artic. a - p		10.5	9.8		8.04 ± 3.012		
Mid-shaft, breadth		14.4	11.2		(7.2 - 8.8)		
Mid-shaft, a - p	22.5	27.5	23.1	22.9-23.7	22.18 ± 3.037		21.35
Distal art., max. br.	17.2	19.5	16.6	13.6 - 13.4	(19.8 - 24.2)		14.7
Distal art., artic. br.							
Distal art., a - p		15.7	12.5				

(1) A. Clot, 1980; (2) Z. Kratochvil, 1976; (3) I.Petrov, 1992; (4) F. C. Howell, SNG, F/M.

Appendix 8.29B. Some dimensions of ulnae attributed to *Felis silvestris* and to *Felis* cf. *chaus*.

Measurement	Yarımburgaz A.71.4 rt. prox.	*FELIS SILVESTRIS* A.71. *indet.* l. prox.	Europe (1) (N = 19)	*F. cf. CHAUS* Untertürkheim (Stuttgart) (2) No. 16788
Maximum length		(61.7 inc.)		173.8
Mid-shaft, breadth	5.5			
Mid-shaft, a - p	10.4			
Olecranon, length		17.5		(17.5)
Olecranon, height		15.6		(14.5)
Semilunar notch, int.1		9.35		(10)
Diaphysial ht., radial artic.	15.9	14.01	ˉ3.16 ± 3.022 (11.7 - 14.6)	(15)
Diameter a - p, radial notch	9.8	10.2		(10.5)

(1) Z. Kratochvil, 1976; (2) H. Hemmer, 1934.

Appendix 8.29C. Some dimensions of radii attributed to *Felis silvestris*.

Measurement	Yarımburgaz A.71.6 l. proximal	Europe (1) (N = 21-25)	Gerde (2)
Maximum length		1ˉ5.96 ± 3.118 (107.0-124.3)	
Prox. artic. breadth	10.5	10.16 ± 3.014 (9.2 - 11.1)	9.0 - 9.3
Prox. artic., a - p	7.65	7.16 ± 3.0(9 (6.6 - 7.9'	6.5 - 6.3
Mid-shaft, breadth	4.85	6.3 ± 3.012 (5.3 - 7.3ˉ	
Mid-shaft, a - p	8.3		
Distal art., max. br.		ˉ5.39 ± 3.C25 (13.7 - 17.1)	

(1) Z. Kratochvil, 1976; (2) A. Clot, 1980.

Appendix 8.29D. Some dimensions of innominate and scapula (glenoid) postcrania attributed to *Felis silvestris* and to *Felis chaus*.

Measurement	*FELIS* Yarımburgaz A. 71. 4, RIGHT	*SILVESTRIS* Europe (1) (N = 20)	*Felis chaus* W/S Asia (2)
INNOMINATE			
Height, ischium-iliac crest	100.2	Ç3.45 ± 3.-106 (85.7 - 101.8)	102.625 (98.8 - 111.9)
Acetabular breadth	14.5	12.92 ± 3.018 (11.9 - 14.6)	
Iliac br., above sac. aur. surf.	21.7		
Ischium length	32.8		
	Yarımburgaz	**Europe (1)**	
SCAPULA - GLENOID	**A.71.4 - 125**	**(N = 19)**	
Glenoid length (ht.)	11.8	14.5 ± 3.(24 (12.3 - 15.9)	
Glenoid breadth	10.8	11.08 ± 3.016 (10.0 - 12.0)	

(1) Z. Kratochvil; (2) F. C. Howell, SNG, F/M.

Appendix 8.33A-B. Dimensions of Mt-III in *Equus caballus mosbachensis* (A), and in some other attributed Pleistocene subspecies of *Equus caballus* (B).

A.	***EQUUS CABALLUS MOSBACHENSIS*** **(v. Reichenau)**								
					Abbeville	**Rhen. Alluv.**	**Vosgien Red Sand**	**Achenheim ****	**Montoussé**
Measurement *	**Mosbach**	**Randersacker (N=3)**			**(N+2)**	**(N=1)**	**(N=1)**	**Loess, anc. imf.**	**(N=1)**
Max. length (1)	(42) 308.5 ± 7.0	(290)	(310)	300	201.5 - 294.0			(8) 297.5 (293.5-302.5)	295.5
Sup. art., transv. (5)	(49) 59.7 ± 1.88 (56 - 64)	63.0	52.5	53.0	55.0	62.7		(4) 59.0 (57.5 - 60.5)	60.0
Sup. art., a-p (6)	(49) 50.1 ± 1.0 (46.0 - 56.0)								
Mid-shaft, transv. (3)	(62) 41.0 ± 1.58 (37.5 - 44.5)	41.5	40.0	39.5	37.5 - 39.0	42.0		(8) 38.9 (29. - 42.5)	
Mid-shaft, a-p (4)	(62) 38.2 ± 1.56 (3.5 - 41.5)	(42.0)	(35.5)	(36.5)					
Distal art., transv. (11)	(27) 54.9 ± 1.76 (53.0 - 59.5)				54.0		61.5	(5) 58.1 (59.0 - 61.0)	
Distal art., a-p (12)	(27) 43.9 ± 1.59 (40.0 - 48.5)				37.0		(42.5)	(2) 41.5 - 46.0	
Robusticity index: 100 x (3) (1)	132.9 (128.4 - 138.2)	143.0	129.0	132.0	127.5; 134.0			133.8 (124.0 - 144.0)	
Caballine index:									
100 x distal sa. br.	[102.44 (100.94				103.7		103.25	(4) 96.165 (93.9 - 100.0)	
distal art. br.	- 103.34)]								

A. continued.	***EQUUS CABALLUS MOSBACHENSIS*** **(v. Reichenau)**								
		MSG. P 105	**Tiraspol**	**NORFOLK FOREST BED LOCS.**					**Yarımburgaz**
Measurement *	**Vergranne**	**St. George-Cariera**	**OPS-95 (N=1)**	**S 82**	**S 299**	**S 72**	**H6753**	**H6754**	**A.71**
Max. length (1)		301	293.5	292	304	305	292	281	308
Sup. art., transv.(5)		60.0	52.4	59.2	59.0	60.7	52.4	49.0	55.0
Sup. art., a-p (6)			45.8						48.0
Mid-shaft, transv. (3)		39.5	38.0	38.3	39.0	35.4	38.4	30.6	33.0
Mid-shaft, a-p (4)		37.5							34.0
Distal art., transv. (11)	(2) 57.5 - 58.0	44.0	54.0	55.5	50.3		54.2	46.6	55.0
Distal art., a-p (12)	(1) 45.0	59.0	41.0						41.0
Robusticity index: 100 x (3) (1)		131.2	126.1	131.2	128.3	116.1	131.5	108.9	107.1
Caballine index: 100 x distal sa. br.	(2) 99.31 - 101.7	129.54	98.15	97.3	113.33		102.03	102.15	101.8

Appendix 8.33A-B. Continued...

B.		***EQUUS CABALLUS SSP.***										
Measure-ment*	***palu-stris*** **(Lunel-Viel)**	***tautavel-ensis*** **(Arago)**	***torralbae*** **(Torr-alba)**	**(Am-brona)**	***pivet-eaui*** **(La Chaise)**	**(Stein-heim)**	***stein-heimensis*** **(Heppen-loch)**	**(Chatillon-St.Jean) (N = 17)**		***insulidens*** **(Tusnov Sat)**		***abeli*** **(Heiligen-stadt) (N= 2)**
Maximum length (1)	(26) 278.34 ± 6.821 (265 - 289)	(2) 295.1 - 294.3	(3) 269.5 (265.5-273)	264.2 - 265.7		291.0 - 292.0	304.	285.9 ± 4.575	285.	296.	290.	325.; 326
Sup. art., transv (5)	(27) 55.87 ± 2.73 (51.3 - 60.4)		(4) 52.375 (49.-56.5)		(1) 63.5	56.5; 57.5	61.0	54.9 ± 1.615	50.	54.5	57.0	61.0
Sup. art., a-p (6)	(28) 47.33 ± 2.218 (34.3 - 44.0)									55.5		
Mid-shaft, transv. (3)	(27) 40.34 ± 2.218 (36.3 -44.0)		(3) 34.67 (32.5 -38.0)			36.0 - 41.	41.	37.9 ± 1.829	36.	37.5	36.	42.; 42.5
Mid-shaft, a-p (4)									34.5	34.5	34.0	
Distal art., transv. (11)	(30) 55.49 ± 2.597 (49.5-99.6)	(12) 55.96 ± 1.99	(7) 49.43 (47.5-51.5)	48.7 49.1	(3) 58.5 (54. - 60.)	54.0; 55.0	61.5	53.7 ± 1.059	39.	40.	42.5	60 - 61.
Distal art., a-p (12)	(34) 41.58 ± 1.436 (39.7 - 45.7)		(4) 38.625 - (36. - 40.5)	38.9 39.0	(6) 43.4 (41.-44.)			41.6 ± 0.020	55	38.0	34.5	
Robusticity index: 100 x (3) / (1)	(21) 144.76 ± 7.489 (1310.- 158.0)		(3) 122.4; 124.1; 139.2	(1) 139.3		130.6; 140.9	134.9	132.53 ± 5.279	133.3	130.2	131.	129.6; 130.0
Caballine index: 100 x distal sa. br. / dist. art. br.	(28) 101.1 ± 3.21 (96.3 - 104.1)	(4) 100.28 ± 2.64 (94.7 - 104.4)	(7) 100.83 (94.7- 104.4)	(1) 105.1	(4) 97.44 (95.7- 98.3)	102.8 - 103.6	101.63	102.64 ± 1.496 (94.2-107.5)	147.44	143.75	129.4	

(*) dimensions after A. von den Driesch (1976); (**) attributed to *Equus achenheimensis* by G. Nobis (1971).

Appendix 8.36. Measurements of ph-1, digit II, pes of *Dicerorhinus*.

	Dicerorhinus	***etruscus***	***Dicerorhinus***	***kirchbergensis***	**Yarımburgaz**
Dimension	Up. Val d'Arno (1)	Voigtstedt (2)	Weimar-Enringsdort (3)	Taubach	U-89, 1, 35
Maximum length	33 (29 - 37)	35 - 36	36.3	29.3	31.8
Proximal breadth	34.7 (34 - 35)	40. - 40.8	38.2	33.4	29.32
Proximal a - p	25.3 (24 - 27)	22.6 - 22.8	26.8	24.3	24.3
Distal breadth	27.7 (23 - 30)	41.2 - 41.0	34.3	30.3	30.0
Distal a - p	- - -	21.0 - 21.3	26.5	27.6	23.3

(1) P. Mazza, 1988; (2) H.-D. Kahlke, 1965b; (3) H.-D. Kahlke, 1975.

Appendix 8.37A-B. Dimensions of (A) postcranial elements of *Dama* cf. *mesopotamica* from Yarımburgaz, and (B) some comparative postcranial *Dama* samples.

Specimen	**Maximum length**	**Proximal Articulation Max. br.**	**Max. a-p**	**Mid-Shaft Max. a-p**	**Max. br.**	**Distal Articulation Max. br.**	**Max. a-p**
A.							
A.71.5, ph-1, manus	44.8	19.7	18.0			19.5	13.3
T-88, 4, 662, ph-1,	47.55	16.71	20.15			15.61	13.34
pes							
U-89, 1, 299, rt.		32.0	33.5				
prox. mt. inc.							
U-89, 1, 91, inc. mt.			(26.13)				
with part. prox.art.							
P-90, 1, 86, distal						34.4	- - -
mt. articulation							
P-90, 1, 70. ph-1							
left, manus	47.92	17.33	21.93			14.92	14.04
B.							
METATARSAL:							
La Fage (1) (N=2)	(+ 252.0)	27.4 - 37.0	28.5 - 37.8			36.9; 36.0	22.0 - 24.8
Riano (2) (N=2)	275 - 278	- - -	41.0			38 - 39	- - -
Petralona (3) (N=1)	- - -	32.4	38.0			- - -	- - -
Swanscombe (2)	256.	- - -	35.25			36.25	- - -
(N = 4/5)	(245 - 266)	- - -	(32 - 37)			(35 - 37)	- - -
D. dama, Recent (2)	210.3	- - -	29.0			30.1	- - -
(N=7)	(192 - 223)		(25. - 33.)			(27. - 34.)	
Ph - 1:							
La Fage (1) (N = 12)	49.725	17.4	21.6				
(not differentiated)	(42.9 - 53.0)	(12.6 - 19.0)	(16. - 24.3)			- - -	- - -
Atapuerca (N = 2)	(42.) - (42.2)	15.2	21.6			14.4 - 14.6	12.5 - 14.0
Riano (2)							
Manus (N = 1)	50.25 (50 - 51)						
Pes (N = 4)	53						
Clacton (Jaywick) (2)							
Manus (N = 2)	52 - 53						

(1) J. Bouchud, 1972; (2) G. Leonardi & C. Petronio, 1976; (3) E. Tsoukala, 1989.

Appendix 8.38A. Dimensions of postcranial elements of *Cervus elaphus* from Yarımburgaz.

		Proximal Articulation		Distal Articulation	
Specimen / Element	Max. length	Max. br	Max. a-p	Max br.	Max. a-p
A.					
HUMERUS					
T-89, 5, 1062, right	(280.0)	93.13	72.36 (head)		
TIBIA					
A.71.4. 231, rt., prox.		(41.8) (*)	(46.8) (*)		
T-89. 3. 230, left,dist.				51.84	38.26
R-90, 1. 228, left,prox.		(82,2)	(49.9)		
METACARPAL					
T-89. 4. 630, left,prox.		(+44.9)			
METATARSAL					
R-90. 1. 1340, prox.		(40.5)	(38.1)		
METAPODIAL					
S-89.4.262. distal,inc.				(+19.9)	(+30.3)
Ph-1					
A.71, - - , 31. manus	60.3	22.93	28.4	19.7	19.2
A.71, 12, 682. manus	58.8	(22.9)	25.41	20.8	16.50
U-89, 1, 48. manus	59.1	22.5	26.74	22.0	16.9
R-90, 1, 399. manus	61.03	22.96	29.3	22.4	18.92
A.71, 4, 233a. pes	59.4	23.0	27.8	- - -	- - -
A.71, 4, 233b. pes	59.3	22.8	25.7	- - -	- - -
	59.65	22.83	27.22	21.225	17.9
	(58.8-61.03)	(22.5-23.0)	(25.4-29.3)	(19.7-22.4)	(16.5-19.2)
Ph-2					
R-90, 1, 316. manus	45.40	23,4	28,30	19.41	24,82
R-90, 1, 850. manus	45.2	22.45	27.04	18.9	25.84
A.71, 4, 234. pes	42.,5	23.0	26.24	17.96	22.63
A.71, 3, ?#. pes	41.0	23.0	27.8	- - -	- - -
Y-88, 1a, 157. pes	44.75	24.4	25.82	19.31	24.5
P-90, 1, 4. pes	43.73	25.3	28.8	20.53	25.83
P-90, 1, 93. pes	43.6	20.15	24.7	16.8	23.3
P-90, 1, 388. pes	47.94	23.85	28.7	19.4	25.7
T-89, 2, 129.	45.7	22.5	27.0	18.05	23.74
U-89, 1, 389.	43.24	22.25	25.3	- - -	- - -
U-89, 1, 11.	43.84	21.13	(20.5)	18.74	- - -
Y-88, 5, ?#.	32.6	23.9	- - -	21.23	20.6
	43.3	22.94	26.4	19.03	24.12
	(32.6-47.9)	(20.1-25.3)	(20.5-28.8)	(16.8-21.23)	(20.6-25.84)

(*) Articulation; measurement across base of tibial crest.

Appendix 8.38B. Some comparative postcranial dimensions of *Cervus elaphus* samples.

Element/ Locality (N)	Max. length	Proximal Articulation Max. br	Max. a-p	Distal Articulation Max br.	Max. a-p
B.					
HUMERUS					
La Fage (1; 4)	273.0	- -	- -	55.0	55.35
				(51.-58.)	(50.4-60.)
** Combe Grenal, lower (5)	- -	- -	- -	54.6	53.4
				(52.0-57.5)	(52.0-55.0)
Riano (5)	275.4	64.25		57.2	
	(259-283)	(56.-74.)		(55.-60.)	
Petralona (1; 2)	266.0	73.0	81.0	58.55	56.25
Burgäschisee	286.33	- -	85.33	60.67	
(Neolithic) (6)	(280-308)		(80.-93.)	(58.-65.)	
TIBIA					
La Fage (1; 2)	379.	82.	76.	51. - 63.	39. - 48.
** Combe Grenal,	- -	- -	- -	51.6	37.6
52-50A (5)				(49.5-53.5)	(36.5-39.0)
Riano (4)	377.75	76.25	77.75	48.25	39.25
	(361.-385.)	(71.-80.)	(77.-79.)	(45. - 51.)	(39.-40.)
Petralona (2)	393 - 393.3	86.5 - 88.	78. - 79.	54.5 - 55.0	41.-41.5
Mosbach (2; 12)	409. - 411.	92.	95.	55.1	39.75
				(45.-61.)	(34.-47.)
Mauer (3)	- -	- -	- -	53.8	40.5
				(50.5-56.0)	(35.-43.5)
Tiraspol (3)	- -	- -	- -	50.6	39.1
				(46.1-55.2)	(38.-41.1)
Burgäschisee	- -	82.2	- -	51.625	- -
(Neolithic) (15; 4)		(76.5-88.5)		(49.-54.5)	
METACARPAL					
La Fage (1; 3)	273.5	40 - 44	29 - 32	41.7 - 46.2	30.9 - 31.0
Chatillon-St.-Jean (6)	295.22	49.23	36.1	49.94	34.6
	(280-315)	(43.5-53.4)	(32.-43.5)	(47.-54.)	(32.-36.)
Riano (6; 8)	263.4	40.8	30.33	41.5	25.64
	(251- 276)	(39.-42.5)	(29.-31)	(38.-49.)	(24.5-27.)
METACARPAL, Cont'd.					
Petralona (2)	255.4-262.5	41.5-42.0	30.-31.7	42.- 42.4	28.5- 29.3
Mosbach (16; 31)	281.	45.4	29.6	45.9	27.3
	(264.-298)	(38.5-52.0)	(29.0-36.0)	(40.0-48.5)	(25.-30.)
Mauer (3; 1)	291.	45.2	34.0	46.	27.5
	(285-300)	(44.5-46.0)	(32.-35.)		(27.-28.)
Hundsheim (2; 4-5)	280.-306	46.95	33.55	48.4	30.7
		(43.4-50.7)	(31.6-36.0)	(47.-50.8)	(26.-34.)
Fontana Ranuccio (1)	268.	44.0	32.0	43.0	29.0
Tiraspol (1; 3)	265.5	46.9	34.0	51.8	33.2
		(44.3-48.3)	(33.6-34.6)		
Atapuerca, TD-3 (2)	337.	53.; 56.6	38.6; 40.5	56.; 58.	37.; 37.
Burgäschisee	264.75	41.1	- -	42.2	- -
(Neolithic) (66;56)	(256.-272.)	(35.-46.5)		(36.5-47.)	
METATARSAL					
La Fage (1; 3)	- -	38.3	40.4		
		(36.-40.)	(39.- 41.2)	46.9- 48.0	29.5-30.8
Chatillon-St. Jean	312. - 327.	44.75	48.1	48.51	32.-36.

Appendix 8.38B. Continued...

(2; 4; 2)		(37.5 - 54 ,5)	(41.-58.0)		
Riano (3)	293.6	39.0	41.67	41.2	30.42
	(282.-300.)	(36.5-41.0)	(39. - 45.)	(40.-43.)	(25.-28.)
Petralona (2)	317.-322.7	50.-52.5	51.7-58.0	60.5-67.0	37.5-41.5
Mosbach (15; 38-26)	314.4	39.1	43.5	48.2	31.35
	(302.-335.)	(34.-48)	(38.-51.)	(42.-58.)	(25.5-34.5)
Mauer (8; 7)	319.75	39.5	42.94	46.3	29.4
	(306.-334.)	(35.-42)	(41.5-46.0)	(44.-48.)	(27.0-30.5)
Hundsheim (2; 4)	312-328	41. - 42.	49.0	48.7	30.5
				(46.5-50.2)	(28.5-32.5)
Tiraspol (3; 1)	- -	42.7	46.3	53.1	33.8
		(41.9-43.5)	(44.4-49.1)		
Burgäschisee	305.65	35.9	- -	42.6	- -
(Neolithic) (43;51)	(301.-311.)	(31.5-40.)		(36.0-49.0)	
Ph-1 (unspecified)					
La Fage (5; 6)	58.1	21.875	27.75	20.9	18.42
	(57.-60.)	(21.-23.)	(25.9-29.)	(20.3-21.2)	(18.-19.1)
** Combe Grenal (20; 17)	- -	20.12± 0.44	25.12 ± 0.62	19.2 ± 0.62	17.2 ± .70
l. 54 - 50		(18.5-22.4)	(23.1-28.6)	(17.5-21.5)	(15.5-19.9)
Petralona (1)	58.	21.5	27.4	20.5	18.4
Atapuerca, TD (2)	71.5-71.6	27.1-27.6	33.1-33.3	24.7-24.9	22. - 23.
Burgäschisee					
(Neolithic) (175; 223)	58.7	20.9	- -	20.0	- -
	(50.5-67.0)	(17.-24.5)		(16.5-23.5)	
Ph-2 (unspecified)					
La Fage (5)	43.52	21.14	26.4	18.84	24.92
	(42.-45.)	(19.-23 2)	(25.2-28.9)	(17.9-19.6)	(23.9-25.9)
** Combe Grenal,	42.09; 42.5	19.8 ± 0.54	26.19 ± 0.71	16.9 ± 0.48	24.37 ± 0.58
(2; 13; 20) l. 52-50A		(18.0-21.5)	(24.0-28.0)	(14.2-18.1)	(23.5-26.0)
Petralona (1)	43.0	20.7	27.0	18.0	26.6
Atapuerca, TD-3 (3)	50.33	25.5	32..8	23.,7	32.3
	(48.7-51.3)	(25.4-25.6)	(32.5-32.9)	(23.4-24.0)	(32.2-32.5)
Burgäschisee	43.2	20.4	- -	- -	- -
(Neolithic) (163; 175)	(37. - 49.)	(17.0 - 22.5)			
Ph-1 MANUS					
Riano (15)	56.5	- -	- -	- -	- -
	(52.-60.)				
Atapuerca, TD-3 (2)	68.; 49.5	23; 24.5	32.; 32.	- -	- -
Ph-2 MANUS					
Riano (16)	40.9	- -	- -	- -	- -
	(41. - 45.)				
Ph-1 PES					
Riano (12)	57.75				
	(54.-61.)				
Süssenborn (1)	69.5	26.0	33.1	23.6	22.5
Voigtstedt (1)	67.7	26.3	34.3	22.8	20.9
Ph-2 PES					
Riano (14)	43.14	- -	- -	- -	- -
	(40.-48.)				

** ***C. e. simplicidens*** (after J.L. Guadelli, 1987). La Fage - J. Bouchud, 1972; Combe Grenal - J. L. Guadelli, 1987; Riano - G. Leonardi & C. Petronio, 1974, G. DiStefano & C. Petronio, 1993; Petralona - E. Tsoukala, 1989; Burgäschisee - H. R. Stampfli, 1963a; Mosbach - G. Di Stefano & C. Petronio, 1993; Mauer - H.-D. Kahlke, 1960; Tiraspol - H.-D. Kahlke, 1971; Chatillon-St.-Jean - C. Chauviré, 1962, C. Mourer-Chauviré, 1972; Hundsheim - W. Freudenberg 1914, G. Di Stefano & C. Petronio, 1993; Fontana Ranuccio - G. DiStefano & C. Petronio, 1993; Atapuerca TD-3 - B. Azanza & B. Sanchez, 1990; Süssenborn - H.-D. Kahlke, 1956; Voigtstedt - H.-D. Kahlke, 1965a.

Appendix 8.39A. Measurements (l x br) of dentitions of Cervidae from Yarımburgaz.

***DAMA* sp.**						
P-90, 1, 21	**dp 4 /**	**M1 /**	**M2 /**			
Maxilla frag. sub-adult	? x 15.17	19.3 x 18.14	20.78 x 18.16			
P-90, 1, 52	**I / 1**	**I / 3**	**dp / 3**	**dp / 4**	**P / 2**	**M / 1**
L mandible, sub-adult	11.73 x - -	5.7 x - -	13.92 x 9.81	20.93 x 8.68	9.17 x 5.9	18.47 x
						10.32
***CERVUS ELAPHUS* ssp.**	**P /**	**M 2 /**	**M 3 /**			
U-89 1, 442	- - -					
T-89, 4, 528	- - -	24.7 x 23.37				
R-90, 1, 917	- - -	26.56 x 26.81	24.78 x 24.23			
	I / 1	I / 2	I / 3	P / 2	P / 4	
P-90, 1, 43	- - -	- - -	- - -	121.86 x 6.39		
R-90, 1, 200	- - -	- - -	- - -	- - -	19.1 x 11.98	
R-90, 1, 223	- - -	- - -	- - -	- - -	- - -	
R-90, 1, 313/314	- - -	6.21 x - -	6.47 x - -			

Appendix 8.39B. Some comparative dental measurements (l x br) of Pleistocene Cervidae, *Dama*.

	M1/	**M2/**	**P/2**	**M/1**
Dama dama	16.1 x 16.3	19.2 x 17.6	9.5 x 6.0	15.5 x 10.5
(extant) (1)				
Dama mesopotamica	15.25 (11-18)	18.75 (15 - 21) x		
(2) (extant)	x 16.25 (15-18)	18.25 (17. - 19)		
D. sp. aff. ***dama*** (1)	17. 1 x 17.8	18.5 - 20.4 x		16.5 x 10.7
Megalopolis basin		18.2 - 20.0		
D. aff. ***mesopotamica***	16.0 x 19.5			
Qafzeh (3)				
D. clactoniana				
Riano (4)	18.7 x 17.8	20.0 x 20.5	11.2 x 6.5	18.3 x 11.5
Swanscombe (4)	18.9 x 19.1	21.0 x 18.9	11.0 x - -	17.0 x - -
Petralona (5)	20.45 (17.5-22.7)	21.4 (20.1-23.1)	10.4 - 12.2 x	20.7 (19.8-22.2)
(***D. dama*** per E.T.)	x 21.8 (21.2-23.0)	x 22.7 (21.4-23.2)	7.0 - 7.8	x 12.0 (11.4-12.6)
Atapuerca (6)	18.8 (17.6-20.8)	20.78 x 21.01	- - -	15.06 (12.65-15.41)
	x 18.3 (16.7-19.01)			x 11.1 (10.9-11.24)

(1) J.K. Melentis, 1964; (2) T. Haltenorth, 1959; (3) J. Bouchud, 1974; (4) G. Leonardi & C. Petronio, 1976: (5) E. Tsoukala, 1989; (6) B. Azanza & B. Sanchez, 1990.

Appendix 8.39C. Some comparative dental measurements (l x br) of Pleistocene Cervidae, *Cervus elaphus*.

	M2/	M3/	P/2	P/4
C. e. elaphus				
Chatillon-St.-Jean (1)	24-28.6 x	24-25 x		
	24.6-26.8	23-23.5		
Grotte Vaufrey (2)	26.0 (24.1-28.7) x	24.6 (23.0 - 27.5) x		19.6 (17.6-21.0) x
	24.55 (23.0-26.8)	24.6 (23.5-26.7)		12.75 (11.3-14.0)
Qafzeh (3)	21.0-24.4 x	22.4 x 23.2		
	21.8 - 26.0			
***Cervus elaphus* ssp. indet.**				
Petralona (7)	25.5 (24.7-26.7)	24.7 (24.4-25.0)	12.6 (12.12-13.0)	19.8 (18.8-21.8)
	x 26.1 (25.2-27.9)	x 24.9 (24.7-25.3)	x 8.5 (8.2-8.8)	13.24 (12.2-14.3)
Atapuerca (9)	24.01 - 25.01 x	23,01 - 24.36 x	10.8 (10.35-11.27)	16.1 (15.28-16.63)
	24.19 - 24.22	23.95 - 24.34	x 6.53 (6.41-6.84)	x 10.04 (9.33-10.79)
C. e. rianensis				
Riano (4)	21.6 (20.3-22.1)	20.9 (20.2-21.3)	11.5 (10.1-12.1)	16.65 (15.9-18.1) x
	x 20.9 (18.9-22.7)	x 19.8 (18.3-21.5)	x 7.05 (6.3-7.9)	11.1 (10.9-11.4)
C. e. simplicidens	24.98 ± 0.21	22.7 ± 0.37	12.58 ± 0.31	18.17 ± 0.34
Combe Grenal, 54-50A,	(21.0- 25.0) x	(19.0 -25.0) x	(11.0 - 14.4) x	(15.1 - 20.0) x
(11)	24.68 ± 0.32	25.07 ± 0.38	7.21 ± 0.15	10.83 ± 0.15
	(23.0 - 27.0)	(20.0 - 26.0)	(6.1 - 8.0)	(9.5 - 12.6)
C. e. eostephanoceros				
Arago (5)	25.1 (22.5-27.3)	24.6 (23.0-27.5)	12.1 (11.4-13.1)	18.35 (15.9-19.7)
	x 24.1 (22.8-25.3)	x 23.8 (22.0-25.0)	x 8.36 (8.0-8.)	x 12.225 (12.-12.6)
Hundsheim (6)	23 x 24	25 x 25		
C. e. acoronatus				
Mosbach (12)	20-.6 (20.3-21.0)	20.9 (20.5-21.5)	10.7 (9.6-11.4)	17.8 (16.0-19.1)
	x 23.33 (23.0-23.9)	x 22.53 (22.5-22.8)	x 5.24 (5.0-7.8)	x 10.6 (8.4-11.5)
Tiraspol (8)			11.1; 15.3 x	19.7 (18.9-20.8) x
			7.0; 10.0	13.2 (13.1-13.2)
Voigtstedt (10)	20.0 x 22.8	20.1 - 20.5 x	15.9 - 16.4 x	
		22.5	9.1 - 9.6	11.63 (10.9-12.4)

(1) C. Chauvire, 1962; (2) F. Delpech, 1988; (3) J. Bouchud, 1974; (4) G. Leonardi & C. Petronio, 1974; (5) E. Cregut, 1979; F. Delpech, 1988; (6) W. Freudenberg, 1914; (7) E. Tsoukala, 1989; (8) H.-D. Kahlke, 1971; (9) B. Azanza & B. Sanchez, 1990; (10) H -D. Kahlke, 1958; (11) J. .. Guadelli, 1987; (12) H.-D. Kahlke, 1960.

Appendix 8.40. Dimensions of postcranial elements of *Capreolus* sp. from Yarımburgaz.

	Maximum	(Prox. Artic.)		Distal Articulation	
SPECIMEN	**length**	**breadth**	**a - p**	**breadth**	**a - p**
***Capreolus* sp.**					
T-89, 2, 575					
ph-1, pes (cf.)	47.86	15.39	18.41	11.65	
T-89, 4, 813					
ph-2, cf. pes	34.52	13.6	16.35	10.0	15.07
EXTANT SPECIES, PH-1					
Females (29)	31.6				
	(29.-34.)				
Males (28)	32.8				
	(30.-36.)				

Extant species sample from Germany, K. Busold, 1968.

Appendix 8.41. Dimensions of postcranial elements of *Megaloceros* sp. from Yarımburgaz.

Megaloceros sp. SPECIMEN	Maximum length	Proximal Articulation br	Mid-shaft a-p	 br	Distal Articulation br	 a-p
A.71.4 Distal mt, partial trochlea					(21.1)*	(29.2)*
A.71.12 Prox. mc, inc.		(48.5)	31.2	29.5		
Y-88, 1a, 130, dist. humerus, rt.,						
medial 1/2 articulation					(62.5)**	
Y-88, 1a, 131 ph-1, pes, incomplete	78.21	---	50.83			38.87

(*) a-p of crest; br posteriorly of incomplete trochlea. (**)Medial trochlea br, from crest to medial margin.

Appendix 8.42. Measurements of astragali of Yarımburgaz Cervidae, and some comparative samples (measurements after A. von den Driesch 1976, p. 88-89).

Samples/Localities	Mid-Axis Length	Length, Lateral Half (GL1)	Length, Medial Half (GLm)	Depth, Lateral Half (D1)	Depth, Medial Half (Dm)	Max. Distal Breadth (Bd)
Cervus elaphus, ssp.						
Yarımburgaz						
T-89, 4, 1050	46.48	(56.15)	(50.9)	(29.53)	(28.1)	(33.0)
T-89, 4, 253	47.11	57.83	54.32	24.43	26.6	36.9
P-90, 1, 364	46.54	58.58	54.50	31.81	--	36.24
R-90, 1, 360	47.62	57.74	52.96	31.43	--	39.63
Burgäschisee-Sud, (1)						
Neolithic (N = 147-148)		56.9(48.-62.5)	53.3 (45.-60.5)	30.8 (26.5-34.)		35.5 (31. - 40.)
C. e. eostephanoceros/rianensis						
Arago (2) (N=3)		59.4 (54.8-61.8)	55.1 (50.-58.8)	32.7 (29.9-34.6)	32.8 (30.8-34.1)	37.4(35.-38.6)
Riano (3) (N=8)		54.1 (53.-56.)				33.875 (32.-35.)
Dama sp. indet.						
Yarımburgaz						
U-89, 1, 458	36.64	44.63	42.58	24.52	25.97	28.91
D. cf. ***clactoniana***						
Abimes de la Fage		46.675				28.775
(Corrèze) (4) (N=4)		(45.5 - 48.1)				(28.2-29.0)
Swanscombe (5) (N=6)	40.0 (38.4-44.)					
Megaloceros sp.						
Yarımburgaz						
S-89, 4, 225	70.42	(87.9)		(48.65)	(47.25)	(52.7)
Chatillon-St.-Jean						
(6) (N=9)	68.7 (65.-72.)	85.5 (83.-91.)	79.5 (75.-84.)	45.2 (43.-49.)	49.2 (47.54)	56.7 (53.-63.)
Megaloceros verticornis						
Süssenborn (7)		80.2			46.5	55.4
Voigtstedt (8)		74.3			43.5	51.0
'Ubeidiya (9)			63.8 (59.-70.)			41.0 (38.-44.)
Capreolus aff süssenbornensis						
Yarımburgaz						
T-89, 4, 656	29.2	34.51	32.42	18.87	20.03	21.63
Süssenborn (7) (N=4-6)		36.15 (35.-37.4)			20.4 (19.1- 22.0)	22.6 (22.4 - 22.8)
Voigtstedt (8) (N=1)		35.2			22.8	21.1
Capreolus capreolus						
Abimes de la Fage						
(Corréze) (4) (N=4)		31.45 (29.-35.5)				20.4 (18.-23.3)
Burgäschisee-Sud,						
Neolithic (1) (N=22)		29.5 (28.- 31.5)	28.43 (27.-30.)	16.4 (15.5-18.)		18.2 (17. - 19.5)

(1) H. R. Stampfli, 1963 ; (2) E. Cregut, 1979; (3) G. Leonardi and C. Petronio, 1974; (4) J. Bouchud, 1972; (5) A. J. Sutcliffe, 1964; (6) C. Mourer-Chauviré, 1972; (7) H. D. Kahlke, 1969; (8) H. D. Kahlke, 1965a; (9) D. Geraads, 1986.

Appendix 8.43A. Dimensions of dental elements from Yarımburgaz cave attributed to Bovidae.

A. DENTITION		**Crown**	**Crown breadth**	
SPECIMEN	**TAXON**	**length**	**Ant. lobe**	**Post. lobe**
A.71.5 right (?) M2/, worn.	***Bos/Bison***	32.5	27.3	
A.71.6 left (?) M/2 distal part.	***Bos/Bison***	(17.7 - dist. lobe)		
Y-88, cleaning, No. 671, left				
M 1- 2/, complete	***Bos***	27.51	18.90	16.91
Y-88, cleaning, No. 671a, M/,				
posterior lobe only	***Bos***	- -	- -	- -
Y-88, cleaning, No. 672,				
left dp/4, complete	***Bos***	29.76		12.02
U-89, 1, 29, left mandible	***Gazella*** sp.	dp/3 = 10.53	3.91	5.98
frag., juvenile with dp/3,		dp/4 = 15.63	6.40	7.81
dp/4 and M/1		M/1 = 12.84	8.18	8.71

Appendix 8.43B. Dimensions of postcranial elements from Yarımburgaz cave attributed to Bovidae.

B. POSTCRANIA	**Taxon**	**(Preserved) length**	**Prox. Articulation breadth**	**a - p**	**Distal Articulation breadth**	**Distal Articulation max a - p**	**min. a-p**
A.71, *indet.*							
(sieving, 1988)							
left distal tibia	***Gazella*** sp.	- -	- -	- -	30.31	21.44	20.14
Y-88, 1, 675,							
prox. half,	cf.						
left mc.	Alcelaphini	(153.7)	36.7	(~ 29.)			

Appendix 8.43C. Dimensions of horn cores from Yarımburgaz cave attributed to Bovidae.

C. HORN CORES					(a)	(a)	**hc section b**		**hc section c**		**hc section d**	
SPECIMEN	**TAXON**	32	33	40	41	42	a - p	m-1	a - p	m - 1	a - p	m - 1
A.71.12 (14) rt. hc,	***Gazella* sp. cf.**				44.7	30.9	39.15	25.4	35.5	18.7	28.6	14.6
lacking distal port.	***subgutturosa***											
A.71,4 Frontlet, with	***Capra* aff.**											
both hc's, tips	***aegagrus***	92.8	72.7	125	43.9	36.1	34.4	27.9	(30.3)	23.0	- -	- -
missing												
Y-88, 0, 7 left hc,	***Capra***				48.9	31.5	42.9	25.7	39.0	22.0	31.4	18.2
missing tip	***aegagrus hircus***											
Y-88, cleaning, 673,	***Capra***				29.8	(13.2)	26.7	14 2	22.0	11.1	16.2	8.3
left hc, missing tip	***aegagrus hircus***											
A.71, *indet.* sieving	***Capra***						(25.0)	(13 6)				
1988, 33. Left	***aegagrus hircus***											
terminal hc seg.												
S-89, 4, 274a, tip of	***Capra* cf. *ibex***										25.72	19.46
(cf.) right hc.												
T-89, 2, 282, much	***Capra* aff. *ibex***						31.44	27.22	27.82	23.1	20.75	17.72
of distal portion,												
left hc.												

* Measurements of horn cores after A.v.d. Driesch (1976, pp. 31-34): 32 = maximum breadth (laterally) between hc bases; 33 = maximum neurocranium breadth (eur-eur); 40 = hc basal circumference (tape); 41 = maximum (a-p) diameter horncore base; 42 = least (m-1) diameter hc base.

Appendix 8.44. Sedimentary phases and level attributions (by excavation squares) in the Yarımburgaz cave infilling.

	Upper Chamber			Lower Chamber					
Phases	A.29/40	Z	R	U	Y	A.71 (V-88)	T	S	P
pre-Eemian									
^^^^^^^^^^^^^	^^^^^^^^^^^	^^^^^^^^^^	^^^^^^^^^^^^^	^^^^^^^^^^^^^^	^^^^^^^^^^^^	^^^^^^^^^^^^^^^^	^^^^^^^^^^^^^^^	^^^^^^^	^^^^^^^^^^^^^^
Y	- - -		0 - i	1	0b	1	1	- - -	
X§		- - -	1	?	1a	2-3	2	1-2-3	1
W§		- - -	2	?	1b	4-7	3-4	4	2
***********	***********	********	***********	************	***********		**************	*************	*************
V§		- - -	3-4	?	1c,d,2a,b(?)	8-9	5	5	0
00000000000	0000000000	00000000	00000000000	000000000000	00000000000	000000000000000	00000000000000	0000000000000	0000000000000
					3a,b				
U		4 - u	5	not reached	4	10-13	6	6	3
***********	***********	********	***********	************	***********	**************	**************	*************	*************
T		4	not reached	not reached	4-i	14-15 (a-b)	? top	not reached	4
^^^^^^^^^^^^^	^^^^^^^^^^^	^^^^^^^^^^	^^^^^^^^^^^^^	^^^^^^^^^^^^^^	^^^^^^^^^^^^	^^^^^^^^^^^^^^^^	^^^^^^^^^^^^^^^	^^^^^^^	^^^^^^^^^^^^^^
S		9-13?	not reached	not reached	5	16+	not reached	not reached	not reached
						travertine (c+d)			
R			not reached	not reached	c. present	d′	not reached	not reached	not reached

Diastems: (^^^) major unconformity; (***) unconformity; (ooo) probable unconformity (non-sequence); (§) artifactual horizon.

Appendix 8.45. Presence of larger mammalian taxa in the Yarımburgaz cave trenches in the Lower Chamber by infilling phases.

Chamber:	UC	LC	LC	LC	LC	LC	LC	LC
Taxon Group		Y	X	W	V	U	T	S
CARNIVORA								
Ursidae								
Ursus deningeri ssp.	A40.14/15	√	√	√	√	√		
Ursus aff. *arctos*			√	√				
Canidae								
Vulpes aff. *vulpes*			√	√	√			
Vulpes sp. indet. cf. *praeglacialis*		√						
Vulpes indet.		√						
Canis lupus mosbachensis		√		√		---	√?	
Canis lupus ssp.			√					
Canida indet.			√	√				√
aff. *Cuon/Xenocyon/Lycaon* sp.	A.40.14							
Felidae								
Panthera leo ssp.		√	√	√				
Panthera sp. indet.		√		√				
Felis silvestris ssp.			√	√	√?			
Felis ex. gr. chaus			√	√				
Felis (Caracal) ex.gr. *caracal*				√	√?			
Felinae indet..								
Hyaenidae								
Crocuta crocuta cf. *praespelaea*		√	√					
Pachycrocuta aff. *brevirostris*				√				
Hyaenidae indet. (coprolite)								√
Mustelidae								
Mustela sp.			√?	√				
PERISSODACTYLA								
Equidae								
E. caballus aff. mosbachensis			√	√	√			
Rhinocerotidae								
Dicerorhinus sp.		√	√	√				
ARTIODACTYLA								
Suidae								
Sus scrofa		√	√	√				
Cervidae								
Dama sp. cf. mesopotamica		√	√	√				√
Cervus elaphus ssp.		√	√	√	√			√
Megalocerini indet.			√	√		√		
Capreolus sp. aff. *süssenbornensis*			√					
Bovidae								
Bovini								
Bos/Bison spp.			√	√				
Caprini								
Capra aegagrus				√				
Capra cf. ibex			√	√				
Antilopini								
Gazella cf. *subgutturosa*	√						√	√
? cf. Alcelaphini indet.,			√?					

(UC) Upper Chamber; (LC) Lower Chamber; (√) taxon present; S through Y refer to specific trenches.

submerged or frequently flooded during the highest interglacial sea stands, and it would have been dry land during the cold glacial periods.

Chapter 2 reports the geological study of the sedimentary sequences in the Upper and Lower Chambers (UC and LC). Information about the UC derives mainly from re-inspection of the profiles of Özdoğan's excavation trenches and re-appraisal of his field notes, whereas geological findings from the LC come mainly from the 1988-1990 excavations. The two chambers have very different sedimentary histories. The UC, no longer an active karst, preserves a long and complex sequence of human occupations from the Lower Paleolithic to the Byzantine period. Extensive modification of the walls and floor of the cave during Classical and Byzantine times, as well as recent looting, have caused extensive disturbance to the older cultural deposits. In the UC, a deposit of cross-bedded marine sand that probably dates to MIS 5e separates the Lower Paleolithic layers from the later human occupations.

The LC preserves a deeper sedimentary sequence than the UC, but it contains a much more limited range of archaeological deposits. The LC sequence has been divided into three main cycles. The earliest of these (Cycle 1) consists of sediments that are mainly fluvial in origin: it contains very few bones and no artifacts. The middle cycle (2) contains more silty, loamy fill with scattered small pebbles. Substantial quantities of bone and stone tools occur in level V at the top of this cycle. The great majority of artifacts and bones come from layers W and X, in the uppermost sedimentary cycle (3). These layers also contain large amounts of angular limestone debris derived from breakdown of the cave's interior vault, in a silty-loamy matrix. The layer of beach sand found in the UC was not encountered in the LC. It may have been been present at one time but was then removed by water flowing from a ingress in the back of the cave.

Chapter 3 summarizes the results obtained by two different radiometric dating techniques, ^{230}Th/^{234}U and Electron Spin Resonance. Numerous samples of speleothems and other potentially dateable materials were collected during the excavations, but most of these proved unsuitable due to excessive detrital contamination or high porosity. One sample of the secondary mineral, collophane, from below the main archaeological and paleontological layers was dated to 160 $^{+}$25/-20 ka. Because this material formed after the bone-bearing layers above, it provides a minimum age for the main deposit. The much larger series of ESR ages were obtained for bear teeth and the results yield relatively consistent age estimates. Using the LU uranium-uptake model, the ESR results indicate that the main deposit of cave bear bones formed between roughly 211 and 226ka BP, which correlates with a cold interval within MIS 7. Although this is the best estimate we have for the age of the main archaeological and fossil layers in the LC, the ESR results alone cannot exclude somewhat earlier or later ages due to uncertainties about uptake models and radiation dose rates.

Important to understanding the environmental context of the hominin occupation at Yarımburgaz Cave are the thirty-eight different taxa of small mammals from the Lower Chamber (Chapter 4). These assemblages include 15 species of bat, 3 species of insectivore, 17 species of rodent, and 3 lagomorph species. The micromammal assemblages from various units are relatively uniform in composition. The evolutionary stage of the steppe lemming, *Lagurus transiens/lagurus*, indicates a Middle Pleistocene age. The rodent fauna includes several taxa that today live far to the northeast of Thrace, indicating that most of the deposits formed under cooler and drier conditions than prevail in the area today. In contrast, the bat fauna includes typical Mediterranean taxa such as *Rhinolophus mehelyi*, and even some species that today live further south. This apparent discordance between the bats and the terrestrial micromammals may be the result of seasonal migrations of the bats, or it could indicate that the bats were present in Turkish Thrace only during warm climate periods.

The large sample of stone artifacts from the 1988-1990 excavations is discussed in Chapter 5. The assemblage of modified pieces is characterized by a high proportion of flake tools--mainly denticulates and notched pieces--and comparatively small numbers of choppers and chopping tools. Bifaces, bifacial shaping debris, and traces of Levallois technology are absent. The frequency of retouched pieces is high, and some artifacts show signs of having been used and reshaped. There are also distinct differences within the assemblages in how the three principal raw materials—flint, quartz and quartzite—were used. Flint and quartz were used primarily for making flakes and flake tools, whereas quartzite more often was used to make choppers and chopping tools. Pleistocene toolmakers at Yarımburgaz Cave also used very distinct methods to produce flakes from flint and quartz. Generally speaking, the lithic assemblages from Yarımburgaz resemble contemporaneous industries from central and Eastern Europe more than they resemble assemblages from the Levant, although all these industries share important features (discussed in greater detail below). These observations support the idea that the Bosporus/Mamara/Dardanelles waterway was a significant barrier for hominins during the Pleistocene.

Chapter 6 examines the circumstances of bone assemblage formation in the Middle Pleistocene deposits. Altogether, these assemblages represent the cumulative actions of three major agents—cave bears, hominins, and a variety non-ursid large carnivores. The episodic actions of these distinct bone collecting or modifying agents cannot be distinguished stratigraphically, but other aspects of the data allow us to distinguish of the contexts and modes of bone accumulation. Species representation, bone damage, body part representation, and mortality data indicate that most or all cave bear remains accumulated as the result of hibernation-related deaths over many generations of den use. Most of the cave bear remains belong to juveniles and sub-adults, along with some very old individuals. Adult bears in the prime years of life are rare. As the paleontologist, Björn Kurtén (1976) noted in his famous study of fossil cave bears, the high frequency of young animals, the complete patterns of skeletal representation, and the occasional presence of bones still in articulation, are typical features of hibernation death assemblages. Small rodents living in or near the cave entrance frequently gnawed the smaller bones of the bears as they lay exposed for long periods on the cave floor. Damage on the bear and ungulate bones indicates that other large carnivores, especially canids such as wolves but also large bears, sometimes scavenged or preyed on hibernating bears. Judging from the patterns of gnawing and scarcity of tool marks on the ungulate bones, it was mainly the non-ursid carnivores that occasionally brought the bones of deer, wild cattle and other large herbivores into the cave. Ungulate bones are quite rare in this site, in marked contrast to other known Lower and Middle Paleolithic archaeofaunas in Southwest Asian sites that preserve bone (e.g., Boëda et al. 1998; Garrard 1983; Monchot and Kolska Horwitz 2007; Rabinovich and Tchernov 1995; Rabinovich et al. 2008; Speth and Tchernov 2007; Stiner 2005; Stiner et al. 2009; Tchernov et al. 1994; Yeshurun et al. 2007). It is interesting that hominins spent enough time in Yarımburgaz Cave to deposit numerous stone artifacts, yet they left few detectable traces on the faunal remains. Based on this evidence, and the absence of traces of fire in the LC, it appears that hominin occupations were quite ephemeral, and hominins focused on resources and activities other than large game hunting while staying at the cave.

Chapter 7 presents a paleontological study of the ursid remains The roughly 5000 bones and bone fragments of bears represent a minimum of 42 individuals. The dominance of bears in the Middle Pleistocene fauna from Yarımburgaz is similar to the situation in certain Late Pleistocene caves such as Pestera cu Oase in Romania (Quilès et al. 2006) and Almopia Speleopark Bear-Cave in Greece (Tsoukala et al. 2006). Two main groups of ursids are represented, cave bears (*Ursus deningeri-spelaeus*), which constitute the great majority of remains, and a small variant of brown bear (*U.* cf. *arctos*). The dimensions of the *U. arctos* lower cheek teeth are remarkably similar to those of *U. etruscus*.

The very large cave bear sample contains specimens that correspond to morphological and metric definitions for both *U. deningeri* and *U. spaelaeus*. The lower limb morphologies of these bears reveal distinct speleoid and arctoid characters, as well as many intermediate forms, showing thus great variability. The presence of these three species together indicates that the deposits formed over an extended interval within the second half of the Middle Pleistocene. The variation gradient in the cave bears from Yarımburgaz remains likely represents a local evolutionary continuum. The *U. spelaeus* specimens in Yarımburgaz Cave may be among the earliest representatives of this species. The *U. deningeri* and *U. spelaeus* in Yarımburgaz are rather robust in comparison to European samples. There is a great similarity between the Yarımburgaz and the *deningeri* populations of Petralona Cave (Tsoukala 1989), Mosbach (Reichenau 1908) and a cave in northern Israel (Tchernov and Tsoukala 1997).

Chapter 8 examines the taxonomy and chronological indications of the non-ursid animals in the Yarımburgaz fauna. The paleontological assemblage is taxonomically diverse in relation to the number of specimens. The assemblage includes nine genera of Carnivora, two of Perissodactyla, and nine or ten of Artiodactyla, as well as a few bird and reptile remains. Proboscideans, *Hippopotamus*, and non-caballine horses are conspicuously absent. The Carnivora show substantial diversity. The feline remains include lion, another probable pantherine, and smaller taxa such as wildcat, swamp (or jungle) cat, and caracal. The Canidae include both vulpines and canines. One large canid shows potential affinities with Asian wild dogs, whereas others are clearly wolves. Although hyena remains are rare, two distinct lineages are represented, the later *Crocuta crocuta praespelaea* and the earlier *Pachycrocuta brevirostris*. Artiodactyl species dominate among the ungulates, perissodactyls are notably rare, and *E. hydruntinus/E. hemionus* seems to be completely absent. Rhinoceros is represented only by a few fragmentary specimens. Cervids, including fallow, red and roe deer, and a form of *Megaloceros*, are the most abundant of the artiodactyls. Suids and bovids are present but uncommon. The possible presence of an alcelaphine antelope in an early sedimentary phase is noteworthy, as is the definite but rare presence of goitered gazelle (*Gazella* cf. *subgutturosa*). This typically central Asian ungulate co-occurs in Yarımburgaz with an unidentified Asian wild dog that definitely is not *Canis*. The taxonomic identities of the cuonoid-like wild dog and the tentatively indentified medium-sized antelope are not sufficient to undermine the dominance of Palaearctic macromammals in the Yarımburgaz faunas; these taxa merely afford links with the central and west Asian region. It is interesting that the gazelle remains occur in three phases in the LC stratigraphy, but only at the base and top of the sequence. The cuonine canid co-occurs with gazelle in a basal part of the UC infill.

Overall, the faunal assemblage indicates a mid-to-late-Middle Pleistocene age, even if certain elements among the carnivores (*P. brevirostris*) could suggest greater antiquity. Unlike some Lower and early Middle Pleistocene occurrences in the Levant, which often include Ethiopian taxa, the large mammal assemblage from Yarımburgaz is wholly Palaearctic in origin. This finding is less surprising, however, in light of recent results from the well-dated (400-200 ka) Acheulo-Yabrudian Levantine site of Qesem Cave (Barkai et al. 2003; Stiner et al. 2009), which also contains an exclusively Palaearctic fauna.

THE AGE OF THE YARIMBURGAZ CAVE PALEOLITHIC DEPOSITS

When F. Clark Howell first examined the material from Özdoğan's sounding in the Lower Chamber at Yarımburgaz, he immediately recognized the potential antiquity of the deposits. The presence of large choppers and chopping tools, the absence of handaxes and Levallois technology, and the seeming informality of the flake tools all spoke to him of great age. Likewise, some elements in the vertebrate fauna suggested that the deposits dated well back into the Middle Pleistocene. The sum total of chronometric, paleontological, and archaeological evidence now at hand indicates that the Paleolithic occupations at Yarımburgaz are indeed Middle Pleistocene in age, but not quite so ancient as originally surmised. In spite of some contradictory indications, the weight of the evidence suggests that hominin and bear presence in the cave dates to the later part of the Middle Pleistocene, specifically MIS 7 or 8.

The ESR results provide relatively consistent indicators of age, but the estimates depend on which uranium uptake model is used. Age estimates from layers W and X, based on the Linear Uptake (LU) model, which Blackwell and colleagues (Chapter 3) believe is the most appropriate in this case, range between 190 and 250 ka BP, roughly the entire span of MIS 7. Averages by stratum provide a more restricted range of ages, between 211 ± 22 (layer Z) and 226 ± 24 ka (layer W). Other uptake models produce significantly different ages, however. Application of an Early Uptake model (EU) would place the hominin and bear occupations of the Lower Cave within MIS 6. Application of a late uptake model (RU) instead yields much older, more dispersed age estimates in the range of 270 to 390 ka BP (± 40-60 ka) and spanning MIS 8-10. Collophane crusts underlying the main archaeological and paleontological levels are composed of phosphates—presumably dissolved from bone—and almost certainly post-date the main accumulation of bear remains. Thus, uranium-series dates obtained from these secondary mineral deposits (160 $^{+25/-20}$) post-date the main ursid and hominin occupations in the Lower Cave.

Paleontological evidence from the micromammal and non-ursid remains provides a few guidelines for narrowing down the ages of the deposits. Von Koeningswald and colleagues (Chapter 4) conclude that the small mammal fauna in Yarımburgaz Cave dates to ..."a cold period in the middle part of the Middle Pleistocene." Likewise, Howell attributes most of the large mammal fauna to the mid-interval of the Middle Pleistocene. While the presence of *Pachycrocuta brevirostris* could point to an earlier Middle or even Lower Pleistocene age for some of the deposits, this species is represented by a single, incomplete tooth, and its origin and relation to the rest of the fauna is uncertain.

Cave bear remains provide some chronological indicators as well. The Yarımburgaz sample contains specimens attributable to both brown bears (*Ursus* arctos) and cave bears. While most of the cave bear specimens possess features indicative of *Ursus deningeri* (the earlier or Middle Pleistocene form), others can be classified as *U. spelaeus*, the later chronospecies. The Yarımburgaz cave bear sample likely represents a local evolutionary continuum between the two forms. Tsoukala's results on the cave bears therefore suggest that the deposits span the period when *U. spelaeus* replaced *U. deningeri* in central and southern Europe. It is of course difficult to pin down the precise time of the transition, since if probably did not occur at precisely the same instant across the breadth of Eurasia. However, most researchers agree that the species replacement took place around MIS 6, and that *U. deningeri* no longer existed by MIS 5 (e.g., Mazza and Rustoni 1994; Orlando et al. 2002; Withalm 2005). The dominance of the *deningeri* form in Yarımburgaz, along with some specimens with distinct *spelaeus*-like characters, strongly suggests that the LC deposits date to the later part of the Middle Pleistocene.

The lithic assemblage from Yarımburgaz does not provide clear chronological indications. The remarkable diversity of late Middle Pleistocene technologies in western Asia and southeastern Europe (see discussion below) suggests that no simple, unilineal chronological scheme can be imposed. However, the lithics do provide indirect evidence that the hominin occupations are older than MIS 5 and 6. Classic Mousterian industries with Levallois technology are present the Levant (Grün and Stringer 2000; Grün et al. 1991; le Tensorer et al. 2007; Mercier and Valladas 2003; Mercier et al. 1995; Schwarcz and Rink 1998; Porat et al. 2002), southern Anatolia (Otte et al. 1995, 1998), and western Bulgaria (Guadelli et al. 2005) by the beginning of MIS 6, if not earlier. In Kozarnika (Bulgaria), the Paleolithic site closest to Yarımburgaz, the earliest Mousterian with Levallois technology is associated with *Ursus deningeri* (Guadelli et al. 2005). The complete absence of Levallois flakes, cores and production debris in the Yarımburgaz assemblage suggests that it predates the spread—whether by demic expansion, cultural transmission or independent invention—of Mousterian technologies with Levallois sometime before MIS 6.

There are further indicators that the occupations in the LC of Yarımburgaz are attributable to a comparatively cold and dry climatic interval within the Middle Pleistocene. One is the presence of steppic micromammals such as *Cricetus, Mesocricetus, Cricetulus, Lagurus, Sicista subtilis,* and *Ochotona pusilla* in the 3rd sedimentary cycle, which contained the majority of archaeological remains (and bear bones). The large mammal fauna also consists (almost) exclusively of Europena and Asian Palaearctic taxa, and contains no African taxa, consistent with comparatively cool climatic conditions. Previous analysis of stable carbon isotopes in cave bear (NISP = 21) and brown bear (NISP = 2) molar enamel from Yarımburgaz by Pigati and Quade (in Stiner et al. 1998) indicates strictly C_3 vegetation in the bear diets. Importantly, there are no indications of C_4 tropical grasses or other C_4 vegetation in the bear diets. C_3 vegetation dominates Europe and the Mediterranean today, and associates with a climate and environmental regime that is driven by winter rains. No significant differences in (unaltered) $\delta\ ^{13}C$ values found between the cave bear and brown bear samples from Yarımburgaz Cave. All of the samples have highly negative $\delta\ ^{13}C$ values and in this regard are similar to cave bear samples from Europe (e.g., Bocherens et al. 1994; Hilderbrand et al. 1996). Highly negative $\delta\ ^{13}C$ values are most consistent with modern bear populations that feed principally or exclusively on terrestrial food sources. No nitrogen is preserved in the Yarımburgaz bear teeth, and thus the isotope study could not evaluate the trophic levels (i.e., degree of carnivory) of the bears. However, the physical condition of adult bear teeth and the patterns of occlusal and interstitial wear testify to habitual chewing and tearing of tough vegetable matter and abrasion from fine sediments (Stiner et al. 1998).

A final indication of the environmental conditions of the human and bear occupations in the LC is the position of the cave itself. The top of the deposit is less than 20 m above modern sea level. Recent work on raised marine terraces along the southwest shore of the Sea of Marmara produced estimates for uplift rates of 0.40 mm/year since the beginning of MIS 7 (Yaltırak et al. 2002). There are no similar studies of terraces in the area of Yarımburgaz, at the eastern end of the Marmara basin. However, even if rates of uplift were much slower in the Yarımburgaz vicinity, the cave would have been flooded or submerged by high sea levels during warmest intervals of the Middle Pleistocene. The presence of a presumed MIS 5e beach deposit in the UC, which clearly postdates the deposits in the LC, is evidence of this sort of marine transgression.

Paleontological evidence, ESR dates, and the site's low elevation and proximity to the sea strongly suggest that hominins and bears occupied Yarımburgaz during a cool climatic interval in the latter part of the Middle Pleistocene. Based on the ESR dates and the lithic assemblage, MIS 6 is probably too recent an attribution. One obvious possibility is that the occupations date to the coldest part of MIS 8, the period between 250ka-270 ka BP. However, ESR dates center on MIS 7. The SPECMAP curve (Imbrie et al. 1984) shows a sharp cooling event, MIS 7d, around 220 ka BP. This cold period, which resolves into two closely-spaced intervals, is associated with a marked decline in oak and other thermophillic plants in the pollen profiles of near-shore marine cores taken off the Atlantic coast of Iberia (Desprat et al. 2006: Fig. 4). The mid-MIS 7 cold interval documented in the marine cores also corresponds with a sharp decline in arboreal pollen in the cores from Tenaghi Philippon, Greece, the deep terrestrial record closest to Yarımburgaz (Tzedakis et al. 2006: 3422). Thus, hominins and bears could have used Yarımburgaz during MIS 7d, between roughly 220ka and 230ka BP. This interval corresponds with the best age estimates based on ranges of ESR dates (Table 3.6).

It is important to remember that the most definitive chronological information for Yarımburgaz comes from the layers (V, W and X) that contain the majority of bear bones and artifacts, and that that all of the ESR dates were obtained from bear teeth. The Upper and Lower Chambers of Yarımburgaz Cave certainly were open and suitable for habitation by animals and Pleistocene hominins before and after these sediments were deposited. Fresh water flowing through the LC, and possible marine incursions, reworked or removed the earlier sediments within the cave (see Chapter 2). Still, remnant deposits or hanging breccias dating to much earlier periods may be preserved in some parts of this large, complex karst system. Thus the presence of a few anomalous chronological indicators among the faunal remains, such as the single specimen of *Pachycrocuta brevirostris*, is not unexpected. Despite the extensive scale of the 1988-1990 excavations, many parts of the Yarımburgaz Cave complex remain unexplored, and the site will likely yield other paleontological and archaeological surprises if the remaining deposits are protected from further destruction by looters and geological forces.

HOMININ AND URSID OCCUPATIONS AT YARIMBURGAZ

The presence of more than 5000 fossil bones and teeth of cave and brown bears is testament to the fact that bears frequently visited the Yarımburgaz caves during the late Middle Pleistocene. A smaller, but still substantial sample of lithic artifacts demonstrates that hominins too were often present in and around the cave. The spatial and stratigraphic distributions of bear remains and stone artifact overlap to a remarkable degree—where one is found, the other also tends to occur (see Chapter 6, Table 6.6). These singular facts raise interesting questions about the nature of the hominin and bear occupations of the cave, and the level of sympatry and perhaps interaction between these two omnivorous predators.

Beyond the demonstrably poor correlations between the spatial distributions of tools and ungulate bones across excavation units in the LC (Chapter 6), spatial analyses have been of limited analytical power for this study of site formation history. Careful examination of taphonomic data, contrasts in the patterns of skeletal completeness, species diversity in relation to sample sizes among mammal categories, age structure analysis, and ethological perspectives have been more useful for sorting out the roles of hominins, bears, and other mammals in site formation processes in this cave.

Skeletal and contextual information provide clear indications of what bears were doing in Yarımburgaz Cave. The age-structure of the bear thanatocenosis, the comparatively unbiased representation of skeletal elements, and evidence for predation of some individuals (see Chapter 6) demonstrate convincingly that most or all of the bear remains were the result of natural deaths, and less commonly predation, within a hibernation den. Over some millennia, bears sometimes overwintered in the cave complex, and some of them perished there. Evidence of a high degree of herbivory in the cave bears (Stiner et al. 1998), implies that both sexes probably hibernated for some period in winter. We should not take the sheer abundance of the bear remains as an indication that large numbers of bears were ever together in the cave. Except in the case of mothers with dependent cubs, bears are solitary hibernators. Rather, the impressive assemblage represents the material remains of uncommon deaths that accumulated over a substantial period. If we assume that the MNI of 42 reported by Tsoukala (Chapter 7) represents only 25% of the individuals that originally died in the Lower Chamber (probably an underestimate based on the area excavated), and that the deposits formed over 10,000 years or more (again, quite possibly an underestimate), bear skeletons accumulated in the deposits at a rate of between one and two bear deaths per century. That does not tell us how often solitary or pregnant female bears actually denned in the cave—they might have been present nearly every year, or there might have been long intervals in which bears chose to hibernate elsewhere.

Explaining the hominin presence in the cave is much more difficult. Normally, we expect that humans used natural shelters as foci for domestic activities. Caves are places where Pleistocene humans often carried meat, where they cooked and ate food, made and repaired artifacts, socialized, and slept. Aside from the stone artifacts, however, there is precious little evidence for domestic activities inside Yarımburgaz Cave. No hearths were observed in the excavation, and there is no unequivocal evidence of burning of bones or stone tools. As discussed in Chapter 6, cut marks and other indications of hominin actions are quite rare on bones of ungulates, and essentially absent from the bones of bears. Natural bear deaths *in situ* and bones transported into the cave by wolves, hyenas and certain other predators are responsible for most of the faunal remains in the LC. Whatever brought hominins to Yarımburgaz, it did not involve transporting or processing hunted game or use of fire to any significant extent.

Several features of the lithic assemblage also suggest that hominin visits to the cave were brief. The high frequencies of retouched tools and heavy reduction of some flake tools suggests that a good deal of the material represents transported artifacts, carried to the cave from elsewhere, used and soon abandoned (Kuhn 1995, 2004; Riel-Salvatore and Barton 2004). These tools may not have been carried far, but they were not necessarily made in place either. In contrast to the tools, cores are not extensively exploited for the most part, giving the appearance of being predominantly local in origin (Table 5.7). Unfortunately, we know little about the geological sources of the lithic raw materials in in Yarımburgaz. Natural flint, quartz and quartzite pebbles all occur in the fluvial sediments inside the cave, suggesting an origin very close by. More than one type of flint may be represented, but heavy patination obscures differences in color and inclusions that might be used to identify rocks from different sources.

Variation in core and flake numbers (Table 9.1) is instructive with respect to the movement of artifacts and hominin frequentation of the cave. More flakes and retouched tools of flint were brought into the cave than were made there, but the opposite is true of the other raw materials. There are 9.3 flakes and retouched tools for every core or core tool of flint material. The ratios for quartz and quartzite are much lower, 2.9 to 1 and 1.1 to 1, respectively. Even considering the more extensive reduction of flint cores (Chapter 5, Table 5.7), these ratios are remarkably divergent. Estimates of the number of flakes produced from each core are equally striking. The total

Table 9.1 Frequencies of cores, flakes and flake tools, by raw material.

	flint	quartz	quartzite
flakes+flake tools/core	9.37	2.87	1.1
flake scars per core	8	7	4
cores + core tools	70	46	83
estimated flakes produced	560	322	332
flake tools+ whole flakes	545	116	72
flake tools+ whole + broken flakes	656	132	92

number of flake tools, whole flakes, and broken flakes of flint is somewhat higher than the sum of flake scars observed on flint cores. In contrast, there are major deficits in flake numbers for quartzite and quartz. For both of these materials, the total number of flakes and flake tools is far fewer than the number of flakes that should have come from the cores and core tools collected during the excavations (Table 9.1).

Beyond the differential use of the three main raw materials, the core and flake numbers provide a glimpse of the behavioral processes leading to the accumulation of the Yarımburgaz lithic assemblages. Hominins arrived at Yarımburgaz carrying artifacts, mainly flake tools, and mostly made of flint. Some of these artifacts were subsequently abandoned in the cave, and replaced by tools made locally from flint, quartz and quartzite. The deficits in flakes of the latter two materials further suggest that hominins left the cave carrying some of the artifacts they had made there. The lithic data do not tell us why the hominins came to the cave, or what they did with the stone tools they brought or made there. What they do indicate is that, in keeping with the zooarchaeological information and the absence of evidence for fire, the Middle Pleistocene hominins did not spend very much time in Yarımburgaz. Their visits were brief, if repeated, and involved a limited array of activities and some tool manufacture on site.

There is no evidence that early hominins were present in the shelter at the same time as live cave bears. The hominins did not prey on the bears. The two species visited the cave at different times. It is likely that they actively avoided one another, or perhaps they visited the cave on different seasonal schedules. Bears were in the cave in winter. Pleistocene hominins visited the cave briefly, perhaps during the warm seasons or at least when no hibernating bears happened to be present. How then can we account for the close spatial associations of artifact and bear remains at Yarımburgaz Cave?

The explanation for the co-occurrences of behaviorally unrelated material can be found in the behavioral tendencies of both species. Hominins and bears would have had generally similar preferences for shelters. In addition to protection from the elements, they both sought out places that were accessible and comparatively dry. During the warmest, wettest periods of the Pleistocene, when sea levels were high, Yarımburgaz was probably inaccessible or unappealingly damp to hominins and bears alike. Likewise, when the karst was most active and fresh water was flowing or pooling up inside the cave, neither species would have found it a congenial shelter. Both hominins and bears would have been attracted to the cave during those periods when it was dry. If their respective tenures alternated at the scale of seasons, years, or even decades, the material remains would be stratigraphically inseparable. Moreover, denning bears tend to disturb the ground on which they sleep, digging out shallow beds and pushing aside the largest, obtrusive objects (see Chapter 6, and Gargett [1996] for a review of bear behavior in modern and Pleistocene cases). So long as sediments accumulate slowly, such small-scale disturbances, repeated year after year, will result in homogenized or "palimpsest" deposits. Any small separation that may once have existed between hominin debris and bones of bears that died in place would be erased as generations of bears prepared their beds. In the end, similar preferences for conditions inside shelters, coupled with sediment-disturbing activities of bears, created an impression of close co-existence between two species that actually had little or nothing to do with one another.

CULTURAL DIVERSITY IN THE LATER MIDDLE PLEISTOCENE

Although the chronological indicators are far from precise, the ESR dating results in combination with the paleontological assessment of the cave bear remains suggest that the main Paleolithic occupation of Yarımburgaz Cave occurred during mid-MIS 7, or possibly late MIS 8. This period between 250,000 and 180,000 years ago was one of extraordinary archaeological diversity in the regions surrounding Yarımburgaz Cave (Porat et al. 2002). Some of this diversity represents diachronic change in archaeological cultures, but a large part of the variation is synchronic. It is instructive to consider how the Yarımburgaz assemblage fits into supra-regional patterns of technological variation.

Late Acheulean assemblages with bifacial handaxes are found from the Near East to Anatolia and the Caucasus, although few of these assemblages are well dated. However, several Acheulean assemblages form the Levant, including Holon and Umm Qatafa, have recently been assigned to MIS 7 based on ESR determinations (Porat et al. 1999, 2002). The components of Jelinek's (1982, 1990) Mugharan Tradition—Acheulo-Yabrudian, Yabrudian and Amudian assemblages—typified by heavily retouched scrapers on thick flakes, handaxes, and sometimes fine blades, show a wide range of ages ranging from MIS 8 through late MIS 6 (Barkai et al. 2003; Mercier et al. 1995, 2000; Mercier and Valladas 1995; Porat et al. 2002). The early Levantine Mousterian and the "Hummalian" industry have also yielded a wide range of ESR and TL ages (Bar-Yosef 1993; Meignen et al. 2001; Mercier et al. 1995; Schwarcz & Rink 1998; Valladas et al. 1998). For the most part these industries center on MIS 7 and 6, although TL dates from Tabun Cave place the initial appearance of the Mousterian and Levallois technology in MIS 8.

Much less is known about the late Middle Pleistocene archaeology of Anatolia. Handaxes are abundant on river terraces in southeastern Turkey (Bostancı 1961; Garrard et al 1996; Minzoni-Deroch and Sanlaville 1988; Taşkıran 2008; Yalçınkaya 1981), but stratified Acheulean assemblages have been documented only at one locality, Kaletepe Deresi 3, on the central Anatolian plateau (Slimak et al. 2007, 2008). The Acheulean levels at Kaletepe Deresi 3 are likely to date to the early Middle Pleistocene (Tryon et al. 2009). Isolated handaxes have even been found along the eastern shores of the Bosporous (Jelinek 1980; Runnels and Özdoğan 2001) but again, none of these occurrences is securely dated.

Only two sites in Turkey have yielded assemblages in good geological context likely to date to MIS 7 and 8. The most extensively-studied of these is from Karain Cave on the Aegean coast near the modern city of Antalya. Archaeological units E and F in Karain chamber E are characterized by denticulates, notches and scrapers made on thick flakes with open platform angles (Otte et al. 1990; Yaçinkaya et al. 1993); bifaces are essentially absent. Termed "Proto-Charentian", these assemblages bear some resemblance to the Yabrudian of the Levant. The overlying archaeological unit G marks the beginning of the true Zagros/Karain Mousterian. Units E and F in Karain have been assigned to the interval between 250 ka and 350 ka BP largely on stratigraphic grounds (Otte et al. 1998). The closest dated horizon, unit H, is attributed to MIS 5, and it is entirely possible that Unit F and E are not as old as has been estimated.

The other site in Anatolia with *in situ* Middle Pleistocene deposits is Kaletepe Deresi 3 (Slimak et al. 2007, 2008). This open air site preserves a sequence of archeological horizons more than 7 meters deep. Levels 2′ and 3, which are sandwiched between Levallois Mousterian levels above and Acheulean levels below, have yielded small assemblages of thick flakes and some tools with irregular retouch. Volcanic tephras attributed to an eruption of the Acıgöl caldera, dating to approximately 160 ka BP, are present above levels 2′ and 3. The lithic assemblages of levels 2' and 3 are therefore likely to date to MIS 6 or 7, although tighter age constraints are not currently available (Tryon et al. 2009).

The later Lower Paleolithic and early Middle Paleolithic of the eastern Balkans and Thrace are also poorly understood. The Acheulean is seldom found in central Europe and the Balkans (e.g., Darlas 1995; Runnels 2001), such that the discovery of a single handaxe in Greece excited considerable interest (Runnels and van Andel 1993). Archaeological reconnaissance survey in Turkish Thrace has so far failed to identify clear evidence for Middle Pleistocene assemblages with bifaces or cleavers. Sites with choppers and heavy flake tools have been documented, unfortunately only as surface occurrences (Dinçer and Slimak 2007; Runnels and Özdoğan 2001).

Earlier Middle Pleistocene assemblages in central and south-central Europe tend to be "core and flake tool" or Tayacian industries. Typical of central Europe are so-called "small tool industries," such as have been described from Bilzingsleben (Mania 1995), Schöningen (Thieme et al. 1993) and Vértesszölös (Kretzoi and Dobosi 1990). These assemblages are characterized by an abundance of steeply retouched flake tools, often with irregular edges: pointed forms predominate at some sites (e.g., Kretzoi and Dobosi 1990). Most of the "small tool industries" date to between 300 and 500 ka BP, although non-Levallois industries with small flake tools continue with the so called "Taubachian," considered to be an early Middle Paleolithic complex (Moncel 2003; Valoch 1984). Farther west, in Moravia, the earliest Mousterian appears in MIS 6 or late MIS 7 (Svoboda et al. 1996: 82).

An important sequence for the eastern Balkans comes from the site of Kozarnika in northwest Bulgaria (Guadelli et al. 2005). Here, the earliest Middle Paleolithic layers (10a and 10b), described as Mousterian with Levallois technology, has been attributed to late MIS 7 and MIS 6 based largely on biostratigraphic indicators, including the presence of *Ursus deningeri*. The underlying Lower Paleolithic assemblage from layer 11a is characterized by a predominance of notches and denticulates among the retouched tools. Tool blanks were produced from discoid and polyhedral/globular cores. It appears that layers 11a and 10b are separated by a significant erosional hiatus (Guadelli et al. 2005: 98).

Taken at face value, the absolute dates currently available for Middle Pleistocene sites in western Asian and southeast Europe suggest considerable chronological overlap between late Acheulean, Yabruduan/Amudian, non-Levallois flake tool and small tool industries, and the earliest Middle Paleolithic. Certainly, these diverse assemblage types are not all contemporaneous. There is little question, for example, that the early Levantine Mousterian succeeds the Yabrudian and Amudian; it is always stratigraphically higher when both occur in the same sequence, such as in the sites of Yabrud I, Tabun, Hayonim and Hummal. It is even possible that the Mousterian appeared and proliferated during or just after MIS 7. However, the transitions need not have been, indeed probably were not, simultaneous throughout this large area.

Based on typological criteria, the Yarımburgaz assemblage most closely resembles the non-Levallois flake tool industries of central and southeastern Europe, as well as the "proto-Charentian" of Karain. However, we suggest that, despite differences in terminologies, there is a deeper and more widespread similarity among these late middle Pleistocene lithic industries. Specifically, a wide range of pre-Mousterian assemblages dating to late MIS 8 and MIS 7, including the Mugharan complex in the Levant, are characterized by 1) heavy reliance on flake tools, 2) comparatively infrequent core tools, 3) relatively simple flake production with little or no "pre-determination", often resulting in large and thick blanks, 4) a strong reliance on local raw materials obtained within 10-20 km of sites, 5) variable but sometimes heavy retouch and reduction, and 6) locally frequent re-use or lateral recycling of artifacts by removal of ventral flakes and so on (see Chapter 5).

We do not propose that this very general set of characteristics represents a "cultural tradition" or complex, in the sense of shared concepts of appropriate behavior maintained by processes of cultural transmission. The criteria outlined above crosscut a great deal of technological and typological variation. In some assemblages, scrapers dominate while in others denticulates and notches are more common, and some assemblages contain handaxes whereas others contain choppers and chopping tools. *Chaînes opératoires* also vary widely: they may include variants of discoid production (cf. Boëda 1993; Terradas 2003), what is called "Quina" type flake production (cf. Bourguignon 1996; Turq 1992) and even simple blade technology. What the diverse and geographically dispersed late Middle Pleistocene assemblages share in common are similar tendencies in "technological provisioning" (*sensu* Kuhn 1991, 1995, 2004), namely an emphasis on flake tools, comparatively little investment in blank production and "predetermination," and heavy reduction of artifacts that is not conditioned by an immediate scarcity of raw materials. Such commonalities as exist should reflect similar foraging adaptations and systems of land use, mediated perhaps by cognitive capacities of the hominins, rather than shared cultural conventions. The important point is that regionally-specific designations such as "Taubachian", "Tayacian", "Yabrudian," and "proto-Charentian" may obscure broadly-shared behavioral tendencies among hominins of MIS 7 and 8.

MIDDLE OR LOWER PALEOLITHIC?

From the outset of the investigation of the Pleistocene deposits in Yarımuburgaz, the lithic materials from the LC have been assigned to the Lower Paleolithic. In light of recent critiques of the traditional classifications of Paleolithic industries, and ongoing uncertainties about the chronology of different Middle Pleistocene industries, it is worthwhile re-examining that terminology.

Like all scientists, archaeologists inherit many of their conceptual tools from previous generations of researchers. Systems of classification show a particular degree of entropy and may continue to be used as a common language long after their relevance declines. The traditional division of the Paleolithic into three phases—Lower, Middle and Upper Paleolithic—is essentially a geological schema for classification. As such it remains useful for placing archaeological assemblages into a rough chronological framework. Nonetheless, as knowledge accumulates over time, such broadly defined categories that were originally formulated on the basis of limited data from a limited geographic area, inevitably lose their clarity and their utility (Bar-Yosef and Kuhn 1999). New data from new areas or new sites and new sources of chronological information eventually mandate that we revise our accounts of long-term technological evolution.

In fact, prehistorians have already begun to critically re-examine the definitions applied to the major Paleolithic subdivisions. Recent studies have cast doubt on the rigidity of the distinctions between the Lower and Middle Paleolithic in Eurasia. As discussed above, assemblages conventionally classified as Middle Paleolithic were apparently contemporaneous at a broad regional level with the late Acheulean, Micoquian, and other nominally Lower Paleolithic assemblage types. The continuing usefulness of rigid distinctions between Middle and Lower Paleolithic based on the relative frequencies of core and flake tools has been thoroughly undermined as well (Monnier 2006).

The lithic assemblage from Yarımburgaz Cave embodies many of the ambiguities surrounding the conventional divide between Lower and Middle Paleolithic. The absolute age of the deposits in Yarımburgaz is uncertain, but based on the results from ESR dating and paleontological assessments of the ursid remains, the site is roughly contemporaneous with a wide range of archaeological assemblage types ranging from late Acheulean to early Mousterian. The Yarımburgaz assemblage also shows a mixture of features conventionally associated with Lower and Middle Paleolithic industries. The abundance of core tools has been considered important to the definition of Lower Paleolithic technologies. However, bifaces are absent from the Yarımburgaz assemblage and chopper/chopping tools account for only about 10% of modified pieces. Based on the predominance of retouched flake tools, one could argue that Yarımburgaz fits more comfortably within the Middle Paleolithic. On the other hand, the complete evidence of Levallois technology would make some researchers hesitate to assign the Yarımburgaz material to either the Mousterian or Mode 3. Yet many Middle Paleolithic assemblages (the Quina Mousterian of France, for example) also lack Levallois more or less completely. It is interesting that this ambiguity extends back to the original definition of the Tayacian, which was initially grouped with the Mousterian "complex" (Bordes and Bourgon 1951) but later was shifted to the Lower Paleolithic.

Looking beyond simple frequencies of different classes of artifact, Goren-Inbar and Belfer-Cohen (1998) have made an intriguing case for the existence of a fundamental difference between Middle and Lower Paleolithic *châines opératoires*. They argue that Lower Paleolithic stone tool production, while often diversified, exhibits a highly linear pattern. A specific raw material is frequently linked to a specific method of core reduction, producing a limited range of tool forms along a relatively independent trajectory. Several independent or quasi-independent trajectories may co-exist within the same assemblage. This pattern has a very great time depth; it can be identified in the Oldowan, for example, and it also is clearly expressed in many Acheulean assemblages. According to these same authors, Middle Paleolithic assemblages typically show a branching or "ramifying" pattern (see also Bourguignon et al. 2004). In the Mousterian, they argue, a few relatively flexible production methods were applied to a range of raw materials to produce comparatively standardized flakes and blades. Levallois technology would be one example of such a flexible technique. The blanks produced were in turn modified by retouch into a whole assortment of edge types and tool forms. In other words, the diversity of Middle Paleolithic tools develops late in the manufacture process, whereas Lower Paleolithic artifact forms are essentially determined from the beginning of the manufacture process, perhaps as early as the selection of raw material. It is important to note that Goren-Inbar and Belfer-Cohen do not propose this model as a framework for assigning assemblages to chrono-stratigraphic units. Rather, they argue that it is a reflection of changing hominid capacities.

Even within the framework proposed by Goren-Inbar and Belfer-Cohen (1998), the Yarımburgaz assemblage exhibits features of both Middle and Upper Paleolithic industries. The strong association between raw material and reduction systems is reminiscent of the "parallel trajectories" which they claim characterizes the Lower Paleolithic *sensu lato*. However, two of the three major production system/raw material pairings at Yarımburgaz—quartz/bipolar and flint/discoid—led to the production of flake tools. Only quartzite may have been used primarily for producing a different kind of end-product, choppers and chopping tools. Flake tools of flint in particular often were extensively modified and perhaps re-sharpened, more reminiscent of the late-stage pattern of "ramification" attributed by Goren-Inbar and Belfer-Cohen to Middle Paleolithic industries. As discussed in Chapter 5, it is difficult to draw clear typological boundaries between scrapers, denticulates, and other tool forms at Yarımburgaz. That does not mean that functionally distinct kinds of artifact did not exist from the hominins' point of view. Rather, it means that these types can no longer be distinguished when the artifacts reached their final state of discard.

Even the model proposed by Goren-Inbar and Belfer-Cohen does not provide a clear solution to the question of whether to assign the Yarımburgaz assemblage to the Lower or Middle Paleolithic. Such systems of categorization cannot, in any case, be expected to provide unambiguous classifications of actual archaeological phenomena. Current evidence indicates evolutionary continuity between Lower Paleolithic and Middle Paleolithic hominid forms in Eurasia (Dean et al. 1998; Harvati 2007; Hublin 1998; Manzi 2004; Rightmire 2007; Tattersall 2007), and we should not be surprised to find continuity in the lithic assemblages as well. It seems likely thatis even possible that the lithic assemblage from Yarımburgaz Cave, and others like it, really form another entity, a "grade" of technology that is neither fully Middle nor Lower Paleolithic, but which exhibits features of both. We might think of these flake-tool-based assemblages from MIS 7 as representing a "Mode 2.5" in Graham Clark's (1969) famous and widely used scheme for classifying Paleolithic industries. We *are not* proposing this fractional value as a new term but rather to make the point that the various industries of MIS 7 may embody the final evolutionary developments in much older behavioral tendencies, at the same time containing the roots of novel forms of behavior. More intensive study of many other lithic assemblages, associated faunas, and ecological contexts is necessary before proposing a fuller account of the common behavioral tendencies that may unite "ambiguous" late Middle Pleistocene industries such as that from Yarımburgaz Cave. It may well turn out that, as in the later Epipaleolithic-Neolithic and Middle-Upper Paleolithic transitions, what lies in the middle is as interesting and important as what came before and after.

REFERENCES CITED

Barkai, R., Gopher, A., Lauritzen, S.E. and Frumkin, A., 2003. Uranium series dates from Qesem Cave, Israel, and the end of the Lower Palaeolithic. *Nature* 423, 977-979.

Bar-Yosef, O., 1993. The role of western Asia in modern human origins, in: Aitken, M., Stringer, C., Mellars, P., (Eds.) *The Origin of Modern Humans and the Impact of Chronometric Dating*. Princeton University Press, Princeton, NJ, pp. 132-147.

Bar-Yosef, O., Kuhn, S., 1999. The big deal about blades: laminar technologies and human evolution. *American Anthropologist* 101, 322-338.

Bocherens, H., Fizet, M., Mariotti, A., 1994. Diet, physiology, and ecology of fossil mammals as inferred from stable carbon and nitrogen isotope biochemistry: implications for Pleistocene bears. *Palaeogeography, Palaeoclimatology, and Palaeoecology* 107, 213-225.

Boëda, E., 1993. Le débitage discoïdes et le débitage Levallois récurrent centrripète. *B.S.P.F.* 90, 392-404.

Boëda, E., Bourguignon, L., Griggo, C., 1998. Activités de subsistence au Paléolithic moyen: Couche V13 b' du gisement d'Umm el Tlel (Syrie), in: Brugal, J.-P., Meignen, L., Patou-Mathis, M. (Eds.), *Économie préhistorique: Les comportements de subsistence au Paléolithique*. Editions du APDCA, Sophia Antipolis, France, pp. 243-258.

Bordes, F., Bourgon, M., 1951. Le complex Moustérien: Moustérien, Levalloisien et Tayacien. *L'Anthropologie* 55, 1-23.

Bostancı, E., 1961. Researches in south-east Anatolia: The Chellan and Acheulean Industry of Dülük and Kartal. *Anatolia* 6, 89-162.

Bourguingnon, L., 1996. La conception de débitage Quina. *Quaternaria Nova* 6, 149-166.

Bourguignon, L., Faivre, J.-P., Turq, A., 2004. Ramification des chaînes opératoires : Une spécificité du Moustérien ? *Paléo* 16, 37-48.

Clark, G., 1969. *World Prehistory: A New Outline*. Cambridge University Press, London.

Darlas, A., 1995. The earliest occupation of Europe: the Balkans, in: Roebroeks, W., Von Kolschoten, T., (Eds.), *The Earliest Occupation of Europe*. University of Leiden, Leiden, pp. 51-59.

Dean, D., Hublin, J.-J., Holloway, R., Ziegler, R., 1998. On the phylogenetic position of the pre-Neanderthal specimen from Reilingen, Germany. *Journal of Human Evolution* 34, 485-508.

Desprat, S., Sánchez Goñi, M.F., Turon, J.-L., Duprat, J., Malaizé, B., Peypouquet, J.-P., 2006. Climatic variability of Marine Isotope Stage 7: direct land–sea–ice correlation from a multiproxy analysis of a north-western Iberian margin deep-sea core. *Quaternary Science Reviews* 25, 1010-1026.

Dinçer, B., Slimak, L., 2007. The Paleolithic of Turkish Thrace: synthesis and recent results. *TÜBA-AR* 10, 49-61.

Gargett, R.H., 1996. *Cave Bears and Modern Human Origins: The Spatial Taphonomy of Pod Hradem Cave, Czech Republic*. University Press of America, Lanham, MD.

Garrard, A.N., 1983. The Palaeolithic faunal remains from Adlun and their ecological context, in: Roe, D.A., (Ed.), *Adlun in the Stone Age: The Excavations of D.A.E. Garrod in the Lebanon, 1958-1963*. British Archaeological Reports, International Series 159(ii), Cambridge, pp. 397-413.

Garrard, A., Conolly, J., Moloney, N., Wright, K., 1996.The early prehistory of the Sakçagözü region, north Levantine rift valley: Report on 1995 survey season. *Anatolian Studies* XLVI, 53-81.

Goren-Inbar, N., Belfer-Cohen, A., 1998. The technological abilities of the Levantine Mousterian: cultural and mental capacities, in: Akazawa, T., Akoi, K., Bar-Yosef, O., (Eds.), *Neandertals and Modern Humans in Western Asia*. Plenum Press, New York, pp. 205-221.

Grün, R., Stringer, C.B., Schwarcz, H.P., 1991. ESR dating of teeth from Garrod's Tabun Cave collection. *Journal of Human Evolution* 20, 231-248.

Grün, R., Stringer, C., 2000. Tabun revisited: revised ESR chronology and new ESR and U-series analyses of dental material from Tabun C1. *Journal of Human Evolution* 39, 601-612.

Guadelli, J.-L., Sirakov, N., Ivanova, S., Sirakova, S., Anastassova, E., Courtaud, P., Dimitrova, I., Djabarska, N., Fernandez, P., Ferrier, C., Fontugne, M., Gambier, D., Guadelli, A., Iordanova, D., Iordanova, N., Kovatcheva, M., Krumov, I., Leblanc, J.-C., Mallye, J.-B., Marinska, M., Miteva, V., Popov, V., Spassov, R., Taneva, S.T., Tisterat-Laborde, N., Tsanova, T., 2005. Une séquence du Paléolithique inférieur au Paléolithique récent dans les Balkans: la grotte Kozarnika à Orechets (nord-ouest de la Bulgarie), in: Molines, N., Moncel, M.-H., Monnier, J.-L., (Eds.), *Les Premiers Peuplements en Europe : Colloque international, données récentes sur les modalités de peuplement et sur le cadre chronostratigraphique, géologique et paléogéographique des industries du paléolithique ancien et moyen en Europe.* British Archaeological Reports International Series 1364. Archaeopress, Oxford, pp. 87-103.

Harvati, K., 2007. 100 years of *Homo heidelbergensis* – life and times of a controversial taxon. *Mitteilungen der Gesellschaft für Urgeschichte* 16, 85-94.

Hilderbrand, G.V., Farley, S.D., Robbins, C.T., Hanley, T.A., Titus, K., Servheen, C., 1996. Use of stable isotopes to determine diets of living and extinct bears. *Canadian Journal of Zoology* 74, 2080-2088.

Hublin, J.-J., 1998. Climatic changes, paleogeography and the evolution of Neanderthals, in: Aoki, K., Bar-Yosef, O.. (Eds), *Neanderthals and Modern Humans in Western Asia.* Plenum Press, New York, pp. 295-310.

Jelinek, A.J., 1980. Collections of Paleolithic materials from valleys on the east side of the Bosporus, in: Çambel, H., Braidwood, R., (Eds.), *Prehistoric Research in Southeastern Anatolia.* Faculty of Letters publication N. 2589. Istanbul University, Istanbul, pp. 319-327.

Imbrie, J., Hays, J.D., Martinson, D.G., McIntyre, A., Mix, A.C., Morley, J.J., Paces, N.G., Prell, W.L., Shackleton, N.J., 1984. The orbital theory of Pleistocene climate: Support from a revised chronology of the marine $\delta^{18}O$ record, in: Berger, A. et. al. (Eds.), *Milankovitch and Climate.* D. Reidel, Norwell, Mass., pp. 269-305.

Jelinek, A.J., 1982. The Tabun cave and Paleolithic man in the Levant. *Science* 216, 1368-1376.

Jelinek, A.J., 1990. The Amudian in the context of the Mugharan Tradition at Tabun Cave (Mount Carmel) Israel, in: Mellars, P., (Ed.), *The Emergence of Modern Humans.* Cornell University Press, Ithaka, NY, pp. 81-90.

Kretzoi, M., Dobosi, V., (Eds.), 1990. *Vértesszölös —Man, Site and Culture.* Akadémiai Kiadó, Budapest.

Kuhn, S., 1991. Unpacking reduction: lithic raw material economy in the Mousterian of west-central Italy. *Journal of Anthropological Archaeology* 10, 76-106.

Kuhn, S., 1995. *Mousterian Lithic Technology: An Ecological Perspective.* Princeton University Press, Princeton, NJ.

Kuhn, S., 2004. Upper Paleolithic raw material economies at Üçağızlı cave, Turkey. *Journal of Anthropological Archaeology* 23, 431-448.

Kurtén, B., 1976. *The Cave Bear Story.* Columbia University Press, NewYork, NY.

Mania, D., 1995. The Elbe-Saale Rgion (Germany), in: Roebroeks, W., van Kolfschoten, T., (Eds.), *The Earliest Occupation of Europe.* University of Leiden, Leiden, pp. 85-101.

Manzi, G., 2004. Human evolution at the Matuyama-Brunhes boundary. *Evolutionary Anthropology* 13, 11-24.

Mazza, P., Rustoni, M., 1994. On the phylogeny of Eurasian bears. *Paleontographica Abt.* A. 230, 1-38.

Meignen, L., Bar-Yosef, O., Mercier, N., Valladas, H., Goldberg, P., Vandermeersch, B. 2001. Apport des datations au problème de l'origin des Hommes modernes au Proche-Orient, in: Barrandon, J.-N., Guibert, P., Michel, V., (Eds.) *Datation, XXIe Rencontres Internationales d'Archéologie et d'Histoire d'Antibes.* Editions APDCA, Antibes, pp. 295-313.

Mercier, N., Valladas, H., 2003. Reassessment of TL age estimates of burnt flints from the Paleolithic site of Tabun Cave, *Israel Journal of Human Evolution* 45, 401-409.

Mercier, N., Valladas, H., Valladas, G., Reyss, J.-L., Jelinek, A., Meignen, L., Joron, J.-L., 1995. TL dates of burnt flints from Jelinek's excavations at Tabun and their implications. *Journal of Archaeological Science* 22, 495-510.

Mercier, N., Valladas, H., Froget, L., Joron, J.-L., Ronen, A., 2000. Datation par thermoluminescence de la base du gisment Paléolithique de Tabun (Mt. Carmel, Israël). *Comptes Rendues Academie di Sciences*, Paris 330, 731-738.

Minzoni-Deroche, A., Sanlaville, P., 1988. Le Paléolithique inférieur de la région de Gaziantep. *Paléorient* 14, 87-98.

Moncel, M.-H., 2003. Some observations on microlithic assemblages in Central Europe during Lower and Middle Palaeolithic: Kůlna and Předmostí (Czech Republic), Vértesszőllős and Tata (Hungary), in: Burdukiewicz, J.M., Ronen, A., (Eds.), *Lower Palaeolithic Small Tools in Europe and the Levant*. B.A.R. International Series 1115. Archaeopress, Oxford, pp. 169-187.

Monchot, H., Kolska Horwitz, L., 2007. Taxon representation and age and sex distribution, in: Chazan, M., Kolska-Horwitz, L., (Eds.), *Holon: A Lower Paleolithic sites in Israel. American School of Prehistoric Research, Bulletin* 50, Peabody Museum Press, Harvard University, Cambridge, MA, pp. 85-109.

Monnier, G., 2006. The Lower/Middle Paleolithic periodization in Western Europe: an evaluation. *Current Anthropology* 47, 709-744.

Orlando, L., Bonjean, D., Bocherens, H., Thenot, A., Argant, A., Otte, M., Hänni, C., 2002. Ancient DNA and the population genetics of cave Bears (*Ursus spelaeus*) through space and time. *Molecular Biology and Evolution* 19, 1920-1933.

Otte, M., Yalçınkaya, I., Kozlowski, J., Bar-Yosef, O., Taşkıran, H , Noiret, P., 1995. Évolution technique au Paléolithique ancien de Karain. *L'Anthropologie* 90, 529-561.

Otte, M., Yalçınkaya, I., Kozlowski, J., Bar-Yosef, O., Lopez Bayon, I., Taşkıran, H., 1998. Long-term technical evolution and human remains in the Anatolian Palaeolithic. *Journal of Human Evolution* 34, 413-431.

Porat, N., Chazan, M., Schwarcz, H., Kolska Horowitz, L., 2002. Toming of the Lower to Middle Paleolithic boundary; New dates form the Levant. *Journal of Human Evolution* 43, 107-122.

Porat, N., Zhou, L.P., Chazan, M., Noy, T., Horwitz, L.K., 1999. Dating the lower Paleolithic open-air site of Holon, Israel, by luminescence and ESR techniques. *Quaternary Research* 51, 328-341.

Quilès, J., Petrea, C., Moldovan, O., Zilhão, J., Rodrigo, R., Rougier, Constantin, H., Milota, Ş., Gherase, M., Sarcină, L., Trinkaus, E., 2006. Cave bears (Ursus spelaeus) from the Peştera cu Oase (Banat, Romania): Paleobiology and Taphonomy. *Palevol* 5, 927-934.

Rabinovich, R., Gaudzinski-Windheuser, S., Goren-Inbar, N., 2008. Systematic butchering of fallow deer (Dama) at the early middle Pleistocene Acheulian site of Gesher Benot Ya'aqov (Israel). *Journal of Human Evolution* 54(1), 134-149.

Rabinovich, R., Tchernov, E., 1995. Chronological, paleoecological and taphonomical aspects of the Middle Paleolithic site of Qafzeh, Israel, in: Buitenhuis, H., Uerpmann, H.-P. (Eds.), *Archaeozoology of the Near East*, II. Backhuys Publishers, Leiden, pp. 5-44.

Reichenau, V.W., 1908. *Der Carnivoren aus den Sanden von Mauer und Mosbach*. Abhandlugen der Grossherx. Hessischen Geologischen Landesanstalt, Band I, Darmstadt.

Riel-Salvatore, J., Barton, C.M., 2004. Late Pleistocene technology, economic behavior, and land-use dynamics in southern Italy. *American Antiquity* 69, 273-290.

Rightmire, G.P., 2007. Later Middle Pleistocene *Homo*, in: Henke, W., Tattersall, I., (Eds.), *Handbook of Paleoanthropology*, Volume 3: Phylogeny of Hominins. Springer Verlag, Berlin,1695-1715.

Runnels, C., 2001. The Stone Age of the Aegean from the Palaeolithic to the Advent of the Neolithic, in: Cullen, T., (Ed.), *Aegean Prehistory: A Review*. Archaeological Institute of America, Princeton, NJ, pp. 225-258.

Runnels, C., van Andel, T., 1993. A handaxe from Kokkinopilos, Epirus, and its implications for the Paleolithic of Greece. *Journal of Field Archaeology* 20, 191-203.

Runnels, C., Özdoğan, M., 2001. The Palaeolithic of the Bosphorus Region, NW Turkey *Journal of Field Archaeology* 28, 69-92.

Schwarcz, H.P., Rink, W.J., 1998. Progress in ESR and U-series chronology of the Levantine Paleolithic, in: Akazawa, T., Aoki, K., Bar-Yosef, O., (Eds.). *Neanderthals and Modern Humans in Western Asia*. Plenum Press, New York, pp. 57-68.

Slimak, L., Kuhn, S., Balkan-Atlı, N., Binder, D., and Dinçer, B., 2007. Kaletepe Deresi 3: de 'l'Acheulleen au Mousterien en Anatolie centrale. *Anatolia Antiqua/Eski Anadolu* 15, 257-273.

Slimak, L., Kuhn, S., Roche, H., Mouralis, D., Buitenhuis, H., Balkan-Atlı, N., Binder, D., Kuzucuoğlu, C., Guillou, H., 2008. Kaletepe Deresi 3 (Turkey): archaeological evidence for early human settlement in Central Anatolia. *Journal of Human Evolution* 54, 99-111.

Speth, J.D., Tchernov, E., 2007. The Middle Paleolithic Occupations at Kebara Cave: a faunal perspective, in: Bar-Yosef, O., Meignen, L., (Eds.), *Kebara Cave, Mt. Carmel, Israel: The Middle and Upper Paleolithic Archaeology*, Part I. American School of Prehistoric Research, Bulletin 49, Peabody Museum Press, Harvard University, Cambridge, MA, pp. 165-260.

Stiner, M.C., 2005 The Faunas of Hayonim Cave (Israel): A 200,000-Year Record of Paleolithic Diet, Demography & Society. American School of Prehistoric Research, Bulletin 48, Peabody Museum Press, Harvard University, Cambridge, MA.

Stiner, M.C., Achyuthan, H., Arsebük, G., Howell, F.C., Josephson, S., Juell, K., Pigati, J., Quade, J., 1998. Reconstructing cave bear paleoecology from skeletons: a cross-disciplinary study of Middle Pleistocene bears from Yarımburgaz Cave, Turkey. *Paleobiology* 24(1), 74-98.

Stiner, M.C., Barkai, R., Gopher, A., 2009. Cooperative hunting and meat sharing 400-200 kya at Qesem Cave, Israel. *Proceedings of the National Academy of Sciences* 106(32), 13207-13212.

Stuart, A.J., 1991. Mammalian extinctions in the late Pleistocene of northern Eurasia and North America. *Biological Review of the Cambridge Philosophical Society* 66, 453-562.

Svoboda, J., Ložek, V., Vlček. E., 1996. *Hunters between East and West: The Paleolithic of Moravia*. Plenum, New York.

Taşkıran, H., 2008. Réflexions sur l'Acheuléen d'Anatolie. *l'Anthropologie* 112, 140-152.

Tattersall, I., 2007. Neanderthals, *Homo sapiens*, and the question of species in paleoanthropology. *Journal of Anthropological Sciences* 85, 139-146.

Le Tensorer, J.M., Jagher, R., Rentzel, P., Hauck, T., Ismail-Meyer, K., Pümpin, K., Wojtczak, D., 2007. Long-term site formation processes at the natural springs Nadaouiyeh and Hummal in the El Kowm Oasis, Central Syria. *Geoarchaeology* 22, 621-640.

Tchernov, E., Kolska-Horwitz, L., Ronen, A., Lister, A., 1994. The faunal remains from Evron Quarry in relation to other Lower Paleolithic hominid sites in the southern Levant. *Quaternary Research* 42, 328-339.

Tchernov, E., Tsoukala, E., 1997. Middle Pleistocene (Early Toringian) carnivore remains from northern Israel. *Quaternary Research* 48, 122-136.

Terradas, X., 2003. Discoid flaking method: conception and technological variability, in: Peresani, M., (Ed.), *Discoid Lithic Technology: Advances and Implications*. BAR International Series 1120. British Archaeological Reports, Oxford, pp. 19-30.

Thieme, H., Urban, B., Mania, D., von Kolfschoten, T., 1993. Schöningen (Nordharzvoorland). Ein altpaläolithische fundstelle aus dem mittleren Eiszeitalter. *Archaeologisches Korrespondenzblatt* 23: 147-163.

Tryon, C., Logan, A., Muralis, D., Kuhn, S., Slimak, L., 2009. Building a tephrochronological framework for the central Anatolian Paleolithic. *Journal of Archaeological Science* 36, 637-652.

Tsoukala, E., 1989. Contribution to the study of the Pleistocene fauna of large mammals (CARNIVORA, PERISSODACTYLA, ARTIODACTYLA) from Petralona Cave (Chalkidiki, N. Greece). Doctoral Thesis, Aristotle University of Thessaloniki, Scientific Annals, School of Geology, 1(8).

Tsoukala, E., Chatzopoulou, K., Rabeder, G., Pappa, S., Nagel, D., Withalm, G., 2006. Paleontological and stratigraphical research in Loutra Arideas Bear Cave (Almopia Speleopark, Pella, Macedonia, Greece), *Proceedings of the 12th International Cave Bear Symposium*, Thessaloniki/Aridea. Scientific Annals, School of Geology, Aristotle University, special volume 98, 41-67.

Turq, A., 1992. Raw material and technological studies of the Quina Mousterian in Perigord, in: Dibble, H., Mellars, P., (Eds.), *The Middle Paleolithic: Adaptation, Behavior and Variability.* University Museum Monograph 78. University of Pennsylvania, Philadephia, pp. 75-85.

Tzedakis, P.C., Hooghiemstrab, H., Pälikec, H., 2006. The last 1.35 million years at Tenaghi Philippon: revised chronostratigraphy and long-term vegetation trends *Quaternary Science Reviews* 25, 3416-3430.

Valoch, K., 1984. Le Taubachien, sa geochronology, paléoécologie et paléthnologie. *l'Anthropologie* 88, 193-208.

Valladas, H., Mercier, N., Joron, J.-L., Reyss, J.-L., 1998. GIF laboratory dates for Middle Paleolithic Levant, in: Akazawa, T., Aoki, K., Bar-Yosef, O., (Eds.). *Neanderthals and Modern Humans in Western Asia.* Plenum Press, New York, pp. 69-75.

Withalm, G., 2005. The development of cave bear metapodial bones in the profile of the Vindija Cave (Croatia). *Mitteilungen der Kommission für Quartärforschung (Wien)*, 14, 241-257.

Yalçınkaya, I., 1981. Le Paléolithique inférieur de Turquie, in: Cauvin, J. Sanlaville, P., (Eds.), *Préhistoire du Levant.* Éditions du CNRS, Paris, pp. 207-248.

Yalçınkaya, I., Otte, M., Bar-Yosef, O., Kozlowski, J., Léotard, J.-M., Taşkıran, H., 1993. The excavations at Karain Cave, southwestern Turkey: an interim report, in Olszewski, D., Dibble, H., (Eds.), *The Paleolithic Prehistory of the Zagros-Taurus.* University Museum Monograph 83. The University of Pennsylvania, Philadelphia, pp. 101-118.

Yaltırak, C., Sakınç, M., Aksu, A.E., Hiscott, R.N., Galleb, B., Ulgen, U.B., 2002. Late Pleistocene uplift history along the southwestern Marmara Sea determined from raised coastal deposits and global sea-level variations. *Marine Geology* 190, 283-305.

Yeshurun, R., Bar-Oz, G., Weinstein-Evron, M., 2007. Modern hunting behavior in the early Middle Paleolithic: Faunal remains from Misliya Cave, Mount Carmel, Israel. *Journal of Human Evolution* 53 (6), 656-677.

CHAPTER ABSTRACTS

CHAPTER 1

THE CAVES OF YARIMBURGAZ: GEOLOGICAL CONTEXT AND HISTORY OF INVESTIGATION

G. Arsebük, F. C. Howell,
S. L. Kuhn, W. R. Farrand, M. Özbaşaran, M. C. Stiner

The Yarımburgaz cave complex is located in eastern Thrace, approximately 22 km west of Istanbul. This cave complex is situated about 1.5 km north of the northern Küçükçekmece lagoon shore, the smaller of the two embayments in the Sea of Marmara. The Yarımburgaz caves comprise an Upper Chamber (UC) and a Lower Chamber (LC), each with its own entrance facing south-southwest. The wider, more open UC extends more that 50 m into the Eocene limestone bedrock, whereas the tunnel-like LC extends as far as 500 m into the bedrock. The two chambers are joined by a passage about 10 m from the entrances. Traces of Roman and Byzantine construction occur within and outside of Yarımburgaz Cave and include wall remnants, columns, niches, and apses carved into the walls in the UC, the passage to the LC, and the rock face outside the entrances. These structures were severely disturbed in recent times. Although Yarımburgaz Cave has been known to science since the second half of the 19th Century, scientific investigations at the site did not begin until the 1960s with the excavations by Ş.A. Kansu and K. Kökten, who recognized Chalcolithic as well as Paleolithic occupations in both chambers. Two decades later, M. Özdoğan turned his attention to the Chalcolithic remains in Yarımburgaz. His excavations in the UC demonstrated the presence of some Neolithic and limited Middle/Upper Paleolithic occupations below the Chalcolithic. Test soundings in the LC revealed deeply buried Lower Paleolithic occupations with abundant artifacts and faunal remains. After examining materials from Özdoğan's excavations, F.C. Howell in collaboration with Prof. G. Arsebük, field director M. Özbaşaran, and colleagues from the University of Istanbul conducted excavations mainly in the LC from 1988 through 1990. These excavations recovered large quantities of vertebrate remains and artifacts, and provided substantial geological and geochronological data on the Lower Pleistocene deposits. This volume reports the results of geological, paleontological, and archaeological work during the 1988-1990 field seasons and subsequent laboratory research of the Lower Pleistocene materials.

CHAPTER 2

THE SEDIMENTARY INFILLING OF YARIMBURGAZ CAVE, TURKEY

W. R. Farrand, J. P. McMahon

Yarımburgaz is a double cave with slightly offset upper and lower entrance chambers, connected laterally inside the cave. The Lower Chamber (LC) is still an active karstic cave, with growing speleothems and pools of standing water. The rounded quartz and chert pebbles and stratified sand on the existing LC cave floor indicate that another opening exists in the cave somewhere far from the current entrances. The Upper Chamber (UC) shows little evidence for ongoing karstic activity. At the time when Middle Pleistocene hominins occupied the cave, both chambers extended out beyond their present mouths, and the Sazlidere Valley was not eroded as deeply as now. The UC was strongly modified in Byzantine and classical times, both inside and outside the cave entrance. Excavation trenches in the LC revealed modern pits as deep as 2 or 3 m, possibly excavated by modern treasure seekers or looters. The two chambers of the Yarımburgaz cave complex have rather different sedimentary sequences. Below the Byzantine and modern debris, the UC contains Chalcolithic to Upper/Middle Paleolithic occupations above a beach sand attributed to the Last Interglacial. Strongly modified sediments beneath the beach sand contain sparse Lower Paleolithic artifacts. The LC has a long (>5 m) and complex sedimentary history, beginning long before humans and bears began to utilize the cave. The lithostratigraphic units in the excavation trenches are divided into six unconformity-bounded sequences. The excavations did not reach bedrock in any excavation trench in either the upper or the lower cave, except in some areas very close to the cave walls. The lowest deposits contain pebbles and sand deposited by rapidly flowing water, which subsequently gave way to the deposition of fine clays under slower flow conditions, presumably at a time when the cave was not open to the exterior, or when the cave entrance was much farther out than it is now. These fluvial sediments do not contain bones or artifacts. The middle levels in the LC deposits are dominated by reddish brown clayey silt with scattered sand and isolated small chert pebbles (1-2 cm), along with increasing numbers of animal bones, especially cave bear, and stone implements imported by humans. The final evidence for Lower Paleolithic occupation in the LC, such as has escaped erosion, occurs in very rocky deposits, suggesting that the cave mouth was rather widely open, perhaps as similar to today. Considerable breakdown of the cave roof occurred, perhaps induced by earthquakes, causing numerous large limestone fragments to drop to the cave floor. Following the Lower Paleolithic occupations, a large quantity of beach sand was deposited in the UC, which has been tentatively correlated with a high sea level stand of the Last Interglacial, MIS 5. Why this sand did not also flood into the LC remains unclear. Either the LC entrance was blocked, or the sand has been completely removed from the LC by subsequent erosion. From the Late Pleistocene to the Holocene, Middle/Upper Paleolithic, Chalcolithic, and finally Byzantine peoples frequented the UC. The LC may also have been used, but if so, subsequent erosion or very recent human disturbance removed any evidence.

CHAPTER 3

ELECTRON SPIN RESONANCE (ESR) AND ^{230}TH/^{34}U DATING IN THE LOWER CHAMBER AT YARIMBURGAZ, TURKEY

B. A. B. Blackwell, H. P. Schwarcz, W. R. Farrand, J. Lundburg, A. R. Skinner, M. N. Divjak, J. I. B. Blickstein

Both ^{230}Th/^{234}U and electron spin resonance (ESR) techniques were used to date the Lower Paleolithic occupation at Yarımburgaz Cave. ^{230}Th/^{234}U dating proved difficult due to the paucity of suitable samples. Poorly constrained isochrons hint that some collophane and fragmentary travertine deposition may have occurred very recently. While the discontinuous collophane layers may be secondary deposits postdating the sedimentary deposition and archaeological occupation, the standard ^{230}Th/^{234}U dates suggest that one collophane unit below the major *Ursus-*

bearing horizons formed at 160^{+25}_{-20} ka. A single radiation-sensitive ESR signal at g = 2.0018 occurs in well crystallized fossil tooth enamel, but not in modern teeth. In dating fossil teeth, the equivalent radiation dose needed to produce the observed ESR signal is the integral with respect to time of the natural, environmental dose rate experienced by the tooth after its deposition. Using standard ESR, 17 *Ursus* teeth from Yarımburgaz have been dated, including two by coupled ESR-^{230}Th/^{234}U dating using dentine ^{230}Th/^{234}U analyses only. Due to the comparatively small tooth size of the bears, no ESR isochron analyses could be attempted. External dose rates were calculated by modeling dose rates from nearby sedimentary units as determined by NAA for both bulk sediment and specific sedimentological components. The tooth analyzed by coupled ESR-^{230}Th/^{234}U dating from Stratum Z dates to 211 ± 22 ka, with uptake parameters of p_{en} = 1.39 ± 0.34 and p_{den} = -0.80 ± 0.08, although, without enamel ^{230}Th/^{234}U data, this, at best, represents its minimum age. The tooth also may have been reworked from an older layer. One tooth from Stratum Y, probably dating at 219 to 236 ka, but possibly as young as 153 ± 32 ka (assuming linear U uptake; LU), may also represent an extraneous tooth resulting from disturbance of the site by later occupants. Alternatively, that tooth may have experienced more recent U uptake associated with the U mobilization during the secondary collophane deposition at about 160 ka. Nine independent standard ESR analyses from Stratum X averaged 222 ± 18 ka (assuming LU). Six independent standard ESR analyses from Stratum W averaged 197 ± 12 ka (assuming LU), but omitting one potentially extraneous tooth from that group yielded an age of 226 ± 24 ka. A tooth from Stratum W gave a minimum coupled age of 154^{-13}_{-12} ka, with uptake parameters of p_{en} = -0.93 ± 0.06 and p_{den} = -0.93 ± 0.06, but 178^{+23}_{-22} ka, with p_{en} = 1.4 ± 0.3. The deposits at Yarımburgaz therefore date to at least 150 ka. The close agreement between the various age determinations, however, suggests that Strata Z to W likely date between 211 and 226 ka, which correlates with Marine Isotope Stage (MIS) 7d. The collophane deposition correlates with a relatively warm interstadial within MIS 6d, a time when stalagmite deposition occurred in several other European cave sites.

CHAPTER 4

ECOLOGICAL SIGNIFICANCE OF THE SMALL MAMMAL FAUNA FROM YARIMBURGAZ CAVE (TURKISH THRACE)

W. v. Koenigswald, C. Lindenau, W. Santel

Approximately 38 different taxa could be distinguished in the small mammal fauna from the Lower Chamber (LC) at Yarımburgaz Cave. These include three lipotyphlan, 15 bat, 17 rodent, and three lagomorph species. The faunal material is unevenly distributed within the various sedimentary cycles, and a large number of specimens cannot be attributed to any single cycle. The assemblages are relatively uniform in composition. Based on the evolutionary stage for the steppe lemming, *Lagurus transiens/lagurus*, found in the site, the fauna dates to the Middle Pleistocene. The fauna also contains several rodent and lagomorph species no longer present in the area. The rodent and lagomorph faunas show a distinct steppic influence with several immigrants from northeastern regions, indicating that most of the sediment in the LC was deposited under somewhat cooler and dryer conditions than prevail in area today. In contrast, the bat fauna does not necessarily indicate different climatic conditions than today. Some of the bat taxa typically inhabit the Mediterranean (e.g., *Rhinolophus mehelyi*), whereas others have ranges that reach much further south. This apparent discordance between the bats and other small mammals in the LC fauna may result from seasonal migrations of the bats, bringing thermophilic species into the area during the summer, or to an absence of bats from Turkish Thrace during the colder periods.

CHAPTER 5

THE YARIMBURGAZ LITHIC ASSEMBLAGE

S. L. Kuhn

The 1988-1990 excavations in the Lower Chamber at Yarımburgaz produced an assemblage of nearly 1700 stone artifacts, ranging from small debris to large cores and choppers. Several different raw materials were used to make stone tools. In order of abundance, these raw materials are flint, quartz, and quartzite. Small quantities of jasper, silicified wood, and unidentified metamorphic rocks also occur in the lithic assemblages. Most or all the raw materials would have been available within the immediate environment. The stone artifacts show little, if any, evidence for transport by water or abrasion generally, although many flint specimens are almost completely desilicified. Flakes and flake tools dominate the Yarımburgaz lithic assemblage; core tools, mainly choppers and chopping tools, are comparatively rare. Most flake tools have irregular edges, corresponding to typological definitions of notches and denticulates. Specimens with smoothly retouched scraper edges are less common. The frequency of modified pieces is high, and many flake tools also show evidence for heavy retouch and resharpening. In contrast, cores tend to show only light exploitation. This apparent inconsistency in raw material exploitation also occurs at other Middle Pleistocene sites. A striking feature in the Yarımburgaz assemblage concerns the differing ways in which specific raw materials were exploited and transformed into tools. Flint was used mainly for making flake tools, and flakes were produced using a distinctive variation on the unifacial discoid method. The occupants also preferred quartz for making flake tools, but for this material they used a bipolar technique. In contrast, quartzite was used mainly for manufacturing choppers and chopping tools. Such differentiated raw material use is a common feature in Lower Paleolithic assemblages. The lithic assemblage from Yarımburgaz falls into the general category of "core and flake tool assemblages" without bifacial handaxes. Similar assemblages are common in the Middle Pleistocene in Central and Eastern Europe, including the Balkans, where the Acheulean is scarce or absent. In contrast, the Acheulean is much more common in Western Europe and Southwest Asia, including Anatolia. The presence of the Yarımburgaz assemblage in eastern Thrace reinforces the notion that the Bosphorous/Mamara/Dardanelles waterway formed a significant barrier to human movement during much of the Pleistocene.

CHAPTER 6

A TAPHONOMIC STUDY OF BONE ASSEMBLAGE FORMATION AND ARTIFACT-BEAR ASSOCIATIONS IN YARIMBURGAZ CAVE

M. C. Stiner

This chapter evaluates the circumstances of bone assemblage formation in the Middle Pleistocene deposits in Yarımburgaz Cave, identifying the agencies responsible for the accumulations and the causes of bone damage before and following deposition. Taphonomic and other evidence indicate that three distinct biological processes contributed to the formation of the Middle Pleistocene faunas: hibernating bears, non-ursid carnivores such as wolves, and hominins in descending order of importance. Species representation, bone damage, body part representation, and mortality data indicate that cave bear remains in Yarımburgaz Cave are unrelated to hominin use of the same site. The bear bone accumulations resulted from hibernation-related mortality over many generations of den use. Other large carnivores, especially canids, also visited the cave occasionally, based on damage observed on bear and ungulate bones, as well as the scant frequencies of the other predators' own remains. The close spatial associations between different components of the fauna, and between lithic artifacts and bear bones, are explained by generally slow or uneven sedimentation rates in the cave, and by post-depositional sedimentary mixing by denning bears. Human-modified bones are surprisingly scarce in the faunal assemblages; hominins spent enough time in Yarımburgaz Cave to generate substantial quantities of stone artifacts, but they did not discard many ungulate bones there. Additionally, no traces of fire were found. The hominin occupations therefore appear to have been quite ephemeral. These observations suggest that hominin foraging efforts focused on resources other than large game while at the cave, and rather different patterns of coexistence among hominins and carnivores than are typical in later prehistoric periods.

CHAPTER 7

MIDDLE PLEISTOCENE BEARS FROM YARIMBURGAZ CAVE (TURKEY)

E. Tsoukala

Yarımburgaz Cave contains a fauna very rich in Pleistocene ursid remainss. Bear bones from the Lower Cave represent 93% of the total mammalian specimens from the 1988-1990 excavations. This proportion is similar to those at Late Pleistocene sites, such as Peştera cu Oase, Romania, and the Loutra Arideas Bear Cave in Greece. The minimum number of bear individuals (MNI) found in the cave is 42. Approximately half of the bear bone and dental specimens from the 1988-1990 excavations were studied intensively from a paleontological point of view. Most of the bear tooth and bone remains belong to juveniles and sub-adults, while fewer belong to very old individuals and adults. Two main ursid taxa co-occur in the cave—cave bears (*Ursus deningeri-spelaeus)* and brown bears (*U.* cf. *arctos*). Brown bear remains are much rarer than those from cave bears. The very large cave bear sample contains specimens that correspond to morphological and metric definitions for both *U. deningeri* and *U. spaelaeus.* The presence of these three species together indicates that the deposits formed over an extended time during the second half of the Middle Pleistocene, an interval in which *U. spaelaeus* replaced *U. deningeri.*

CHAPTER 8

THE LARGER MAMMAL FAUNA (EXCLUSIVE OF URSIDAE) FROM YARIMBURGAZ CAVE

F. C. Howell

The Lower Cave at Yarımburgaz cave provides one of the largest vertebrate faunal collections from the Middle Pleistocene in Turkey, and for extreme southeastern Europe more generally. The large mammal assemblage includes nine genera of Carnivora, two Perissodactyla, and nine or ten Artiodactyla. Proboscidea, *Hippopotamus*, and non-caballine Equidae are conspicuously absent. *U. deningeri*, the Middle Pleistocene cave bear, numerically dominates the much rarer brown bear, *U. arctos*, by two orders of magnitude. Remains of other carnivores and ungulates occur in much lower numbers yet show both groups show substantial taxonomic diversity. Felines include lion, another probable pantherine, but mostly smaller taxa such as wildcat, swamp (or jungle) cat, and caracal. The Canidae include both vulpines and canines, and, from both the size and some morphological features, several taxa are represented within each group. The foxes appear to represent the extant *V. vulpes*, another smaller, and perhaps, even extinct, species. The canine group appears to constitute wolf-like forms, both smaller and larger species, but jackal can not be excluded. Another large canid with cuonine affinities is also represented. Mustelids and hyaenids are extremely rare, yet two distinct hyaena lineages are represented, the later *C. c. praespelaea* and the earlier *P. brevirostris*. Artiodactyls dominate among the ungulates and perissodactyls are rare. Cervids are the most abundant and ubiquitous of the artiodactyls, including fallow, red and roe deer, and a form of Megalocerini. Suids and bovids are uncommon, but the presence of a possible alcelaphine in an early sedimentary phase is noteworthy. A large caballine horse is represented, but no evidence exists for *E. hydruntinus/E. hemionus*. Rhino is represented by a few fragmentary specimens. Although biostratigraphic indicators are represented from among each of the three larger mammal orders, their implications are neither straightforward, due to fragmentation and/or lack of critical diagnostic elements, nor entirely consistent with respect to chronology. Broadly considered, the composite assemblage indicates a mid-Middle Pleistocene age, although certain elements among the carnivores in particular could suggest a greater age. Strong links to Soutwest Asian biota are apparent, but unlike Lower or even mid-Pleistocene occurrences in the Levant, which often include several African taxa, the biogeographic affinities for the large mammal assemblage from Yarımburgaz are wholly Palaearctic. The presence of a cuonoid-like canid and a presumptive medium-sized antelope taxonomic affiliation could point to African affinities, but the taxonomic attributions for these fragmentary specimens are not sufficiently clear to alter the general conclusion.

CHAPTER 9

CULTURE AND BIOLOGY IN TRANSITION: SUMMARY OF FINDINGS AND CONCLUSION

S. L. Kuhn, M. C. Stiner

The site of Yarımburgaz Cave, and its Pleistocene assemblages and faunas, are geographically, paleontologically and culturally transitional. The site is located at the geographic interface between Europe and Asia. The cave bear fossils represent an evolving population, from *Ursus deningeri* and *U. spelaeus*. The lithic assemblage bridges two broad chronological and taxonomic units, the Lower and Middle Paleolithic. This concluding section discusses the geological age and environmental context of the human and bear occupations of Yarımburgaz Cave, probably ecological relationships between humans and bears, the behavioral implications of the archaeological and taphonomic findings. A range of geological, faunal and archaeological evidence indicates that the cultural deposits and bear remains in the Lower Chamber (LC) of Yarımburgaz Cave date to the latter part of the Middle Pleistocene. The ESR dates suggest that the occupations probably date to Marine Isotope Stage (MIS) 7, an age estimate generally consistent with the evolutionary status of the cave bears. The presence of steppic elements among the rodent remains indicates that the fauna accumulated during a cold interval. The most likely period of occupation is the brief cold interval, MIS 7d, between 211 and 226 kya, although we cannot exclude the possibility that the occupations date to earlier in MIS 8. The spatial associations between the cave bear remains and lithic artifacts in the LC raises questions about possible interactions between bears and hominins at this site. The bears used the cave as a hibernation den and occasionally died there. The nature of the hominin occupation is less clear. It appears that hominins and bears interacted little, if at all, and visits to the site by hominins were quite brief. In fact there is no evidence of fire, nor of extensive butchering of animal remains. Hominins and bears used the cave at disparate, limited intervals, at times when the cave was dry and accessible. The spatial associations between the cave bear remains and stone tools in the LC sediments is the result of slow sedimentation rates and post-depositional disturbances by bears and other animals. Although the Middle Pleistocene artifacts from Yarımburgaz Cave have been described as Lower Paleolithic industries in the past, the assemblage does not fit easily into conventional definitions of either the Lower or the Middle Paleolithic. Comparisons with other assemblages from the Levant, Anatolia, and southeastern Europe dating to MIS 7 reveal some similarities that cross geographic, typological and technological boundaries. We suggest that the common features of these MIS 7 assemblages do not represent a shared cultural heritage, but instead reflect broadly similar patterns of foraging and mobility. The widespread behavioral tendencies of the MIS 7 Paleolithic record may represent the evolutionary foundation for later Mousterian technologies and subsistence adaptations.

GENERAL PLATES

Pl. 0.1. The upper and lower entrances to Yarımburgaz cave.

Pl. 0.2. View of the Sazlıdere Valley.

Pl. 0.3. View of the Sazlıdere River.

Pl. 0.4. Ruins of Byzantine church in the Upper Cave of Yarımburgaz.

Pl. 0.5. Sectioned Pleistocene deposits in foreground and ruins of Byzantine church on surface in background.

Pl. 0.6. View outward through the entrance of the Lower Cave of Yarımburgaz.

Pl. 0.11. View of part of the section exposed in Square V, 1989.

Pl. 0.12. Square V viewed from the north, 1988 (and A.71 from the 1986 excavation also visible).

Pl. 0.13. Large block in the wall of Square V, 1988, with "reaction rim".

Pl. 0.14. West section of Square R, 1990.

Pl. 0.15. Squares T and S, 1989.

Pl. 0.16. General view of Square P, 1990.

Pl. 0.17. Square P, level 2 (plan 9) under excavation, 1990.

Pl. 0.18. Cave bear mandible and other bones in situ in Square P, 1990.

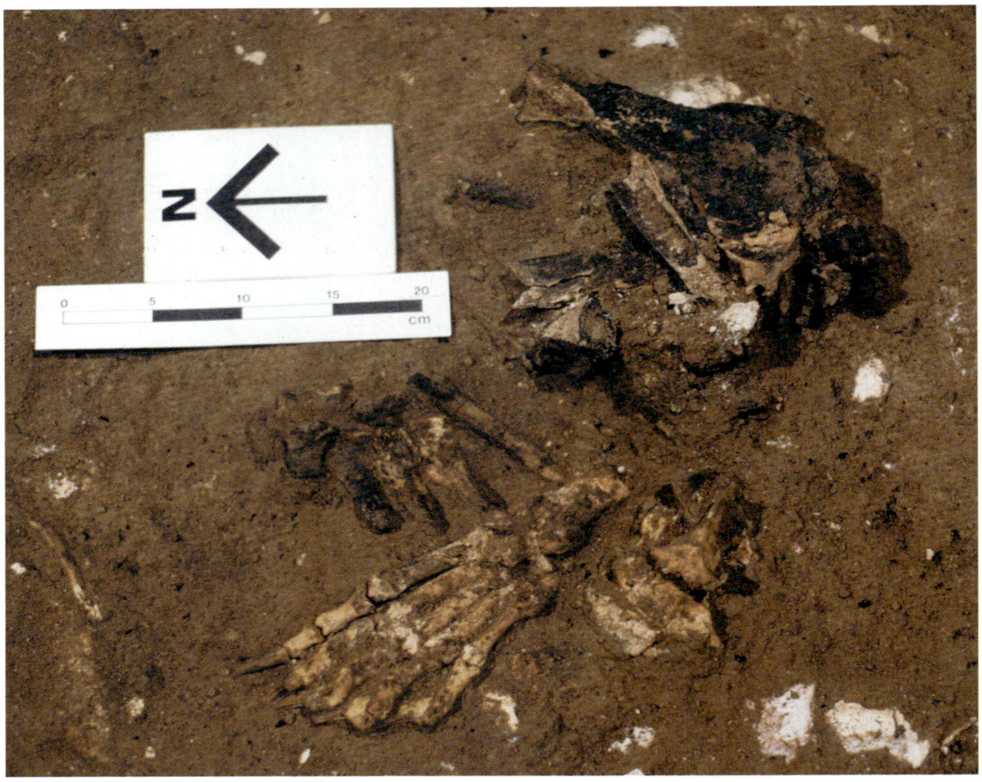

Pl. 0.19.
Cave bear articulated hind food bones in situ in Square P. 1990.

Pl. 0.20.
Concentration of cave bear bones in the northeast corner of Square P, 1990.

Pl. 0.21. Large cave bear bones in Square P, 1990.

Pl. 0.22. East section of Square P, 1990.

Pl. 0.23. West section of Square P, 1990.

Pl. 0.24. Complete cave bear cranium (specimen no. 220) in Square Y, level 1 (plan 7a), 1988.

Pl. 0.25. F. Clark Howell and student with Paleolithic artifacts from the Lower Cave of Yarımburgaz, 1990.

Pl. 0.26. Chopping tools from the Lower Cave excavations of Yarımburgaz.

Pl. 0.27. Flake tools from the Lower Cave excavations of Yarımburgaz.